Applications Manual

Fundamentals
of
Anatomy and Physiology

F O U R T H   E D I T I O N

Frederic H. Martini, Ph.D.

Kathleen Welch, M.D.

with

William C. Ober, M.D.
Art coordinator and illustrator

Claire W. Garrison, R.N.
Illustrator

Ralph T. Hutchings
Biomedical photographer

Prentice Hall, Upper Saddle River, New Jersey 07458

Notice: Our knowledge in clinical sciences is constantly changing. The author and the publisher of this volume have taken care that the information contained herein is accurate and compatible with the standards generally accepted at the time of publication. Nevertheless, it is difficult to ensure that all the information given is entirely accurate for all circumstances. The author and publisher disclaim any liability, loss, or damage incurred as a consequence, directly or indirectly, of the use and application of any of the contents of this volume.

Library of Congress Cataloging-in-Publications
Martini, Frederic.
    Fundamentals of anatomy & physiology. Applications manual /
Frederic H. Martini, Kathleen Welch.—4th ed.
       p.   cm.
    Companion v. to: Fundamentals of anatomy and physiology / Frederic
H. Martini, 4th ed. 1997.
    Includes index.
    ISBN 0-13-751868-4 (alk. paper)
    1. Human physiology. 2. Human anatomy   I. Welch, Kathleen, M.D.
II. Martini, Frederic   Fundamentals of anatomy and physiology.
III. Title
    [DNLM: 1. Anatomy—handbooks. 2. Anatomy—examination questions.
3. Physiology—handbooks. 4. Physiology—examination questions.
QS 4 M3855f   1998 Suppl.]
QP34.5.M4615   1997 Suppl.
612—dc21
DNLM/DLC
for Library of Congress

Executive Editor: *David Kendric Brake*
Development Editor: *Karen Karlin*
Production Editor: *James Buckley*
Managing Editor: *Linda Schreiber*
Special Projects Manager: *Barbara A. Murray*
Assistant Vice President of Production
 & Manufacturing: *David W. Riccardi*
Editor in Chief: *Paul F. Corey*
Editorial Director: *Tim Bozik*
Editor in Chief, Development: *Ray Mullaney*
Executive Managing Editor, Production: *Kathleen Schiaparelli*
Director of Marketing: *Kelly McDonald*
Marketing Manager: *Jennifer Welchans*

Copy Editor: *Margo Quinto*
Manufacturing Manager: *Trudy Pisciotti*
Cover Designer: *Paul Gourhan*
Creative Director: *Paula Maylahn*
Art Director: *Heather Scott*
Art Manager: *Gus Vibal*
Art Editor: *Warren Ramezzana*
Page Layout: *Black Dot Group*
Photo Editor: *Carolyn Gauntt*
Illustrators: *William C. Ober, M.D., Claire Garrison, R.N.*
Biomedical Photographer: *Ralph T. Hutchings*
Photo Research: *Stuart Kenter Associates*
Editorial Assistance: *Byron Smith, David Stack*

Cover Photograph: *Andrea Weber and Andrew Pacho Photo, © 1993 Lois Greenfield*

Printed in the United States of America
10 9 8 7 6 5 4 3 2

ISBN 0-13-751868-4

Prentice-Hall International (UK) Limited, *London*
Prentice-Hall of Australia Pty. Limited, *Sydney*
Prentice-Hall Canada, Inc., *Toronto*
Prentice-Hall Hispanoamericana, S.A., *Mexico*
Prentice-Hall of India Private Limited, *New Delhi*
Prentice-Hall of Japan, Inc., *Tokyo*
Simon & Schuster Asia Pte. Ltd., *Singapore*
Editora Prentice-Hall do Brasil, Ltda., *Rio de Janiero*

Contents

Preface

This *Applications Manual*, which accompanies the fourth edition of *Fundamentals of Anatomy and Physiology* (FAP), has one goal: to help students apply the concepts introduced in the main text to their personal lives. Together with the boxed essays and Clinical Discussions in the text, the discussions in the *Applications Manual* complete an introduction to the major pathological conditions and diagnostic procedures you may encounter in the "real world." Because most of the detailed clinical and applied topics are located in the *Applications Manual*, the applied information can be complete enough to provide a valuable reference without interrupting the text narrative that deals with normal anatomy and physiology. In *Fundamentals*, you will find the icon $\boxed{\text{AM}}$, followed by the discussion title, wherever a topic in the *Applications Manual* is linked to the main text.

The *Applications Manual* has been reorganized to enhance its value as a separate but integrated reference. This edition also contains several features not found in previous editions. For example, it contains optional background material and supplemental material that instructors can use to increase their depth of coverage on specific topics. These discussions are referenced in *Fundamentals* by the title "$\boxed{\text{AM}}$ A Closer Look." Such topics include enzyme regulation and control mechanisms, pathogens, cancer, pain pathways, autonomic pharmacology, and AIDS. This edition also has a new section, called "Origins and Insertions," that will help students visualize and appreciate the anatomical relationships between the skeletal and muscular systems. In addition, the "Surface Anatomy and Cadaver Atlas" contains many new dissection photographs.

The pedagogical framework of this edition of the *Applications Manual* complements that of the main text. *Fundamentals* uses a three-level review system in which Level 3 consists of problem-solving questions and clinical or situational exercises. The *Applications Manual* in turn has three types of problem-solving questions; pages containing these questions are marked by a black band along the margin. The first type, "Critical-Thinking Questions," deals with disorders or situations that involve only one system. Those questions are therefore comparable to the Level 3 questions in *Fundamentals*. The second type, "Clinical Problems," requires students to integrate and interpret information about two or three systems. The third type, "Case Studies," offers multisystem problems that students must approach in a specific sequence. Each Case Study is based on a real patient's history. By completing the Case Study exercises, students can practice working through complex problems by using a logical, stepwise approach.

Few instructors are likely to cover all the material in the *Applications Manual*. Indeed, some instructors may choose not to cover all the boxed material in the text. Because courses differ in their emphases and students differ in their interests and backgrounds, the goal in designing the *Applications Manual* has been to provide maximum flexibility of use. The diversity of applied topics in the text discussions and boxes, the *Applications Manual*, and *The New York Times* articles provides instructors with a wide variety of ways in which to integrate the treatment of normal function, pathology, and other clinical or health-related topics. Boxed material and topics in the *Applications Manual* that are not covered in class can be assigned, recommended, used for reference, or left to the individual student. Experience indicates that each student will read those selections that deal with disorders that affect friends or family members, address topics of current interest and concern, or include information relevant to a chosen career path.

TO THE STUDENT

This *Applications Manual* is organized in units that deal with a wide variety of applied topics:

- **An Introduction to Diagnostics** discusses the basic principles involved in the clinical diagnosis of disease states.

- **Applied Research Topics** considers principles of chemistry and cellular biology that are important to understanding, diagnosing, or treating homeostatic disorders.

- **The Body Systems: Clinical and Applied Topics** is organized to parallel the text chapter by chapter and system by system. This section includes more detailed discussions of many clinical topics introduced in the main text, along with discussions of diseases and diagnostic techniques not covered in the text. Each discussion is cross-referenced to the text by page number; relevant chapter numbers in *Fundamentals* are indicated by black thumb tabs that appear in the margins of right-hand pages.

- The **Critical-Thinking Questions** at the end of each system help you sharpen your ability to think analytically.

- The **Clinical Problems,** located after each group of related systems, assist you in integrating information about several body systems and give you a chance to practice making reasonable inferences in realistic situations.

- The **Case Studies,** which follow The Body Systems section, provide further opportunities for you to develop your powers of analysis, integration, and problem solving. The studies presented, based on actual case histories, draw on material from the entire text (as they would in real life). The questions, keyed to crucial points in the presentation, help you identify the relevant facts and form plausible hypotheses. You can use the case studies as the basis for discussion with other students or tackle them yourself to hone your reasoning skills.

- The **Origins and Insertions** section consists of images of the bones of the skeleton that show the locations of the origins and insertions of the major muscles and muscle groups that are discussed in Chapter 11 of *Fundamentals*. This section will help you understand the relationship between muscle placement and muscle action and will help you remember the names and functions of the bone surface markings introduced in Chapters 6–8 of the text.

- A **Scanning Atlas** of photographs produced by various modern imaging techniques lets you view the interior of the human body section by section. These images will help you develop an understanding of three-dimensional relationships within the body. The Scanning Atlas includes a number of unlabeled images. By labeling them yourself, you can test your knowledge of anatomical structure while you develop your powers of visualization in three dimensions.

- A full-color **Surface Anatomy and Cadaver Atlas** of live-model and cadaver-dissection pho-

tographs allows you to visualize the superficial and internal structures of all major body regions and organs.

You can use the material in the *Applications Manual* in several ways. For example:

- You can read the *Applications Manual* simultaneously with *Fundamentals of Anatomy and Physiology*, referring to topics as each is referenced in the main text.

- You can read the *Applications Manual* separately, referring to *Fundamentals* for relevant background information as needed.

- You can use the *Applications Manual* as a reference, reading only those discussions that are of personal interest to you, that are relevant to your intended career, or that you need to research for some special purpose. You can locate information about specific topics, such as diagnostic procedures or drugs, by referring to the icons that appear next to each heading that is cross-referenced to *Fundamentals*. The following icons are used:

= Reference material, including discussions preceded by *A Closer Look* (as in "A Closer Look: The Nature of Pathogens," p. 21)

= Discussions relating to the diagnosis of disease (as in "Blood Tests and RBCs," p. 116)

= Information about specific diseases and their treatment (as in "Heart Failure," p. 128)

= Discussions about topics in pharmacology and treatment methods (as in "Pharmacology and the Autonomic Nervous System," p. 89)

= Exercise and sports-related topics (as in "Sports Injuries," p. 67)

This manual was written to help you see the relevance of the text material and to give you information that you can use in your daily life. When a family member becomes ill or a medical crisis develops on a prime-time TV show, we hope that this manual will help you make sense of the situation. The organization and coverage of the *Applications Manual* have been greatly influenced by student feedback. If you have suggestions or comments about this edition, please do not hesitate to contact us at the Prentice Hall A & P web site or at the address on page ix.

ACKNOWLEDGMENTS

This was a complex project, and we thank everyone who helped bring it to completion. Linda Schreiber coordinated the assembly of all the components and still found time to manage the review process. We acknowledge Dr. Eugene C. Wasson, III, the staff of Maui Radiology Consultants, and the Radiology Department of Maui Memorial Hospital for providing many of the MRI and CT scans used in the Scanning Atlas; Ralph Hutchings, who contributed the photographs for the Surface Anatomy and Cadaver Atlas; and Bill Ober, M.D., and Claire Garrison, R.N., who created the new Origins and Insertions section. The diagnostic tables for each system, which have been updated for this edition, were prepared for the first edition of the *Applications Manual* by Martha Newsome, D.D.S. Her hard work in assembling those complex tables and her continued interest in the project are deeply appreciated. Finally, we also express our thanks to David Brake, Executive Editor for Biology, for supporting this project; to Karen Karlin, Senior Development Editor, for her attention to detail; to James Buckley, Supplements Production Editor, for coordinating the mechanics of production; to Stuart Kenter Associates for assisting with the photo research; and to the Prentice Hall production staff who worked on the design, layout, and assembly of this manual.

Frederic Martini, Ph.D.
Kathleen Welch, M.D.

e-mail address:
martini@maui.net

 A & P Web site address:
www.prenhall.com/martini/fap

mailing address:
c/o Prentice Hall, Inc.
1208 East Broadway Road
Tempe, AZ 85282

Photo Credits

Unless noted otherwise below, all photographs are courtesy of Ralph Hutchings.

Elizabeth M. Abel, M.D., from the Leonard C. Winograd Memorial Slide Collection, Stanford University School of Medicine:
A-12a, A-12b

AP/Wide World Photos:
A-47

Christopher J. Bodin, M.D., Tulane University Medical Center:
SA 14, SA 17, SA 18

Centers for Disease Control and Prevention:
A-41

Dr. Harold Chen:
A-17a, left and right

Custom Medical Stock Photo:
A-18a, A-18b, A-20a, A-29a, A-33b, SA 16, CA 5-3a, CA 6-4b, CA 7-2b

Hewlett-Packard Company:
A-36b, left

Medcom:
A-19a, A-19c

Medichrome:
SA 10

Medtronic, Inc.:
A-33c

Mentor Networks Inc.:
CA 4-1a, CA 4-1b, CA 4-2a, CA 4-2b, CA 5-1a, CA 5-1b, CA 5-1c, CA 5-2b, CA 5-3c, CA 6-1a, CA 6-1b, CA 6-2a, CA 6-2b, CA 7-2e, CA 7-4a, CA 7-4d

Monkmeyer Press:
A-9c, A-17b

National Heart Institute/World Health Organization:
A-33d

Photo Researchers Inc.:
A-3b, A-8a, A-8b, A-17c, A-18c, A-33a, A-36a, SA 11, SA 15, SA 19, CA 3-5b

Picker International, Inc.:
A-36c, A-36d

Martin M. Rotker:
A-9b, SA 12

Stuart Kenter Associates:
A-19b

Patrick M. Timmons/Michael J. Timmons:
CA 3-6, CA 5-4, CA 5-5b, CA 7-3c

United Nations:
A-51

Visuals Unlimited:
A-45, SA 13

Dr. Eugene C.Wasson, III, and staff of Maui Radiology Consultants, Maui Memorial Hospital:
A-20b, A-20c, SA 1a, SA 1b, SA 1c, SA 1d, SA 1e, SA 2a, SA 2b, SA 2c, SA 2d, SA 3a, SA 3b, SA 4, SA 5a, SA 5b, SA 6a, SA 6b, SA 7a, SA 7b, SA 8a, SA 8b, SA 9a, SA 9b, SA 9c, SA 9d, SA 9e, SA 9f, SA 9g, SA 9h, SA 9i

Kathleen Welch, M.D.
SA 20

An Introduction
to Diagnostics

A **diagnosis** is a conclusion or decision based on a careful examination of relevant information. Each of us has made simple diagnoses in our everyday experiences. When the car won't start, the kitchen faucet leaks, or the checkbook doesn't balance, most of us will try to determine the nature of the problem. Sometimes the diagnosis is simple: The car battery is dead, the faucet is not completely turned off, or the amount of a check was recorded incorrectly. Once we make the diagnosis, we can take steps to remedy the situation.

Most of us use similar observational skills to diagnose simple medical conditions. For example, imagine that you awaken with a headache, feeling weak and miserable. Your face is flushed, your forehead is hot to the touch, and swallowing is painful. You know that these are the general symptoms of the flu, and you know also that your lab partner missed Tuesday's class because of the flu. You diagnose yourself as sick with the flu, and you open the medicine cabinet in search of appropriate medication.

The steps you took in arriving at the conclusion "I probably have the flu" were quite straightforward: (1) You made observations about your condition; (2) you compared your observations with available data; and (3) you determined the probable nature of the problem. **Clinical diagnosis**, or the identification of a disease, can be much more complicated, but these same steps are always required. In this section, we will examine the basic principles of diagnosis. The goal is not to train you to be a clinician but rather to demonstrate how these basic steps can be followed in a clinical setting.

Any diagnosis—of a disease or of a leaky kitchen faucet—requires an analysis that proceeds in a series of logical steps. Logical analysis, a process often called *critical thinking*, does not come naturally; it is too easy to become distracted or misled and then make a hasty or incorrect decision. Critical thinking is a learned skill that follows rules designed to minimize the chances of error. Nowhere is critical thinking more important today than in the sciences, especially the medical sciences. In applying critical thinking to scientific investigation, we follow what has been called the *scientific method*.

📖 The Scientific Method

Your course in anatomy and physiology should do more than simply teach you the names and functions of different body parts. It should provide you with a frame of reference that will enable you to understand new information, draw logical conclusions, and make intelligent decisions. A great deal of confusion and misinformation exists about just how medical science "works," and people make unwise and even dangerous decisions as a result. Nowhere is this more apparent than when a discussion drifts to health, nutrition, and cancer. If you are going to be working in a health-related profession or are just trying to make sound decisions about your own life, you must learn how to organize information, evaluate evidence, and draw logical conclusions.

FORMING A HYPOTHESIS

There is a lot more to science than just the collection of information. You could spend the rest of your life carefully observing the world around you, but such a task won't reveal very much unless you can see some kind of pattern and come up with an idea, or *hypothesis*, that explains your observations.

Hypotheses are ideas that may be correct or incorrect. To evaluate one, you must have relevant data and a reliable method of data analysis. For example, you could propose the hypothesis that radiation emitted by planet X confers immortality. Could anyone prove you wrong? Not very likely, particularly if you didn't specify the location of the planet or the type of radiation. Would anyone believe you? If you were a "leading authority" on something (anything), a few probably would.

That's not as ridiculous as it might seem. For almost 1500 years "everyone knew" that inhaled air was transported from the lungs through blood vessels to the heart. They knew this because the famous Roman physician Galen had said so. Because he was right in several other respects, all his statements were accepted as true, and contrary opinions were held in low esteem. To avoid making this kind of error, you must always remember to evaluate the hypothesis, not the individual who proposed it!

In the evaluation process, we must examine the hypothesis to see if it makes correct predictions about the real world. The steps in this process are diagrammed in Figure A-1. A valid hypothesis will have three characteristics: It will be (1) testable, (2) unbiased, (3) and repeatable.

A testable hypothesis is one that can be studied by experimentation or data collection. Your assertion concerning planet X qualifies as a hypothesis, but it cannot be tested unless we find the planet and detect the radiation. An example of a testable hypothesis would be "left-handed airplane pilots have fewer crashes than do right-handed pilots." That is testable because it makes a prediction about the world that can be checked—in this case by collecting and analyzing data.

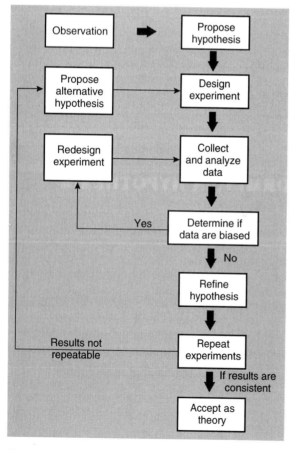

Figure A-1   The Scientific Method.
The basic sequence of steps involved in the development and acceptance of a scientific theory.

AVOIDING BIAS

Suppose, then, that you collected information about all the plane crashes in the world and discovered that 80 percent of all crashed airplanes were flown by right-handed pilots. "Aha!" you might shout, "The hypothesis is correct!" The implications are obvious: Ban all right-handed airline pilots, eliminate four-fifths of all crashes, and sit back and wait for your prize from the Air Traffic Safety Association.

Unfortunately, you would be acting prematurely, for your data collection was biased. To test your hypothesis adequately, you need to know not only how many crashes involved right-handed or left-handed pilots but how many right-handed and left-handed pilots were flying. If 90 percent of the pilots were right-handed, but they accounted for only 80 percent of the crashes, then left-handed pilots are the ones to watch out for! Eliminating bias in this case is relatively easy, but health studies may have all kinds of complicating factors. Because 25 percent of us will probably develop cancer at some point in our lives, we will use cancer studies to exemplify the problems encountered.

The first example of bias in action concerns cancer statistics, which indicate that there are definite regional variations in cancer rates in the United States and abroad. For example, although the estimated yearly cancer death rate in the United States was 173 per 100,000 population in 1997, the rate in Utah was only 126 per 100,000, whereas the rate in the District of Columbia was 221 per 100,000. It would be very easy to assume that this difference is the direct result of rural versus urban living. But these data alone should not convince you that moving from the District of Columbia to Utah will lower your risk of developing cancer. To draw that conclusion, you would have to be sure that the observed rates were the direct result of just a single factor, the difference in physical location. As you will find in later chapters, many different factors can promote cancer development. To exclude all possibilities other than geography, you would have to be certain that the populations were alike in all other respects. Here are a few possible sources of variation that could affect that conclusion:

- *Different population profiles.* Cancer rates vary between males and females, among racial groups, and among age groups. Therefore, we need to know how the populations of Utah and the District of Columbia differ in each respect.

- *Different occupations.* Because chemicals used in the workplace are implicated in many cancers, we need to know how the populations of each region are employed and what occupational hazards they face.

- *Different mobilities.* Because the region in which a person dies may not be the region in which he or she lived and developed cancer, we need to know whether people with cancer in Utah stay in the state or go elsewhere for critical care and whether people with cancer travel to the District of Columbia to seek treatment at special clinics.

- *Different health care.* Because cancer death rates reflect differences in patterns of health care, we need to know whether residents of Utah pay more attention to preventive health care and have more regular checkups, whether their medical facilities are better, and whether they devote a larger proportion of their annual income to health services than do residents in the District of Columbia.

You can probably think of additional factors, but the point is that avoiding experimental bias can be quite difficult!

A second example of the problem of bias comes from the collection of "miracle cures" that continue to appear and disappear at regular intervals. Pyramid power, pendulum power, crystals, magnetic energy fields, and psychic healers come and go in the news. Wonder drugs are equally common, whether they are "secret formulas" or South American plant extracts discovered by colonists from other planets. The proponents of each new procedure or drug report glowing successes with patients who would otherwise have surely succumbed to the disease. And all these remedies are said to have been suppressed or willfully ignored by the "medical establishment."

Even accepting that the claims aren't exaggerated, does the fact that 1, or 100, or even 1000 patients have been cured prove anything? No, it doesn't, for a list of successes doesn't mean very much. To understand why, consider the questions you might pose to an instructor who announced on the first day of class that he or she had given 20 A's last semester. You would want to know how many students were in the class: only 20, or several hundred? You would also want to find out how the rest of the class performed—20 A's and 200 D's might be rather discouraging. You could also check on how the students were selected. If only students with A averages in other courses were allowed to enroll, your opinion should change accordingly. Finally, you might check with the students and compare their grades with those given by other instructors who teach the same course.

With just a couple of modifications, the same questions could be asked about a potential cancer cure:

- How many patients were treated, how many were cured, and how many died?

- How were the patients selected? If selection depended on wealth, degree of illness, or previous exposure to other therapeutic techniques, then the experimental procedure was biased from the start.

- How many might have recovered regardless of the treatment? Even "terminal" cancers sometimes simply disappear for no apparent reason. Such occurrences are rare, but they do happen. Thus, any treatment, however bizarre, will in some cases appear to work. If the frequency of recovery is lower than that among other patient groups, the treatment may actually be harmful despite the reported "cures."

- How do the foregoing statistics compare with those of more traditional therapies when subjected to the same unbiased tests?

THE NEED FOR REPEATABILITY

Finally, let's examine the criterion of repeatability. It's not enough to develop a reasonable, testable hypothesis and collect unbiased data. Consider the hypothesis that every time a coin is tossed, it will come up heads. You could build a coin-tossing machine, turn it on, and find that in the first experiment of 10 tosses, the coin came up heads every time. Does this result prove the hypothesis?

No, it doesn't, despite the fact that it was an honest experiment and the data supported the hypothesis. The problem here is one of statistics, sample size, and luck. The odds that a coin will come up heads on any given toss are 50 percent, or 1 in 2—the same as the odds that it will come up tails. The odds that it will come up heads 10 times in a row are about 1 in 2000—small but certainly not inconceivable. If that coin is tossed 50 times, however, the chance of getting 50 heads drops to 1 in 4,500,000,000,000,000, a figure that most people would accept as vanishingly small. Proving that the hypothesis "a tossed coin always lands heads up" is false requires that the coin come up tails only once. So the truth could be revealed by running the experiment with more coin tosses or by letting other people set up identical experiments and toss their own coins.

The point is that if a hypothesis is correct, anyone and everyone will get the same results when the experiment is performed. If it isn't repeatable, you have to doubt the conclusions even when you have complete confidence in the abilities and integrity of the original investigator.

If a hypothesis satisfies all these criteria—that is, it is testable, unbiased, and repeatable—it can be accepted as a scientific *theory.* The scientific use of this term differs from its use in general conversation. When people discuss "wild-eyed theories," they are usually referring to untested hypotheses. Hypotheses may be true or false, but by definition theories describe real phenomena and make accurate predictions about the world. Examples of scientific theories include the theory of gravity and the theory of evolution. The "fact" of gravity is not in question, and the theory of gravity accounts for the available data. But this does not mean that theories cannot change over time. Newton's original theory of gravity, though used successfully for more than two centuries, was profoundly modified and extended by Albert Einstein. Similarly, the theory of evolution has been greatly elaborated since it was first proposed by Charles Darwin in the middle of the nineteenth century. No one theory can tell the whole story, and all theories are continuously being modified and improved as we learn more about our universe.

⚕ HOMEOSTASIS AND DISEASE
FAP p. 16

The ability to maintain homeostasis depends on two interacting factors: (1) the status of the physiological systems involved and (2) the nature of the stress imposed. Homeostasis is a balancing act, and each of us is like a tightrope walker. Homeostatic systems must adapt to sudden or gradual changes in our environment, the arrival of pathogens, accidental injuries, and many other factors, just as a tightrope walker must make allowances for gusts of wind, frayed segments of the rope, and thrown popcorn.

The ability to maintain homeostatic balance varies with age, general health, and genetic makeup. The geriatric patient or young infant with the flu is in much greater danger than an otherwise healthy young adult with the same viral infection. If homeostatic mechanisms cannot cope with a particular stress, physiological values will drift outside the normal range. This change can ultimately affect all other systems, with potentially fatal results. After all, a person unable to maintain balance will eventually fall off the tightrope.

Consider a specific example. A young adult exercising heavily may have a heart rate of 180 bpm for several minutes. Such a heart rate can be disastrous for an older person with cardiovascular and respiratory problems. If the heart rate cannot be reduced, due to problems with the pacemaking or conducting systems of the heart, the cardiac muscle tissue will be damaged, leading to decreased pumping efficiency and a drop in blood pressure.

These changes represent a serious threat to homeostasis. Other systems soon become involved, and the situation worsens. For example, the kidneys stop working when the blood pressure falls too far, so waste products begin accumulating in the blood. The reduced blood flow in other tissues soon leads to a generalized *hypoxia*, or low tissue oxygen level. Cells throughout the body then begin to suffer from oxygen starvation. The person is now in serious trouble. Unless steps are taken to correct the situation, survival is threatened.

Disease is the failure to maintain homeostatic conditions. The disease process may initially affect a tissue, an organ, or a system, but it will ultimately lead to changes in the function or structure of cells throughout the body. A disease can often be

overcome through appropriate, automatic adjustments in physiological systems. In a case of the flu, the disease develops because the immune system cannot defeat the flu virus before that virus has infected cells of the respiratory passageways. For most people, the physiological adjustments made in response to the presence of this disease will lead to the elimination of the virus and the restoration of homeostasis. Some diseases cannot easily be overcome. In the case of the person with acute cardiovascular problems, some outside intervention must be provided to restore homeostasis and prevent fatal complications.

Diseases may result from the following:

- *Pathogens or parasites that invade the body.* Examples include the viruses that cause flu, mumps, or measles; the bacteria responsible for dysentery or tetanus; and pinworms, flukes, and tapeworms. The invasion process is called **infection**. Some parasites do not enter the body but instead attach themselves to the body surface. This process is called **infestation**.

- *Inherited genetic conditions that disrupt normal physiological mechanisms.* These conditions make normal homeostatic control difficult or impossible. Examples (discussed in later sections) include the *lysosomal storage diseases*, *cystic fibrosis*, and *sickle cell anemia*.

- *The loss of normal regulatory control mechanisms.* For example, cancer involves the rapid, unregulated multiplication of abnormal cells. Many cancers have been linked to abnormalities in genes responsible for controlling the rates of cell division. A variety of other diseases, called *autoimmune disorders*, result when regulatory mechanisms of the immune system fail and normal tissues are attacked.

- *Degenerative changes in vital physiological systems.* Many systems become less adaptable and less efficient as part of the aging process. For example, we experience significant reductions in bone mass, respiratory capacity, cardiac efficiency, and kidney filtration as we age. If the elderly are exposed to stresses that their weakened systems cannot tolerate, disease results.

- *Trauma, toxins, or other environmental hazards.* Accidents may damage internal organs, impairing their function. Toxins consumed in the diet or absorbed through the skin may disrupt normal metabolic activities.

- *Nutritional factors.* Diseases may result from diets inadequate in proteins, essential amino acids, essential fatty acids, vitamins, minerals, or water. Kwashiorkor, a protein deficiency disease, and scurvy, a disease caused by vitamin C deficiency, are two examples. Excessive consumption of high-calorie foods, fats, or fat-soluble vitamins can also cause disease.

Pathology is the study of disease, and *pathophysiology* is the study of functional changes caused by disease. Different diseases may result in the same alteration of function and produce the same symptoms. For instance, a patient who has paler-than-normal skin and complains of a lack of energy and breathlessness may have (1) respiratory problems that prevent normal oxygen transfer to the blood, as in *emphysema*, or (2) cardiovascular problems that interfere with normal oxygen transport (*anemia*) or circulation (heart failure). Clinicians must ask questions and collect appropriate information to make a proper diagnosis. This process often involves eliminating possible causes until a specific diagnosis is reached.

For example, if tests indicate that anemia is responsible for the symptoms, the specific type of anemia must then be determined before treatment can begin. After all, the treatment for anemia due to a dietary iron deficiency is very different from the treatment for anemia due to internal bleeding. Of course, you could not hope to identify the probable cause of the anemia unless you were already familiar with the physical and biochemical structure of red blood cells and with their role in the transport of oxygen. This realization brings us to a key concept: *All diagnostic procedures assume an understanding of normal anatomy and physiology.*

SYMPTOMS AND SIGNS

When disease processes affect normal functions, the alterations are the *symptoms* or *signs* of the disease. An accurate diagnosis, or identification of the disease, is accomplished through the observation and evaluation of signs and symptoms.

A **symptom** is the *patient's perception* of a change in normal body function. Examples of symptoms include nausea, malaise, and pain. Symptoms are difficult to measure, and a clinician must ask appropriate questions. The following are typical questions:

"When did you first notice this symptom?"

"What does it feel like?"

"Does it come and go, or does it always feel the same?"

"Are there things you can do to make it feel better or worse?"

The answers provide information about the duration, sensations, recurrence, and potential triggering mechanisms of the symptoms important to the patient.

Pain, an important symptom of many illnesses, is often an indication of tissue injury. The flow chart in Figure A-2 demonstrates the types of pain and introduces important related terminology. Pain sensations and pathways are detailed in Chapter 17 of the text, and we shall consider the control of pain in related sections of the *Applications Manual* (p. 91).

A **sign** is a physical manifestation of the disease. Unlike symptoms, signs can be measured and observed through sight, hearing, or touch. The yellow color of the skin caused by liver dysfunction or a detectable breast lump are signs of disease. An observable change due to a disturbance in the structure of tissue or cells is called a **lesion**. We shall consider lesions of the skin in detail in a later section dealing with the integumentary system (p. 37).

Steps in Diagnosis

A person experiencing serious symptoms usually seeks professional help and thereby becomes a patient. The clinician, whether a nurse, physician, or emergency medical technician, must determine the need for medical care on the basis of observation and assessment of the patient's symptoms and signs. This is the process of diagnosis: the identification of a pathological process by its characteristic symptoms and signs.

Diagnosis is a lot like assembling a jigsaw puzzle. The more pieces (clues) available, the more complete the picture will be. The process of diagnosis is one of deduction and follows an orderly sequence of steps:

1. *Obtain the patient's medical history.* The medical history is a concise summary of past medical disorders, general factors that may affect the function of body systems, and the health of the patient's family. This information provides a framework for considering the individual's current problem. The examiner gains information about the person's concerns by asking specific questions and using good listening skills. Physical assessment begins here, and this is the time for unspoken questions such as, "Is this person moving, speaking, and thinking normally?" The answers will later be integrated with the results of more-precise observations. Other components of the medical history may include the following:

   - *Chief complaint.* The person, now a patient, is asked to specify the primary problem that requires attention. This is recorded as the *chief complaint.* An example would be the entry "Patient complains of pain in the right lower quadrant."

   - *History of current illness.* Which areas of the body are affected? What kind of functional problems have developed? When did the patient first notice the symptoms? The duration of the disease process is an important

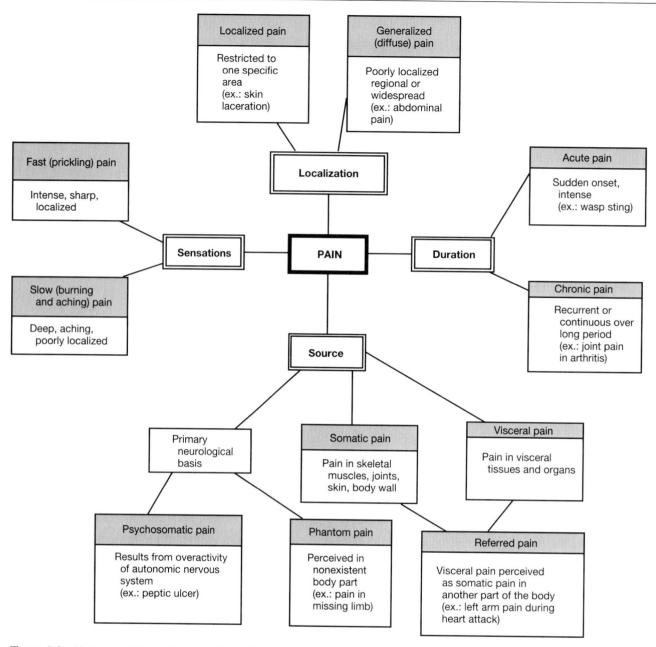

Figure A-2   Methods of Classifying and Describing Pain

factor. For example, an infection may have been present for months, only gradually increasing in severity. This would be called a *chronic infection.* A disease process may have been underway for some time before the person recognizes that a problem exists. Over the initial period, the individual experiences mild *subclinical symptoms* that are usually ignored. Chronic infections have different causes and treatments from *acute infections,* which produce sudden, intense symptoms.

- *Review of systems.* The patient is asked questions that focus on the general status of each body system. This process may detect related problems or causative fac-

tors. For example, a chief complaint of headache pain may be *related* to visual problems (stars, spots, blurs, or blanks seen in the field of vision) or *caused* by visual problems (eyeglasses poorly fitted or the wrong prescription).

2. *Perform a physical examination.* The physical examination is a basic but vital part of the diagnostic process. The common techniques used in physical examination include *inspection* (vision), *palpation* (touch), *percussion* (tapping and listening), and *auscultation* (listening):

- **Inspection** is careful observation. A general inspection involves examining body proportions, posture, and patterns of movements. Local inspection is the examination of sites or regions of suspected disease. Of the four components of the physical exam, inspection is often the most important because it provides the largest amount of useful information. Many diagnostic conclusions can be made on the basis of inspection alone; most skin conditions, for example, are identified in this way. A number of endocrine problems and inherited metabolic disorders can produce subtle changes in body proportions that can be detected by the trained eye.

- In **palpation**, the clinician uses hands and fingers to feel the body. This procedure provides information on skin texture and temperature, the presence of abnormal tissue masses, the pattern of the pulse, and the location of tender spots. Once again, the procedure relies on an understanding of normal anatomy. In one spot, a small, soft, lumpy mass is a salivary gland; in another location it could be a tumor. A tender spot is important in diagnosis only if the observer knows what organs lie beneath it.

- **Percussion** is tapping with the fingers or hand to obtain information about the densities of underlying tissues. For example, when tapped, the chest normally produces a hollow sound, because the lungs are filled with air. That sound changes in pneumonia, when the lungs contain large amounts of fluid. To get the clearest chest percussions, the fingers must of course be placed in the right spots.

- **Auscultation** (aws-kul-TĀ-shun; *auscultare*, to listen) is listening to body sounds, typically using a stethoscope. This technique is particularly useful for checking the condition of the lungs during breathing. The wheezing sound heard in people with asthma is caused by constriction of the airways, and pneumonia produces a gurgling sound, indicating that fluid has accumulated in the lungs. Auscultation is also important in diagnosing heart conditions. Many cardiac problems affect the sound of the heartbeat or produce abnormal swirling sounds during blood flow.

Every examination also includes measurements of certain vital body functions, including body temperature, blood pressure, respiratory rate, and heart (pulse) rate. The results, called **vital signs**, are recorded on the patient's chart. As we noted earlier, each of these values can vary over a normal range that differs according to the age, gender, and general con-

Table A-1 Normal Range of Values for Resting Individuals by Age Group

Vital sign	Infant (3 months)	Child (10 years)	Adult
Blood pressure (mm/Hg)	90/50	125/60	95/60 to 140/90
Respiratory rate (per minute)	30–50	18–30	8–18
Pulse rate (per minute)	70–170	70–110	50–95

dition of the individual. Table A-1 indicates the representative ranges for vital signs in infants, children, and adults.

3. *If necessary, perform diagnostic procedures.* The physical examination alone may not provide enough information to permit a precise diagnosis. Diagnostic procedures can then be used to focus on abnormalities revealed by the physical examination. For example, if the chief complaint is knee pain after a fall, and the physical examination reveals swelling and localized, acute pain on palpation, the *preliminary diagnosis* may be a torn cartilage. An X-ray or MRI scan or both may be performed to ensure that there are no other problems, such as broken bones or torn ligaments. With the information the diagnostic procedure provides, the *final diagnosis* can be made with reasonable confidence. Diagnostic procedures thus extend, rather than replace, the physical examination.

There are two general categories of diagnostic procedures:

1. Tests performed on the individual, generally within a hospital facility. Information on representative tests of this type is summarized in Table A-2. These procedures allow the clinician to:

   - Visualize internal structures (endoscopy, X-rays, scanning procedures, ultrasonography, mammography)

   - Monitor physiological processes (EEG, ECG, PET, RAI, pulmonary function tests)

   - Assess the patient's homeostatic responses (stress testing, skin tests)

2. Tests performed in a clinical laboratory on tissue samples, body fluids, or other materials collected from the patient. Table A-3 (p. 10) includes details on a representative sample of such tests.

Table A-2   Representative Diagnostic Tests, Their Principles and Uses

Procedure	*Principle*	*Examples of Uses*
Endoscopy	Insertion of fiber-optic tubing into a body orifice or through a small incision (laparoscopy and arthrosopy); permits visualization of a body cavity or organ interior	Allows direct visualization of internal structures and detection of abnormalities of surrounding soft tissues Bronchoscopy: bronchi and lungs Laparoscopy: abdominopelvic organs Cystoscopy: urinary bladder Esophagoscopy: esophagus Gastroscopy: stomach Colonoscopy: colon Arthroscopy: joint cavity
Standard X-rays	A beam of X-rays passes through the body and then strikes a photographic plate. Radiodense tissues block X-ray penetration, leaving unexposed (white) areas on the film negative. (FAP, *p. 25*)	Long bones: to detect fracture, tumor, growth patterns Chest: to detect tumors, pneumonia, atelectasis, tuberculosis Skull: to detect fractures, sinusitis, metastatic tumors
Contrast X-rays	X-rays taken after infusion or ingestion of radiodense solutions (FAP, *p. 25*)	Barium swallow (upper GI): series of X-rays after the ingestion of barium, to detect abnormalities of esophagus, stomach, and duodenum Barium enema: series of X-rays after barium enema is given to detect abnormalities of colon IV pyelography: series of X-rays after intravenous injection of radiopaque dye filtered by kidneys; reveals abnormalities of kidneys, ureters, and bladder; allows assessment of renal function Mammogram: X-rays of each breast taken at different angles for early detection of breast cancer and other masses, such as cysts
Computerized tomography (CT or CAT)	Produces cross-sectional images of area to be viewed; together, all sections can produce a three-dimensional image for detailed examination. (FAP, *p. 26*)	CT scans of the head, abdominal region, (liver, pancreas, kidney), chest, and spine, to assess organ size and position, to determine progression of a disease, and to detect abnormal masses
Nuclear scans	Radioisotope ingested or injected into the body becomes concentrated in the organ to be viewed; gamma radiation camera records image on film. Area should appear uniformly shaded; dark or light areas suggest hyperactivity or hypoactivity of the organ.	Bone scan: to detect tumors, infections, and degenerative diseases Scans of the brain, heart, thyroid, liver, spleen, and kidney, to assess organ function and the extent of disease
Radioactive iodine uptake test (RAI)	Radioactive iodine is given orally; scans are taken at 3 different times to determine thyroid percentage uptake of radioiodine	Aids in the determination of a hyperthyroid or hypothyroid condition
Positron emission tomography (PET)	Radioisotopes are given by injection or inhalation; gamma detectors absorb energy and transmit information to computers to generate cross-sectional images	Used to measure metabolic activity of heart and brain and to analyze blood flow through organs

Table A-2 *(continued)*

Procedure	Principle	Examples of Uses
Magnetic resonance imaging (MRI)	A magnetic field is produced to align hydrogen protons and then exposed to radio waves that cause the aligned atoms to absorb energy. The energy is later emitted and captured to produce an image. (FAP, *p. 26*)	MRI gives excellent contrast of normal and abnormal tissue, tumor progression, demyelination, obstructions in arteries, aneurysms, and the extent of organ disease
Ultrasonography	A transducer contacting the skin or other body surface sends out sound waves and then picks up the echoes. (FAP *p. 26*)	Used in obstetrics, to detect ectopic pregnancy, determine fetal size, check fetal rate of growth; upper abdominal ultrasound detects gallstones, visceral abnormalities, and measures kidneys
Echocardiography	Ultrasonagraphy of the heart	Used to assess the structure and function of the heart
Electrocardiography (ECG)	Graphed record of the electrical activity of the heart, using electrodes on the skin surface	Useful in detection of arrhythmias, such as premature ventricular contractions (PVCs) and fibrillation, and to assess damage after myocardial infarction
Electroencephalo-graphy (EEG)	Graphed record of electrical activity in the brain through the use of electrodes on the surface of the scalp	Analysis of brain wave frequency and amplitude aids in the diagnosis of tumors, seizure disorders, and strokes
Electromyography (EMG)	Graphed record of electrical activity resulting from skeletal muscle contraction, using electrodes inserted into the muscles	Determination of neural or muscular origin or muscle disorder; aids in the diagnosis of muscular dystrophy, myasthenia gravis, and other neuromuscular disorders
Pulmonary function tests	Measurement of lung volumes and capacities by a spirometer or other device	Aids in the differentiation between obstructive and restrictive lung diseases
Pap smears	Removal of cells for laboratory analysis	Detection of precancerous cells or infections; most often used to assess mucosal cells of cervix
Stress testing	Monitoring of blood pressure, pulse rate, and ECG during exercise; may include intravenous injection of radioisotopes	Aids in the determination of the extent of coronary artery disease, which may not be apparent while the individual is at rest
Skin tests	Injection of a substance under the skin, or placement of a substance on the skin surface, to determine the response of the immune system	Tuberculin test: injection of tuberculin protein under skin Allergen test: injection of allergen or application of a patch containing allergen

THE PURPOSE OF DIAGNOSIS

Several hundred years ago, a physician would arrive at a final diagnosis and consider the job virtually done. Once the diagnosis was made, the patient and family would know what to expect. In effect, the physician was more of an oracle than a healer. Wounds could be closed and limbs amputated, but few effective treatment options were available. Therapy consisted largely of some combination of cupping and bleeding, typically performed by barbers rather than by surgeons. In *cupping*, a glass cup created suction at a portion of the body surface. This procedure was done to bring blood to an area of intact skin (*dry cupping*) or to bring blood to a cut (*wet cupping*).

We have an incredible variety of treatment options today, and a final diagnosis is vital because it determines which treatment options will be selected. A modern physician with a new patient follows the *SOAP* protocol:

S is for *subjective*. The clinician obtains subjective information from the patient and completes the medical history.

O is for *objective*. The clinician now performs the physical examination and obtains objective information about the physical condition of the patient. This may include the use of diagnostic procedures.

A is for *assessment*. The clinician arrives at a diagnosis and, if necessary, reviews the literature on the condition. A preliminary con-

Table A-3 Laboratory Tests Performed on Samples Taken from the Body

BLOOD TESTS: Serum, plasma, or whole blood samples can be evaluated. Depending on the blood constituent or chemical being monitored, venous or arterial blood will be taken.

Laboratory Test	Significance	Notes
Complete blood count (CBC) RBC count Hemoglogin (Hb, Hgb) Hematocrit (Hct)	Data from this test series inform the practitioner about a change in the number of red blood cells; changes may indicate the presence of disease, hemorrhaging, or other problems. A CBC is generally performed during a normal physical exam to give the practitioner more information about the patient's general health. Changes may indicate blood loss, infections, or other problems.	For more information see Table A-19c, p. 112.
RBC indices: Mean corpuscular hemoglobin (MCH) Mean corpuscular hemoglobin concentration (MCHC)	Provide information about the status of hemoglobin production and RBC maturation	See Table A-19c, p. 112.
WBC count Differential WBC count	The white blood cell count reflects the body's immune system and the ability to fight infection. Increased WBC count could indicate the presence of infection.	See Table A-19c, p. 112.
Hemostasis tests: Platelet count Bleeding time Factors assay Plasma fibrinogen Plasma prothrombin time (PT) Plasminogen	A decreased number of platelets could result in uncontrolled bleeding. Other constituents, such as fibrinogen, clotting factors, and prothrombin, also contribute to the clotting process, and these can be assessed separately.	See Table A-19c, p. 113.
Serum electrolytes: Sodium Potassium Chloride Bicarbonate	Sodium, potassium, and chloride levels are important because these electrolytes function in nerve transmission, skeletal muscle contraction, and cardiac rhythm. Abnormal levels of bicarbonate indicate problems with acid–base balance.	See Tables A-19b, A-27, A-28, pp. 111, 168, 186.
Iron (serum)	Decreased levels cause iron deficiency anemia; increased levels may cause liver damage.	
Arterial blood gases and pH: pH P_{CO_2} P_{O_2}	Respiratory acidosis and alkalosis can be monitored with these values. Decreased oxygen levels occur in respiratory system dysfunction.	See Table A-25, p. 152.
Hemoglobin electrophoresis: Hemoglobin A Hemoglobin F Hemoglobin S	Electrophoresis separates the different types of hemoglobin for quantitative measurement. Abnormal levels of hemoglobin occur in many anemias. Abnormal types of hemoglobin occur in sickle cell anemias.	See Table A-19c, p. 113.
ABO and Rh typing	Blood typing is critical for correct matching of blood types prior to transfusion. Rh typing during pregnancy is important to determine risk of fetal–maternal Rh incompatibility.	See Table A-30, p. 209.
Serum cholesterol	Elevated cholesterol levels reflect the potential for atherosclerosis and coronary artery disease.	See Table 19a, p. 109.
Serum lipoproteins	Electrophoresis is used to separate the LDL fraction to determine the HDL and LDL levels in the assessment of the potential for coronary artery disease.	See Table 19a, p. 109.

Table A-3 *(continued)*

Laboratory Test	Significance	Notes
Enzymes: Creatine phosphokinase (CPK or CK) Isoenzymes (CPK-MM, CPK-MB, CPK-BB)	Abnormal enzyme levels in the blood are generally due to cellular damage. CPK-MM is useful in the diagnosis of muscle disease; CPK-MB is used in the diagnosis of myocardial infarction.	See Table A-19b, p. 111.
Aspatate aminotransferase (AST)	AST levels are important to assess following a myocardial infarction or liver damage	
Lactate dehydrogenase (LDH)	Different isoenzymes of LDH can be useful in the detection of heart damage, liver problems, and pulmonary dysfunction	
Rheumatoid factor	Measures presence of antibodies characteristic of rheumatoid arthritis and (less often) systemic lupus erythematosus and other autoimmune diseases	See Table A-10, p. 49.
Hormones	Increased or decreased levels reflect endocrine system disorders	See Table A-18, p. 99.
Blood urea nitrogen (BUN)	Used to assess kidney function	See Table A-28, p. 187.
Immunoglobin electrophoresis (IgA, IgG, IgD, IgE, IgM)	Monitoring of infections and allergic response	See Table A-21, p. 134.
Alcohol	To determine level of intoxication or detect metabolic problems or poisoning	
Human chorionic gonadotropin (hCG)	To determine pregnancy	See Table A-30, p. 208.
Phenylketonuria	To detect a genetic disorder of amino acid metabolism, phenylketonuria (PKU)	
Alpha fetoprotein	To identify probability of fetal defects or presence of twins	
Glucose tolerance test	To detect hyperglycemia (diabetes mellitus) and other abnormalities	See Table A-18, p. 101.
Blood culture	Presence of pathogen occurs in septicemia or other infectious disorders	See Tables A-19c, A-21, A-27, pp. 112, 134.

URINE TESTS: A single urine sample may be tested, or urine may be collected over a period of time and tested (usually from 2 to 24 hours). A routine urinalysis aids in the detection of kidney dysfunction as well as metabolic imbalances and other disorders. Presence of abnormal cellular constituents in urine indicates urinary system disorder, including infection, inflammation, or the existence of a tumor.

Laboratory Test	Significance	Notes
Creatine clearance	Abnormal values indicate reduced renal function	
Urine electrolytes: Sodium Potassium	Abnormal levels reflect fluid or electrolyte imbalances and the effects of hormones on renal function	See Table A-28, p. 186.
Uric acid	Increased levels occur with gout	See Table A-28, p. 187.
Human chorionic gonadotropin (hCG)	Determination of pregnancy	See Table A-30, p. 208.
Urine culture	Pathogens may be present in urinary tract infections	

Table A-3 *(continued)*

OTHER LABORATORY TESTS: Additional laboratory tests can be used to monitor other body fluids, excretory products, or tissues. Here are several examples.

Laboratory Test	*Significance*	*Notes*
Cerebrospinal fluid	Analysis of CSF pressure, color, sugar and protein content, and the presence of antibodies or pathogens	See Table A-12, p. 71.
Stool sample	Culturing of sample to identify microorganisms and determine their antibiotic sensitivities	See Table A-27, p. 168.
Semen analysis	Useful in diagnosis of male infertility	See Table A-29, p. 196.
Tissue biopsy	Removal of tissue for microscopic examination	See Table A-11, p. 61.

clusion as to the **prognosis** (probable outcome) is made.

P is for *plan.* A treatment plan is designed. This can be very simple (take two aspirin) or very complex (radiation, chemotherapy, or surgery). If the treatment is complex, one or more treatment options are usually prepared for review by the patient and, in many cases, the patient's family. The options are discussed and the treatment plan finalized.

As you may have noticed, these are precisely the steps you followed in diagnosing the flu at the very beginning of this section: subjective (you felt ill), objective (flushed face, fever), assessment (flu-like symptoms), and plan (take medicine). The SOAP protocol is both simple to remember and remarkably effective.

The primary goal of an introductory anatomy and physiology course is to provide you with the foundation for other, more specialized courses. In the sections of this manual that deal with body systems, you will be introduced to clinical conditions that demonstrate the relationships between normal and pathological anatomy and physiology. The sections on diagnostic procedures and the Case Study exercises are intended to demonstrate how information can be extracted and organized to reach a reasonable tentative diagnosis. The goal is to acquaint you with the mechanics of the process. This knowledge will not enable you to make accurate clinical diagnoses, for situations in the real world are much more complicated and variable than the examples provided here. Making an accurate clinical diagnosis is generally a complex process that demands a far greater level of experience and training than this course can provide.

For similar reasons, we will not discuss detailed treatment plans; the treatment of serious diseases requires specialized training and competence in advanced biochemistry, pharmacology, microbiology, pathology, and other clinical disciplines. However, many of the discussions in later sections include information about the use of specific drugs and other therapeutic procedures in the treatment of disease. These are representative examples intended to show potential treatment strategies rather than to endorse specific protocols and therapies.

Applied Research Topics

The Application of Principles in Chemistry

Cells, tissues, organs, and organ systems are composed of chemicals. The survival of cells, tissues, organs, and systems depends on the control of chemical reactions, both within individual cells and in the extracellular fluids of the body. It is therefore not surprising that you cannot understand physiological principles without a familiarity with basic chemistry. As our understanding of physiological mechanisms has improved, physicians have become relatively adept at using chemical tests to diagnose disease. Physicians have also developed ways of manipulating intracellular and extracellular chemical reactions to help restore homeostasis. In this section, we will consider the practical application of some basic chemical principles introduced in Chapter 2 of the text.

▢ MEDICAL IMPORTANCE OF RADIOISOTOPES   FAP *p. 32*

Many recent technological advances in medicine have involved the use of radioisotopes for diagnosis and the visualization of internal structures. In this section, we will focus on two examples:

1. *The use of radioactive tracers in diagnosis.* Radioisotopes can be attached to organic or inorganic molecules and injected into the body. Within the body, these labeled compounds emit radiation energy that can be used to create images. These images provide information about tissue structure, tumorous growths, blocked or weakened blood vessels, and other abnormalities in the body.

2. *The use of radioactive compounds to treat disease.* If a suitable radiation source can be accurately delivered to a target site, the radioactivity can be used to destroy abnormal cells or tissues.

Radioisotopes and Clinical Testing

A **radioisotope,** or radioactive isotope, is an isotope whose nucleus is unstable—that is, the nucleus spontaneously decays, or emits subatomic particles, in measurable amounts. **Alpha particles** are generally emitted by the nuclei of large radioactive atoms, such as uranium. Each alpha particle consists of a helium nucleus: two protons and two neutrons. **Beta particles** are electrons, more typically released by radioisotopes of lighter atoms. **Gamma rays** are very-high-energy electromagnetic waves comparable to the X-rays used in clinical diagnosis.

The **half-life** of any radioactive isotope is the time required for half of a given amount of the isotope to decay. The half-lives of radioisotopes range from fractions of a second to billions of years.

Gamma rays, beta particles, and alpha particles—like X-rays—can damage or destroy living tissues. The danger posed by radiation exposure varies with the nature of the emission and the duration of exposure. But radiation also has a variety of beneficial uses in medical research and clinical diagnosis. Weakly radioactive isotopes with short half-lives can sometimes be used to check the structural and functional state of an organ without surgery.

Radioisotopes can be incorporated into specific compounds normally found within the body. These compounds, called **tracers,** are said to be labeled: When introduced into the body, labeled compounds can be tracked by the radiation they release. After a labeled compound is swallowed, its uptake, distribution, and excretion can be determined by monitoring the radioactivity of samples taken from the digestive tract, body fluids, and waste products. For example, compounds labeled with radioisotopes of cobalt are used to monitor the intestinal absorption of vitamin B_{12}. Normally, cobalt-58, a radioisotope with a half-life of 71 days, is used.

Radioisotopes can also be injected into the blood or other body fluids to provide information on circulatory anatomy and the anatomy and function of specific target organs. In **nuclear imaging,** the radiation emitted by injected radioisotopes creates an image on a special photographic plate. Such a procedure is used to identify regions where particular radioactive materials are concentrated or to check the circulation through vital organs. Radioisotopes can also produce pictures of specific organs, such as the liver, spleen, or thyroid, where labeled compounds are removed from the circulation.

The thyroid gland (Figure A-3a) sits below the larynx (voicebox) on the anterior portion of the neck. A normal thyroid gland absorbs iodine, which is then used to produce thyroid hormones. As a result, the thyroid gland will actively absorb and concentrate radioactive iodine. The **thyroid scan** in Figure A-3b was taken following the injection of iodine-131, a radioisotope with an 8-day half-life. This procedure, called a *thyroid radioactive iodine uptake measurement,* or **RAIU,** can provide information about (1) the size and shape of the gland and (2) the amount of absorptive activity under way. Comparing the rate of iodine uptake with the level of circulating hormones makes possible the evaluation of the functional state of the gland.

Radioactive iodine is an obvious choice for imaging the thyroid gland. For most other tissues and organs, a radioactive label must be attached to another compound. *Technetium* (^{99m}Tc), a versatile label, is the primary radioisotope used in nuclear imaging today. The isotope is artificially produced and has a half-life of 6 hours. This brief half-life significantly reduces the radiation exposure of the patient. Technetium is used in more than 80 percent of all scanning procedures. The nature of the technetium-labeled compound varies with the identity of the target organ. Technetium scans are performed to examine the thyroid gland, spleen, liver, kidneys, digestive tract, bone marrow, and a variety of other organs.

PET (**P**ositron **E**mission **T**omography) scans utilize the same principles as standard radioisotope scans, but the analyses are performed by computer. The scans are much more sensitive, and the computers can reconstruct sections through the body that permit extremely precise localization. Among other things, this procedure can analyze blood flow through organs and assess the metabolic activity within specific portions of an organ as complex as the human brain.

Figure A-3c is a PET scan of the brain showing activity at a single moment in time. The scan is dynamic, however, and changing patterns of activity can be followed in real time. PET scans can be used to analyze normal brain function as well as to diagnose brain disorders. To date, the technique has served primarily as a research tool. Because the equipment is expensive and bulky, it is unlikely to be available anywhere except in large, regional medical centers or universities. The research

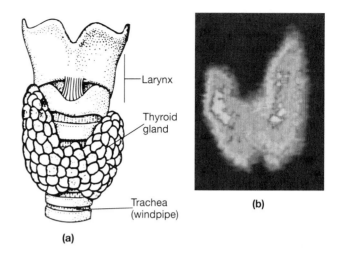

(a)

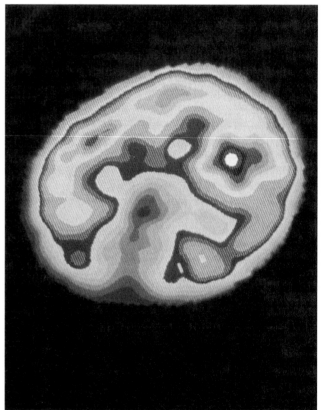

(b)

(c)

Figure A-3   Imaging Techniques.
(a) The position and contours of the normal thyroid gland as seen in dissection. (b) After it has been labeled with radioactive iodine, the thyroid can be examined by special imaging techniques. In this computer-enhanced image, different intensities indicate differing concentrations of the radioactive tracer. (c) A PET scan of the left cerebral hemisphere in lateral view. The light areas indicate regions of increased metabolic activity.

advantages of PET scans have diminished considerably since the advent of real-time CT analysis (*cine-CT*) and the realization that MRI can be used to monitor small changes in blood flow and tissue activity without the use of radioactive tracers.

Radiopharmaceuticals

Nuclear medicine involving injected radioisotopes has been far more successful in producing useful images than in treating specific disorders. The problem is that relatively large doses of radiation must be used to destroy abnormal or cancerous tissues, and it is very difficult to control the distribution of these radioisotopes in the body with sufficient precision. As a result, radiation exposure may damage both normal and abnormal tissues. For the same reason, it is difficult to control the radiation dosage administered to the target tissues. This is a problem because underexposure can have very little effect, whereas overexposure can destroy adjacent normal tissues.

Radioactive drugs, or *radiopharmaceuticals*, can be effective only if they are delivered precisely and selectively. One success story has been the treatment of *hyperthyroidism*, or thyroid over-secretion. As we noted earlier, the thyroid gland selectively concentrates iodine. Large doses of radioactive iodine (^{131}I) can be administered to treat hyperthyroidism. The radiation released destroys the abnormal thyroid tissue and stops the excessive production of thyroid hormones. (Most individuals then become *hypothyroid*—deficient in thyroid hormone—but this condition can be treated by taking thyroid hormones in tablet form.) This is now the preferred treatment method for hyperthyroid patients over 40 years of age.

A relatively new application of nuclear medicine involves the attachment of a radioactive isotope to a *monoclonal antibody (MoAb)*. Antibodies are proteins produced in the body to provide a selective defense against foreign proteins, toxins, or pathogens. A substance that triggers antibody production is called an *antigen*. Monoclonal antibodies are produced by culturing immune cells that are sensitized to a particular antigen. (The process is detailed on p. 146.) The antibodies these cells manufacture can be extracted, labeled with radioactive materials, and concentrated. If injected into the body, the antibodies will bind to their target antigens and expose the surrounding tissues to radiation.

MoAbs specific for the antigens on certain types of tumor cells have already been approved by the Food and Drug Administration (FDA). When injected into the body, the radiolabeled MoAbs travel to the tumor site and attach to the specific antigen displayed on the cancer cell surface. Upon attachment, radiation is emitted from the target area. The amount of radiation emitted is low, however, and the procedure is used to produce images rather than to treat the disease. This technique is very sensitive and can detect small tumors for early diagnosis and treatment. Experiments continue, with the eventual goal of using radiolabeled MoAbs to destroy tumor cells.

📖 SOLUTIONS AND CONCENTRATIONS FAP *pp. 33, 43*

Physiologists and clinicians pay particular attention to ionic distributions across membranes and the electrolyte composition of body fluids. Standard values for physiological tests are provided throughout the text and are summarized in Appendix VI. Data must be analyzed from several different perspectives, and physiological values may be reported in several different ways. One method is to report the concentration of atoms, ions, or molecules in terms of weight per unit volume of solution. Although grams per liter (g/l) may be used, values are most often expressed in terms of grams (g), milligrams (mg), or micrograms (μg) per 100 ml. Because 100 ml is 0.1 liter, or 1 deciliter **(dl)**, the abbreviations most often used in this text are **g/dl** and **mg/dl**.

Osmolarity depends on the total number of individual atoms, ions, and molecules in solution, without regard to molecular weight, electrical charge or molecular identity. As a result, if fluid balance and osmolarity are being monitored, concentrations are usually reported in terms of moles per liter (mol/l or M) or millimoles per liter (mmol/l or mM) rather than in terms of g/dl or mg/dl. To convert from g/dl to mol/l, multiply by 10 and divide by the atomic weight of the element. For example, a sample of plasma (blood with the cells removed) contains sodium ions at a concentration of roughly 0.32 g/dl (320 mg/dl). We convert this value to mmol/l as follows:

$$\frac{\text{g/dl} \times 10}{\text{atomic weight}} = \frac{0.32 \times 10}{22.99}$$
$$= 0.14 \text{ mol/l (or 140 mmol/l)}$$

Moles or millimoles per liter can also be used to indicate the concentration of molecules in solution. We can perform the same conversion by substituting molecular weight for atomic weight in the above equation. The total solute concentration of a solution can be determined by adding the concentrations of individual solutes, expressed in moles per liter or millimoles per liter. The resulting value is reported in **milliosmoles per liter (mOsm/l)**. The use of mOsm rather than mmol indicates that multiple solutes are present, each contributing to the total osmolarity.

Because electrolyte concentrations have profound effects on living cells, it is often important to know how many positive and negative charges are present in a biological solution. In this case, the important question is not just how many ions or molecules are present, but how many positive or negative charges they bear. For example, a single calcium ion (Ca^{2+}) has twice the electrical charge of a single sodium ion (Na^+), although the two are identical in terms of their effects on osmolarity. One **equivalent (Eq)** is a mole of positive or negative charges, and physiological concentrations are often reported in terms of **milliequivalents per liter (mEq/l).** You should become familiar with both methods of expression. Fortunately, the conversion from millimoles to milliequivalents is relatively easy to perform. For **monovalent ions,** those with a +1 or –1 charge, millimole and milliequivalent values are identical, and no calculation is needed. For **divalent ions,** with +2 or –2 charges, the number of charges (mEq) is twice the number of ions (mmol). For an ion with a +3 or –3 charge, the number of milliequivalents is three times the number of millimoles. To convert mEq to mmol, simply divide by the ionic valence (number of charges).

Table A-4 compares the different methods of reporting the concentration of major electrolytes in plasma in terms of weight, moles, and equivalents; the notation "nr" indicates that concentrations are not reported in those terms. The tables included in Appendix VI of the text provide data in terms currently accepted for clinical laboratory reports.

There is no doubt that physiologists and clinicians would benefit from the use of standardized reporting procedures. It can be very frustrating to consult three references and find that the first reports electrolyte concentrations in mg/dl, the second in mmol/l, and the third in mEq/l. In 1984, the American Medical Association House of Delegates endorsed a plan to standardize clinical test results through the use of **SI** (Système Internationale) units, with a target date of July 1, 1987, for the switchover. Unfortunately, there was no mechanism for enforcing compliance, and the standardization attempt ultimately failed. As of 1997, all scientific and medical journals around the world report data in SI units, but most clinical laboratories continue to use their traditional reporting methods.

The major problem is that the relationships to values currently in use are difficult to remember.

Electrolyte concentrations, now most often given in mEq/l, will be reported in mmol/l in the SI. That means the values for sodium and potassium concentrations remain unchanged, but the normal values for calcium and magnesium are reduced by 50 percent. The situation becomes more confusing in terms of metabolite concentrations. Cholesterol and glucose concentrations are now most often reported in mg/dl, but the SI units are mmol/l. However, total lipid concentrations, also currently reported as mg/dl, and total protein concentrations, now given as g/dl, are reported in terms of g/l under the SI. For these units to be useful in a clinical setting, physicians must not only remember the definition of each SI unit but must also convert and relearn the normal ranges. As a result, it appears unlikely that the conversion to SI units will be completed in the immediate future.

℞ THE PHARMACEUTICAL USE OF ISOMERS
FAP p. 46

A chemical compound is a combination of atoms bonded together in a particular arrangement. The chemical formula specifies the *number* and *types* of atoms that combine to form the compound. The *arrangement* of the atoms, which determines the specific shape of each molecule, is shown by the molecule's structural formula.

Isomers are chemical compounds that have the same chemical formula but different structural formulas. Some isomers are mirror images of each other. These are called *stereoisomers*. Stereoisomers are analogous to the left hand and right hand of the human body. The two hands contain the same palm and finger bones (metacarpals and phalanges), but each hand is a mirror image of the other. A glove designed for the left hand will not fit the right hand, and vice versa. Chemical compounds are also said to be left-handed or right-handed, depending on their structural configuration. For example, there are left *(levo)* isomers and

---

Table A-4   A Comparison of Methods for Reporting Concentrations of Solutes in the Blood*

Solute	mg/dl	mmol/l	mEq/l	SI Units
Electrolytes				
Sodium (Na$^+$)	320	140	140	140 mmol/l
Potassium (K$^+$)	16.4	4.2	4.2	4.2 mmol/l
Calcium (Ca^{2+})	9.5	2.4	4.8	2.4 mmol/l
Chloride (Cl$^-$)	354	100	100	100 mmol/l
Metabolites				
Glucose	90	5	nr	5 mmol/l
Lipids, total	600	nr	nr	0.6 g/l
Proteins, total	7 g/dl	nr	nr	70 g/l

***nr = not reported in these units.**

right (dextro) isomers of glucose. Just as the right hand cannot fit into a left glove, receptors and enzymes within our cells cannot bind the levo-isomer of glucose. Our cells are therefore unable to metabolize levo-glucose as an energy source.

This is a common pattern: Our cells and tissues will typically respond to only one structural form—either levo or dextro—but not to both. This feature can pose a problem for pharmaceutical chemists, because many of the chemical reactions used to synthesize a drug produce a mixture of levo- and dextro-isomers. In some cases, the inactive isomer is simply ignored; in others, the inactive isomer is removed. For example,

- The antibiotic *chloramphenicol* contains both levo- and dextro-isomers, but only the levo form is effective in killing bacterial pathogens.

- *Ephedrine* is a drug used to produce bronchiolar dilation; it is often administered to treat asthmatic attacks. The popular tablet *Primatene®* contains only the levo form of ephedrine, which is the active form.

In some cases, both forms of an isomer are biologically active but have strikingly different effects. The drug *Thalidomide* was given to pregnant women in the 1960s to alleviate symptoms of morning sickness. The sedative effect of one isomer was well documented, but the medication sold contained both forms. Unfortunately, the other isomer caused tragic abnormalities in fetal limb development. (We discuss the mechanisms that underlie Thalidomide's effects on fetal development on p. 204.)

TOPICS IN METABOLISM

Metabolism is the sum of all the biochemical reactions that proceed in the body. Hundreds of thousands of reactions occur in each cell. At any moment, biochemical pathways may be producing phospholipids for the cell membrane or peptide hormones for secretion, while breaking down carbohydrates to generate ATP. In this section, we will consider three aspects of metabolism:

1. Many disease processes are the result of a faulty biochemical pathway. For example, an enzyme may be missing or nonfunctional, or the necessary enzymatic substrates may be unavailable. *Phenylketonuria, galactosemia, albinism,* and *hypercholesterolemia* are metabolic disorders that we will consider.

2. Enzymes play a pivotal role in controlling metabolic processes in our cells. The mechanisms responsible for controlling enzymatic reactions are therefore important, and problems with enzymatic regulation can cause severe metabolic disorders.

3. Diet and nutrition have an obvious impact on metabolic operations within the body. A substantial research effort is under way to manipulate metabolic operations by dietary changes. The control of cholesterol in the diet is only one of several pertinent examples.

® Artificial Sweeteners   FAP *p. 46*

Some people cannot tolerate sugar for medical reasons; others avoid it to comply with recent dietary guidelines that call for reduced sugar consumption or to lose weight. Thus, many people today use artificial sweeteners in their foods and beverages.

Artificial sweeteners are organic molecules that can stimulate taste buds and provide a sweet taste to foods without adding substantial amounts of calories to the diet. These molecules have a much greater effect on the taste receptors than do natural sweeteners, such as fructose or sucrose, so they can be used in minute quantities. For example, *saccharin* is about 300 times as sweet as sucrose. The popularity of this sweetener has declined since it was reported that saccharin may promote bladder cancer in rats. The risk is very small, however, and saccharin continues to be used. Several other artificial sweeteners are currently on the market, including *aspartame (NutraSweet®)*, *sucralose*, and *acesulfame potassium (Ace-K, or Surette®)*. The market success of an artificial sweetener ultimately depends on its taste and its chemical properties. Stability in high temperatures (as in baking) and resistance to breakdown in an acidic pH (as in carbonated drinks) are important properties for any artificial sweetener.

Molecules of artificial sweeteners do not resemble those of natural sugars. Saccharin, acesulfame potassium, and sucralose cannot be broken down by the body and have no nutritional value. Aspartame consists of a pair of amino acids. Amino acids are the building blocks of proteins (as we will discuss later in this chapter), and they can be broken down in the body to provide energy. However, because aspartame is 200 times as sweet as sucrose, very small quantities are needed, so the sweetener adds few calories to a meal. Aspartame does not produce the bitter aftertaste sometimes attributed to saccharin, so aspartame is used in many diet drinks and low-calorie desserts.

Two recent entries into the market for artificial sweeteners, *thaumatin-1* and *monellin*, are proteins extracted from African berries. Thaumatin, roughly 100,000 times as sweet as sucrose, has been approved by the Food and Drug Administration for use in chewing gums. Another artificial sweetener, *cyclamate*, was banned in 1970 after experiments suggested that it caused bladder tumors in laboratory rats. Those conclusions have been shown to be incorrect, so cyclamates may soon be reapproved. However, because this sweetener is only about 30 times as sweet as sucrose, it may not have much impact on the marketplace.

℞ Fat Substitutes   FAP *p. 47*

Although the average diet in the United States is not as rich in fats as that of Eskimos, Americans consume more fat than do people in many other parts of the world. Diets high in fat have been linked not only to heart disease but also to certain forms of cancer. Recent recommendations suggest that lowering the percentage of calories we derive from fat would benefit our health. This suggestion has led to an increased interest in the development of fat substitutes.

Fat substitutes provide the texture, taste, and cooking properties of natural fats. Two fat substitutes, *Simplesse®* and *Olestra®*, have been approved by the Food and Drug Administration; a third, *Trailblazer®*, is currently under review. Both Simplesse and Trailblazer are made from proteins of egg white and skim milk or whey. The heated proteins are treated to form small spherical masses that have the taste and texture of fats. Simplesse can be used in place of fats in any application other than baking; it is found in low-calorie "ice creams" under the trade name *Simple Pleasures®*. These fat substitutes can be broken down in the body, but they provide less energy than do natural fats. For example, ice cream with Simplesse has half the calories of ice cream that contains natural fats.

Olestra is made by chemically combining sucrose and fatty acids. The resulting compounds cannot be used by the body and so contribute no calories. Olestra has been approved as an ingredient in margarines, baked goods, and other snack foods; its use as a shortening and cooking oil remains under review. One of the problems is that Olestra droplets within the digestive tract collect lipid-soluble materials, including fat-soluble vitamins, and prevent their absorption. In addition, if eaten in large quantities, Olestra can cause diarrhea. These side effects pose a serious threat—a combination of fluid loss and vitamin deficiency. To prevent vitamin deficiencies among consumers, manufacturers of snack foods prepared with Olestra now fortify them with fat-soluble vitamins.

Olestra is not the only approved fat substitute derived from carbohydrates. *Oatrim*, a fat substitute derived from soluble fiber and complex carbohydrates, is now used in muffins, cookies, fat-free cheeses, and lean hot dogs and luncheon meats.

Fat substitutes provide the texture of natural fats, but the flavor is generally quite different. For this reason, efforts are underway to develop "designer fats" using fatty acids that provide fewer calories than do the lipids normally found in the diet. For example, the fats in *SALATRIM®* contain long-chain fatty acids that are poorly absorbed and short-chain fatty acids that provide relatively few calories when broken down. On a gram-for-gram basis, the fats in SALATRIM provide only 60 percent of the calories of natural triglycerides.

Fat substitutes must be tested and approved by the FDA. To receive FDA approval, a substance must be proven harmless to laboratory animals when it is consumed at levels 100 times the expected dosage in humans. (This requirement is currently under review, as even otherwise harmless compounds can cause adverse effects at such high doses.)

The use of fat substitutes may also pose secondary metabolic problems. Fat-soluble vitamins (A, D, E, and K) are normally absorbed by the intestinal tract in company with dietary lipids. A drastic reduction in the lipid content of the diet may therefore lead to deficiencies of these vitamins.

📖 A CLOSER LOOK: THE CONTROL OF ENZYME ACTIVITY   FAP *p. 56*

Four major factors determine the rate at which a particular enzymatic reaction occurs: (1) *substrate and product concentrations,* (2) *enzyme concentration,* (3) *competitive inhibition,* and (4) *enzyme activation states.* We will examine each of these factors individually.

Substrate and Product Concentrations

Consider a situation in which there are many enzymes but no substrate molecules. No reaction occurs. Figure A-4a plots the effect of increasing substrate concentrations on the reaction rate. At low concentrations, substrate availability limits the reaction rate. The higher the substrate concentration, the faster the reaction proceeds. This acceleration does not continue. As product concentration rises, the chances increase that a product molecule will contact the active site instead of substrate molecules. When all the active sites are bound to substrate or product molecules, the enzyme system is **saturated.** Any further increase in substrate concentration will have no effect on the reaction rate. The graph in Figure A-4a is a typical enzyme **saturation curve.**

Enzyme Concentration

The concentration of enzymes has a direct effect on the rate of the reaction (Figure A-4b). The higher the enzyme concentration, the faster the initial reaction rate. Varying rates of enzyme synthesis or destruction may change enzyme concentrations in a cell. Changing the cytoplasmic concentration of enzymes is a slow process. During that time, turning enzymes ON or OFF regulates the concentration of functional enzymes.

Competitive Inhibition

Enzyme specificity results from the shape and charge characteristics of the active site. That specificity is not perfect. Molecules that closely resemble the normal substrate can bind to the active site and interfere with substrate binding.

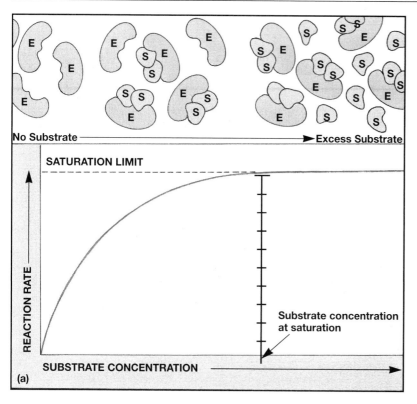

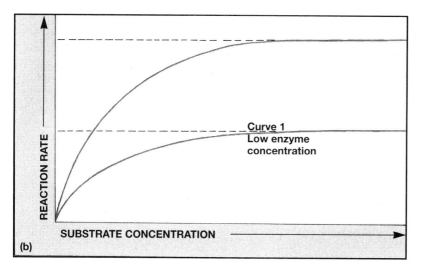

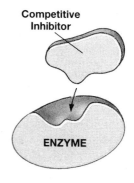

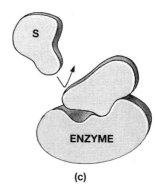

(c)

Figure A-4 Factors That Affect Enzymatic Reaction Rates.

(a) A saturation curve, showing the effects of changing substrate concentrations. Here E = enzyme and S = substrate. (b) The effect of altering the concentration of enzymes on a reaction rate. Curve 2 has a higher enzyme concentration and therefore a faster initial reaction rate than curve 1 has. (c) A competitve inhibitor blocks the enzyme's active site, so the substrate cannot bind. (d) The effects of competitive inhibitors on reaction rates.

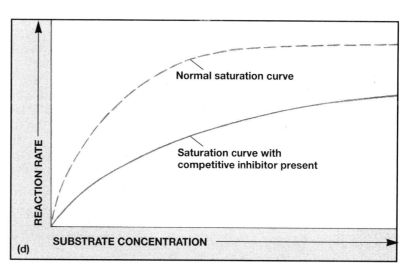

This process is called **competitive inhibition,** because substrate molecules must compete for a space on the active site. The higher the concentration of competitive inhibitors, the lower the rate of reaction. Figure A-4c,d indicates the effects of competitive inhibition on reaction rates.

Enzyme Activation States

An activated enzyme will catalyze a particular reaction; an inactivated enzyme will not. In an inactive enzyme, the active site cannot interact with substrate molecules. During activation, the shape of the active site changes. Four types of factors may activate or inactivate a specific enzyme:

1. *Physical factors.* Environmental factors such as high or low temperature or a change in pH can alter the tertiary or quaternary structure of the enzyme. Variations outside normal limits will temporarily or permanently inactivate enzymes. For example, body temperature regulation is essential because temperature directly affects enzyme activity. At high body temperatures (over 40°C), enzymes begin to denature, becoming permanently nonfunctional. Enzymes are equally sensitive to pH changes. Each enzyme works best at an optimal combination of pH and temperature. For example, *pepsin,* an enzyme that breaks down proteins in the stomach contents, works best at a pH of 2.0 (strongly acidic). The small intestine contains *trypsin,* another enzyme that attacks proteins. Trypsin works only in an alkaline environment, with an optimum pH of 9.5.

2. *The presence or absence of cofactors.* **Cofactors** are ions or molecules that must attach to the active site before substrate binding can occur (Figure A-5). A **holoenzyme** is an enzyme activated by an appropriate cofactor. An enzyme without its cofactor is an inactive **apoenzyme** (ap-ō-EN-zīm). Examples of cofactors include mineral ions, such as calcium (Ca^{2+}) and magnesium (Mg^{2+}), and several vitamins.

3. *Allosteric effects.* An **allosteric** (*allos,* "other" + *stereo,* "solid") **effect** is a change in the shape of the active site caused by interaction between an enzyme and some other molecule. The molecule involved is called a **modulator. Activators** are modulators that turn an enzyme ON. This process is **allosteric activation. Inhibitors** turn an enzyme OFF, and they perform **allosteric inhibition.** Complex proteins or small molecules may have powerful allosteric effects. Many hormones, such as *adrenaline,* activate enzymes in cell membranes throughout the body. The hormonal effects vary with the nature of the activated enzyme.

4. *Phosphorylation.* **Phosphorylation** is the enzymatic attachment of a phosphate group to a molecule. Some enzymes are activated or inactivated by phosphorylation. For example, phosphorylation of a muscle enzyme results in glycogen breakdown and glucose release.

Chemicals that affect enzyme activity can have powerful effects on cells throughout the body. For instance, several deadly compounds, such as hydrogen cyanide (HCN) and hydrogen sulfide (H_2S), kill cells by inhibiting mitochondrial enzymes involved in ATP production. Many inhibitors with less drastic effects are in clinical use. For example, *warfarin* slows blood clotting by inhibiting liver enzymes responsible for the synthesis of clotting factors. Several of the beneficial effects of *aspirin*—notably, the reduction of inflammation—are related to the inhibition of enzymes involved with prostaglandin synthesis. Many important antibiotics, such as penicillin, kill bacteria by inhibiting enzymes that are essential to bacteria but are absent from our cells.

⚕ METABOLIC ANOMALIES

FAP p. 61

The following metabolic disorders are examples of conditions that result from the absence of an

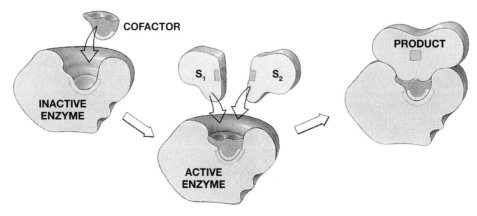

Figure A-5 Cofactors and Enzyme Activity.
In some cases, a cofactor must bind to the active site before the enzyme can bind substrate molecules and function normally.

enzyme necessary for normal cellular function. Without these specific proteins, growth and development are impaired, and vital tissues are damaged or destroyed. You will find additional information about many of these conditions elsewhere in the *Applications Manual* and in the text.

Anomalies in Amino Acid Metabolism

Phenylketonuria. Patients diagnosed with *phenylketonuria (PKU)* lack the enzyme that converts the amino acid *phenylalanine* to the amino acid *tyrosine*. Without this enzyme, phenylalanine accumulates in the blood and tissues, and large quantities are excreted in the urine. If this condition is not detected shortly after birth, mental retardation may occur due to damage to the developing nervous system. Because milk is a major source of phenylalanine, newborns generally undergo a blood test for PKU 48 hours after nursing begins. Abnormally high circulating levels of phenylalanine may indicate PKU. Once a diagnosis of PKU is made, the diet is controlled to avoid foods containing high levels of phenylalanine.

Albinism. Albinism is a genetic disorder that results in a lack of pigment production in the skin. The cause is a defective enzyme involved in the metabolism of the amino acid tyrosine. Because this enzyme is abnormal, the protein pigment *melanin* cannot be synthesized. The skin is white, and the hair and eyes are also affected. Among its other functions, melanin helps protect the skin from the effects of ultraviolet (UV) radiation. When outdoors, individuals with albinism must be careful to avoid skin damage from the UV radiation in sunlight.

Anomalies in Lipid Metabolism

Hypercholesterolemia. Familial hypercholesterolemia is a genetic disorder resulting in a reduced ability to remove cholesterol from the circulation. As circulating levels rise, cholesterol accumulates around tendons, creating yellow deposits called *xanthomas* beneath the skin. The worst aspect of this disorder is the deposition of cholesterol within the walls of blood vessels. This condition, a form of *atherosclerosis*, can restrict the circulation through vital organs such as the heart and brain. Atherosclerosis can develop in individuals with normal cholesterol metabolism, but clinical symptoms do not ordinarily appear until age 40 or older. Individuals with hypercholesterolemia may develop acute coronary artery disease or even suffer a heart attack at or before 20 years of age.

Anomalies in Carbohydrate Metabolism

Galactosemia. Milk contains a monosaccharide called *galactose* that can be converted to glucose within cells. The genetic disorder *galactosemia* is caused by the absence of the enzyme that catalyzes this reaction. Affected individuals have elevated levels of galactose in the blood and urine. Chronically high levels of galactose during childhood can cause abnormalities in nervous system development, jaundice, liver enlargement, and cataracts. Preventive treatment involves early detection of galactosemia and restriction of dietary intake of this monosaccharide.

Topics in Cellular Biology

Cells are the smallest living units in the body, but they are not the only forms of life. Discussions throughout the text and the *Applications Manual* assume that you are already familiar with the basic properties of cells and with the characteristics of potential *pathogens* (disease-causing organisms). This section begins with a review of the important distinctions among various types of pathogens and then proceeds to a discussion of other topics at the cellular level of organization.

A CLOSER LOOK: THE NATURE OF PATHOGENS

FAP p. 66

Chapter 3 of the text presents the structure of a "typical" cell. The cellular organization depicted in Figure 3-2 and described in that chapter is that of a *eukaryotic* cell (ū-kar-ē-OT-ik; *eu*, "true" + *karyon*, "nucleus"). The defining characteristic of eukaryotic cells is the presence of a nucleus. All eukaryotic cells have similar membranes, organelles, and methods of cell division. All multicellular animals, plants, and fungi (plus some single-celled organisms) are composed of eukaryotic cells.

The eukaryotic plan of organization is not the only one found in the living world, however. There are organisms that do not consist of eukaryotic cells. These organisms are of great interest to us because they include most of the pathogens that can cause human diseases.

Bacteria

Prokaryotic cells do not have nuclei or other membranous organelles. They do not have a cytoskeleton, and typically their cell membranes are surrounded by a semirigid cell wall made of carbohydrate and protein. Figure A-6a shows the structure of a representative **bacterium.**

Bacteria are generally less than 2 µm in diameter. Many bacteria are quite harmless, and many more—including some that live within our bodies—are actually beneficial to us in a variety of ways. Other bacteria are dangerous pathogens that, given the opportunity, will destroy body tissues. These bacteria are dangerous because they absorb nutrients and release enzymes that damage cells and tissues. A few pathogenic bacteria also release toxic chemicals. Bacterial infections are responsible for

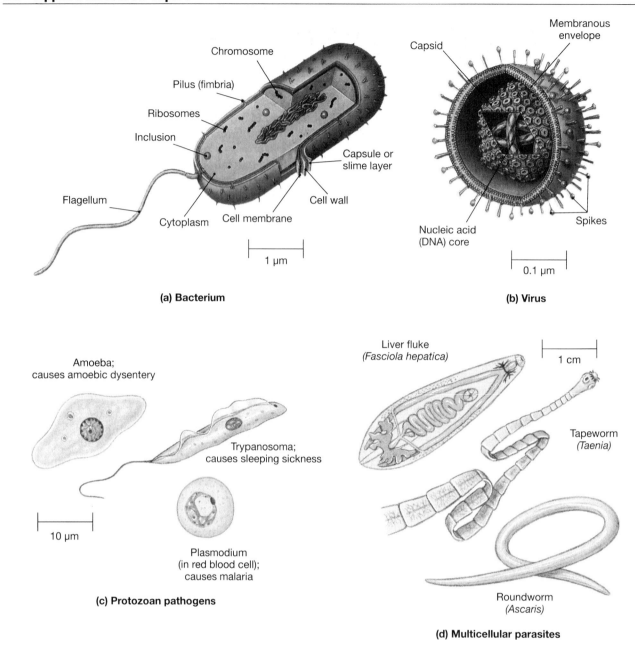

Figure A-6   Representative Pathogens.
(a) A bacterium, with prokaryotic characteristics indicated. Compare with Figure 3-2 (FAP *p. 67*), which shows a representative eukaryotic cell. (b) A typical virus. Each virus has an inner chamber containing nucleic acid, surrounded by a protein capsid or an inner capsid and an outer membranous envelope. The herpes viruses are enveloped DNA viruses; they cause chickenpox, shgles, and herpes. (c) Protozoan pathogens. Protozoa are eukaryotic single-celled organisms, common in soil and water. (d) Multicellular parasites. Several different groups of organisms are human pathogens, and many have complex life cycles.

many serious diseases, including *tetanus, cholera, gangrene, pneumonia, meningitis, syphilis, gonorrhea, typhus, plague, tuberculosis, typhoid fever,* and *leprosy.* We consider these and other bacterial infections in various chapters of the text and in other sections of the *Applications Manual.*

Viruses

Another type of pathogen conforms neither to the prokaryotic nor to the eukaryotic organizational plan. These tiny pathogens, called **viruses,** are not cellular. In fact, when free in the environment, they do not show any of the characteristics of living organisms. They are classified as **infectious agents** because they can enter cells (either prokaryotic or eukaryotic) and replicate themselves.

Viruses consist of a core of nucleic acid (DNA or RNA) surrounded by a protein coat called a *capsid.* (Some varieties have an *envelope,* a membranous outer covering, as well.) The structure of a representative virus is shown in Figure A-6b. Important viral diseases include influenza (flu), yellow fever, some leukemias, AIDS, hepatitis, polio, measles, mumps, rabies, and the common cold.

To enter a cell, a virus must first attach to the cell membrane. Attachment occurs at one of the normal membrane proteins. Once the virus has penetrated the cell membrane, the viral nucleic acid takes over the cell's metabolic machinery. In the case of a DNA virus (Figure A-7a), the viral DNA enters the cell nucleus, where transcription begins. The mRNA produced then enters the cytoplasm and is used in translation, in which the cell's ribosomes begin synthesizing viral proteins. The viral DNA replicates within the nucleus, "stealing" the cell's nucleotides. The replicated viral DNA and the new viral proteins then form new viruses that pass through the cell membrane.

In the case of an RNA virus, the situation is somewhat more complicated (Figure A-7b). In the simplest RNA viruses, the viral RNA entering the cell functions as an mRNA strand that carries the information needed to direct the cell's ribosomes to synthesize viral proteins. These proteins include

enzymes essential to the duplication of viral RNA. When the cell is packed with new viruses, the cell membrane ruptures and the RNA viruses are released into the interstitial fluid.

In the *retroviruses,* a group that includes HIV (the virus responsible for AIDS), the replication process is more complex. These viruses carry an enzyme called *reverse transcriptase,* which directs "reverse transcription"—the assembly of DNA according to the nucleotide sequence of an RNA strand. The DNA created in this way is then inserted into the infected cell's chromosomes. The viral genes are then activated, and the cell begins producing RNA through normal transcription. The RNA produced includes viral RNA, mRNA carrying the information for the synthesis of reverse transcriptase, and mRNA controlling the synthesis of viral proteins. These components then combine within the cytoplasm, which gradually becomes filled with viruses. The new RNA viruses are then shed at the cell surface.

Even if the host cell is not destroyed outright by these events, normal cell function is usually disrupted. In effect, the metabolic activity of the cell is

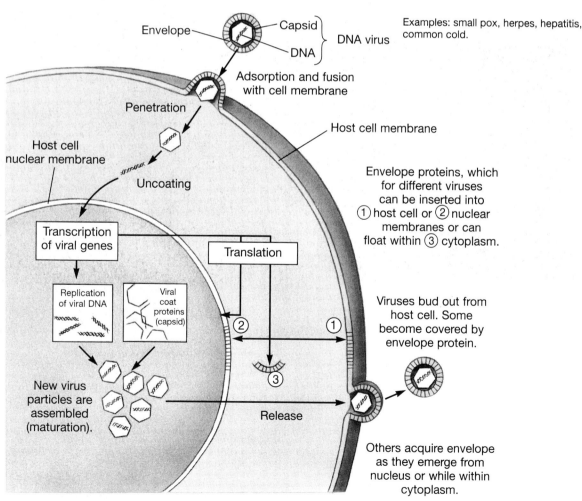

(a)

Figure A-7   Viral Replication.
(a) Replication of a DNA virus.

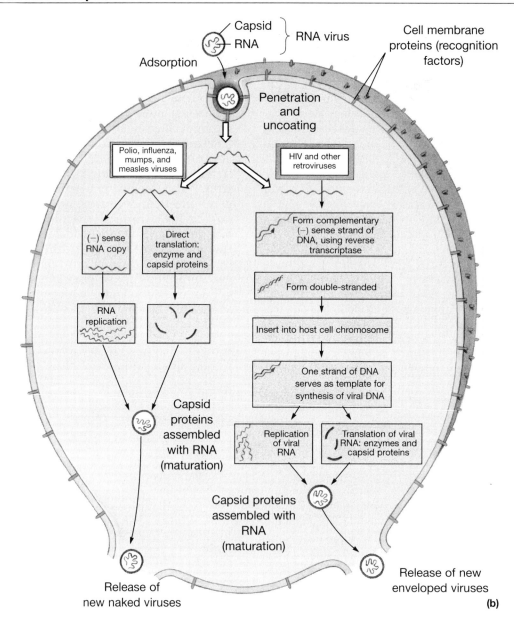

Figure A-7   (continued).
(b) Replication of RNA viruses.

diverted to create viral components rather than performing tasks needed for cell maintenance and survival. Some viruses, however, can lie dormant within infected cells for long periods of time before initiating this process of replication.

Viruses are now becoming important as benefactors as well as adversaries. In genetic engineering procedures, viruses whose nucleic acid structure has been intentionally altered can be used to transfer copies of normal human genes into the cells of individuals with inherited enzymatic disorders. This was the method used to insert the gene for the enzyme missing in adenosine deaminase deficiency patients (p. 31). Attempts are now planned to treat *cystic fibrosis (CF)* in the same way.

Cystic fibrosis is a debilitating genetic defect whose most obvious—and potentially deadly—symptoms involve the respiratory system. The underlying problem is an abnormal gene that carries instructions for a chloride ion channel that occurs in cell membranes throughout the body. Researchers have recently treated CF in laboratory animals by inserting the normal gene into a virus that infects cells lining the respiratory passageways. The virus could be given to human patients via an inhalant.

Prion Diseases

Prions (PRĒ-onz) are controversial proteins that are either infectious agents or the result of infection by an as-yet unidentified virus. (Current evidence suggests that the proteins are themselves infectious.) Normal neurons produce a very similar protein that exists as individual molecules. They are located in the cell membrane and exposed to the extracellular fluid. When exposed to a prion, this protein changes shape, and the individual molecules interact to form large insoluble complexes that are released into the extracellular fluid. These complexes are called *amyloid plaques.* Either the presence of the amyloid plaques or the absence of the normal protein then disrupts neural function.

The first prion disease described was *kuru,* a deadly disease affecting members of a cannibalistic society in New Guinea. The prions were passed from person to person when uninfected individuals ate infected brains. The infection, which led to death within a year, caused half of all childhood and adult deaths in New Guinea. Other known prion diseases include *Cruetzfeldt–Jakob disease* and *fatal familial insomnia.* Most cases of prion disease (kuru excepted) are the result of mutations in the normal gene that produce amyloid plaques. However, if a normal person becomes exposed to these abnormal proteins, a prion disease will result.

Prion infection also occurs in domesticated animals. In sheep, the condition is called *scrapie;* in cows, it is called *bovine spongiform encephalopathy.* Infected cows ultimately develop an assortment of strange neurological symptoms (such as pawing at the ground and difficulty in walking), giving the condition the common name of "mad cow disease." In 1995, European researchers attributed an upsurge in the incidence of Cruetzfeldt–Jakob disease in humans to the consumption of meat from prion-infected cows. Although it remains unproven that the human infections resulted from infected beef, the incident has focused significant public interest and research attention on prion diseases.

Unicellular Pathogens and Multicellular Parasites

Bacteria and viruses are the best-known human pathogens, but there are eukaryotic pathogens as well. Examples of the most important types are included in Figure A-6c. **Protozoa** are unicellular eukaryotes that are abundant in soil and water. They are responsible for a variety of serious human diseases, including *amoebic dysentery* and *malaria.* **Fungi** (singular, *fungus*) are eukaryotic organisms that absorb organic materials from the remains of dead cells. Mushrooms are familiar examples of very large fungi. In a fungal infection, or *mycosis,* a microscopic fungus spreads through living tissues, killing cells and absorbing nutrients. Several relatively common skin conditions (including *athlete's foot*) and a few more-serious

diseases (including *histoplasmosis*) are the result of fungal infections.

Larger multicellular organisms (Figure A-6d), generally referred to as *parasites,* such as *flatworms* and *nematodes,* can also invade the human body and cause diseases. These organisms, which range in size from microscopic flukes to tapeworms a meter or more in length, typically cause weakness and discomfort but do not *by themselves* kill their host. However, complications resulting from the parasitic infection, such as malnutrition, chronic bleeding, or secondary infections by bacterial or viral pathogens, can ultimately prove fatal.

METHODS OF MICROANATOMY   FAP *p. 66*

Over the last 30 years, our technological gadgetry has improved remarkably, enabling us to view the insides and outsides of cells in new ways. Sophisticated equipment has permitted the detailed analysis of physiological processes within cells. The basic problems facing cytologists stem from the considerable size difference between the investigator and the object of interest. Cytologists and histologists measure intracellular structures in terms of *micrometers* (µm), also known as *microns.* Although the range of cell sizes is considerable, an "average cell" is a cube roughly 10 µm × 10 µm × 10 µm. To fill a cubic millimeter, we would need a million cells. Because the human eye cannot recognize details smaller than about 0.1 mm, cytologists rely on special equipment that magnifies cells and their contents.

Light Microscopy

Historically, most information has been provided by **light microscopy,** in which a beam of light is passed through the object to be viewed. A light microscope can magnify cellular structures about 1000 times and can show details as fine as 0.25 µm. A camera can be attached to the microscope and used to produce a photograph called a **light micrograph (LM).** Unfortunately, you cannot simply pick up a cell, slap it onto a microscope slide, and take a photograph. Because individual cells are so small, you must work with large numbers of cells. Most tissues have a three-dimensional structure, and small pieces of tissue can be removed for examination. The component cells are prevented from dying and decomposing by first exposing the tissue sample to a poison that will stop metabolic operations but will not alter cellular structures.

Even then, you still cannot look at the tissue sample through a light microscope, for a cube only 2 mm (0.078 in.) on a side will contain several million cells. You must slice the sample into thin sections. Living cells are relatively thick, and cellular contents are not transparent. Light can pass through the section only if the slices are thinner than the individual cells. Making a section that slender poses interesting technical problems. Most tissues are not very sturdy, so an attempt to slice a fresh piece will destroy the sample. (To appreciate the problem, try to slice a marshmallow into thin sections.) Thus, before you can make sections, you must embed the tissue sample in something that will make it more stable, like wax, plastic, or epoxy. These materials will not interact with water molecules, so your sample must first be dehydrated (typically by immersion in 30 percent, 70 percent, 95 percent, and finally 100 percent alcohol). If you are embedding in wax, the wax must be hot enough to melt; if you are using plastic or epoxy, the hardening process generates heat on its own.

After embedding the sample, you can section the block with a machine called a *microtome,* which uses a metal, glass, or diamond knife. For viewing by light microscopy, a typical section is about 5 μm (1/5000 in.) thick. The thin sections are then placed on microscope slides. If the sample was embedded in wax, you can now remove the wax with a solvent, such as xylene. But you are not done yet. In thin sections, the cell contents are almost transparent; you cannot yet distinguish intracellular details by using an ordinary light microscope. You must first add color to the internal structures by adding special dyes called *stains* to the slides. Some stains are dissolved in water, and others in alcohol. All types of cells do not pick up a given stain equally, if they pick it up at all; nor do all types of cellular organelles. For example, in a sample scraped from the inside of the cheek, one stain may dye only certain types of bacteria; in a semen sample, another stain may dye only the flagella of the sperm. If you try too many stains at one time, they all run together, and you must start over. After the staining is completed, you can put coverslips over the sections (generally after you have dehydrated them again) and can see what your labors have accomplished.

Any single section can show you only a part of a cell or tissue. To reconstruct the tissue structure, you must look at a series of sections made one after the other. (See the related discussion in Chapter 1 of FAP, *p. 21.*) After examining dozens or hundreds of sections, you can understand the structure of the cells and the organization of your tissue sample—or can you? Your reconstruction has left you with an understanding of what these cells look like after they have (1) died an unnatural death; (2) been dehydrated; (3) been impregnated with wax or plastic; (4) been sliced into thin sections; (5) been rehydrated, dehydrated, and stained with various chemicals; and (6) been viewed with the limitations of your equipment. A good cytologist or histologist is extremely careful, cautious, and self-critical, and much of the laboratory preparation is an art as well as a science.

Electron Microscopy

More-elaborate procedures can allow for the examination of finer details. In **electron microscopy,** a beam of electrons is passed through or reflected off the surface of a suitably prepared object. In **transmission electron microscopy,** the electrons pass through an ultrathin section. Once through the section, they strike a photographic plate and produce an image known as a **transmission electron micrograph (TEM).** Transmission electron microscopy can magnify structures up to approximately 500,000 times, revealing details less than a nanometer in size; a transmission electron microscope can permit you to visualize large organic molecules. In **scanning electron microscopy,** a beam of electrons reflects off the surface of an object such as a cell, a broken portion of a cell, or extracellular structures. (The surfaces are specially coated to enhance reflectivity.) After bouncing off the surface, the electrons strike a photographic plate to produce an image known as a **scanning electron micrograph (SEM).** Scanning electron microscopy can magnify structures up to about 50,000 times but provides a three-dimensional perspective on cellular anatomy that cannot be obtained by other methods.

Such detail poses problems of its own. At the level of the light microscope, if you were to slice a large cell as you would slice a loaf of bread, you might produce 10 sections from the one cell. You could review the entire series under a light microscope in a few minutes. If you sliced the same cell for examination under an electron microscope, you would have 1000 sections, each of which could take several hours to inspect! Figure A-8a,b compares SEM and TEM views of cells that line the intestinal tract, and Figure A-8c shows a diagrammatic representation of the intact cell.

CELL STRUCTURE AND FUNCTION
FAP *p. 20*

Each cell in the body has a particular role to play in maintaining the integrity of the individual as a whole. Some conduct nerve impulses, and others manufacture hormones, build bones, or contract to produce body movements. When any of these cells malfunction, whether due to genetic abnormalities affecting enzyme function, a viral or bacterial infection, trauma, or cancer, homeostasis is threatened. This section introduces clinical and practical applications of basic principles of cellular function and discusses representative disorders resulting from problems at the cellular level of organization.

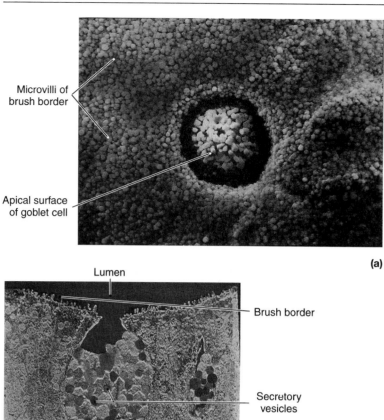

Microvilli of
brush border

Apical surface
of goblet cell

(a)

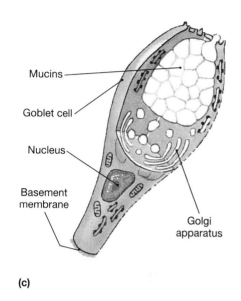

Mucins

Goblet cell

Nucleus

Basement
membrane

Golgi
apparatus

(c)

Lumen

Brush border

Secretory
vesicles

Columnar
epithelial cells

Nucleus of
goblet cell

(b)

Figure A-8 A Comparison of Histological Techniques.
(a) Cell surfaces can be seen with a scanning electron
microscope. (b) Similar cells as viewed with a transmission electron microscope. (c) A composite drawing
that summarizes the information provided by both scanning and transmission electron microscopy.

℞ Drugs and the Cell Membrane

FAP p. 73

Many clinically important drugs affect cell membranes. Although the mechanism behind the action
of general anesthetics, such as *ether, chloroform,*
and *nitrous oxide,* has yet to be determined, most
are lipid-soluble hydrophobic molecules. There is a
direct correlation between the potency of an anesthetic and its lipid solubility. Lipid solubility may
speed the drug's entry into cells and enhance its
ability to block ion channels or alter other properties of cell membranes. The most important clinical result is a reduction in the sensitivity and
responsiveness of neurons and muscle cells.

Local anesthetics such as *procaine* and *lido-
caine,* as well as *alcohol* and *barbiturate* drugs, are
also lipid-soluble. These compounds affect membrane properties by blocking sodium channels in
the cell membranes of neurons. This blockage

reduces or eliminates the responsiveness of neurons to painful (or any other) stimuli. The very powerful toxin *tetrodotoxin (TTX)* is found in some
species of puffer-fish (family Tetraodontidae). Eating the internal organs of these fish causes a severe
and potentially fatal form of food poisoning, marked
by the disruption of normal neural and muscular
activities. (Nevertheless, the flesh is considered a
delicacy in Japan, where it is prepared by specially licensed chefs and served under the name *fugu.*)

Other drugs interfere with membrane receptors
for hormones or chemicals that stimulate muscle or
nerve cells. *Curare* is a plant extract that interferes
with the chemical stimulation of muscle cell membranes. South American Indians use it to coat their
hunting arrows so that wounded prey cannot run
away. To prevent reflexive muscle contractions or
twitches while surgery is being performed, anes-
thesiologists may administer a curare derivative (*d-
tubocurarine* or related drugs) preoperatively to
patients.

📖 Mitochondrial DNA, Disease, and Evolution
FAP *p. 88*

Several inheritable disorders result from abnormal mitochondrial activity. The mitochondria involved have defective enzymes that reduce their ability to generate ATP. Cells throughout the body may be affected, but symptoms involving muscle cells, neurons, and the receptor cells in the eye are most commonly seen, because these cells have especially high energy demands. Disorders caused by defective mitochondria are called *mitochondrial cytopathies.* In several instances, the disorders have been linked to inherited abnormalities in mitochondrial DNA. In some cases, the problem appears in one population of cells only. For example, abnormal mitochondrial DNA has been found in the motor neurons whose degeneration is responsible for the condition of *Parkinson's disease,* a neurological disorder characterized by a shuffling gait and uncontrollable tremors.

More commonly, mitochondria throughout the body are involved. Examples of conditions caused by mitochondrial dysfunction include one class of epilepsies *(myoclonic epilepsy)* and a type of blindness *(Leber's hereditary optic neuropathy).* These are inherited conditions, but the pattern of inheritance is very unusual. Although men or women may have the disease, only affected women can pass the condition on to their children. The explanation for this pattern is that the disorder results from an abnormality in the DNA of mitochondria, not in the DNA of cell nuclei. All the mitochondria in the body are produced through the replication of mitochondria present in the fertilized ovum. Few if any of those mitochondria were provided by the father; most of the mitochondria of the sperm do not remain intact after fertilization takes place. As a result, children can generally inherit these conditions only from their mother.

This brings us to an interesting concept: Virtually all your mitochondria were inherited from your mother, and hers from her mother, and so on back through time. The same is true for every other human being. Now, it is known that small changes in DNA nucleotide sequences accumulate over long periods of time. Mitochondrial DNA, or mDNA, can therefore be used to estimate the degree of relatedness between individuals. The greater the difference between the mDNA of two individuals, the more time has passed since the lifetime of their most recent common ancestor, and the more distant their relationship. On this basis, it has been estimated that all human beings now alive shared a common female ancestor roughly 350,000 years ago. Appropriately, that individual has been called "Mitochondrial Eve." The existence and history of Mitochondrial Eve remain controversial.

⚕ Lysosomal Storage Diseases
FAP *p. 90*

Problems with lysosomal enzyme production cause more than 30 storage diseases that affect children. In these conditions, the lack of a specific lysosomal enzyme results in the buildup of materials normally removed and recycled by lysosomes. Eventually the cell cannot continue to function. We will consider three important examples here: Gaucher's disease, Tay-Sachs disease, and glycogen storage disease.

Gaucher's disease is caused by the buildup of *cerebrosides,* glycolipids found in cell membranes. This is probably the most common type of lysosomal storage disease. There are two forms of this disease: (1) an infantile form, marked by severe neurological symptoms ending in death, and (2) a juvenile form, with enlargement of the spleen, anemia, pain, and relatively mild neurological symptoms. Gaucher's disease is most common among the Ashkenazi Jewish population, where it occurs at a frequency of approximately 1 in 1000 births.

Tay-Sachs disease is another hereditary disorder caused by the inability to break down glycolipids. In this case, the glycolipids are *gangliosides,* which are most abundant in neural tissue. Individuals with this condition develop seizures, blindness, dementia, and death, generally by age 3–4. Tay-Sachs disease is most common among the Ashkenazi Jewish population, where it occurs at a frequency of 0.3 per 1000 births.

Glycogen storage disease (Type II) affects primarily skeletal muscle, cardiac muscle, and liver cells—the cells that synthesize and store glycogen. In this condition, the cells are unable to mobilize glycogen normally, and large numbers of insoluble glycogen granules accumulate in the cytoplasm. These granules disrupt the organization of the cytoskeleton, interfering with transport operations and the synthesis of materials. In skeletal and heart muscle cells, the buildup leads to muscular weakness and potentially fatal heart problems.

Topics in Molecular Biology

Molecular biology is the study of the synthesis, structure, and function of macromolecules important to life, such as proteins and nucleic acids. Deciphering the genetic code and relating the intricate structure of a protein to its particular functions are major goals of molecular biology. Research in this area has greatly enhanced our understanding of normal functions as well as disease processes.

The field of molecular biology has revolutionized the study of medicine by providing a clear biochemical basis for many complex diseases. For example, in *sickle cell anemia* red blood cells undergo changes in shape that result in blocked vessels and tissue damage due to oxygen starvation. It is now known that this condition results when an individual carries two copies of a defective gene that determines the structure of *hemoglobin,* the oxygen-binding protein found within red blood cells. The genetic defect changes just 2 of the 574 amino acids in this protein. That is enough to alter the functional properties of the hemoglobin molecule, leading to changes in the properties of the red blood cells. This type of disorder is often called a *molecular disease,* because it results from abnormalities at the molecular level of organization.

Roughly 2500 inherited disorders have now been identified, and researchers have located the defective genes responsible for cystic fibrosis, Duchenne's muscular dystrophy, and Tay-Sachs disease. Identifying the genetic defect is, of course, the vital first step toward the development of an effective gene therapy or other treatment. The treatments that are now evolving make use of the principles of *genetic engineering* (p. 146).

Ⓡ GENETIC ENGINEERING AND GENE THERAPY   FAP *p. 99*

Once the mechanics of the genetic code were understood, everyone realized that it would be theoretically possible to change the genetic makeup of organisms—perhaps even of a human being. The popular term for activities related to this goal is *genetic engineering.*

What are some of the key problems confronting genetic engineers? Genes code for proteins; the makeup of each protein is determined by the sequence of codons (nucleotide triplets) in a stretch of DNA. A human cell has 46 chromosomes, 2 meters of DNA, and roughly 10^9 triplets. If all the DNA in the human body were extracted and strung together, the resulting strand would be long enough to make several hundred round-trips between Earth and the sun. Simply finding a particular gene among the estimated 100,000 that each of us carries is an imposing task. Yet, before a specific gene can be modified, its location must be determined with great precision. Locating a gene involves preparing a map of the appropriate chromosome.

Mapping the Human Genome

Several techniques can be used to create a general map of the chromosomes. **Karyotyping** (KAR-ē-ō-tī-ping; *karyon,* nucleus + *typos,* a mark) is the determination of an individual's chromosome complement. Figure A-9a shows a set of normal human chromosomes. Each chromosome has characteristic banding patterns, and segments can be stained with special dyes. Unusual banding patterns can indicate structural abnormalities. These abnormalities are sometimes linked to specific inherited conditions, including a form of

leukemia. *Down syndrome* (Figure A-9b,c) results from the presence of an extra chromosome, a copy of chromosome 21.

In December 1993, French researchers at the Centre d'Etude de Polymorphism Humaine (CEPH; Center for the Study of Human Polymorphism) completed the first preliminary mapping of the entire human genome. This provided the landmarks and reference points needed to make more precise maps that indicate the locations of specific genes. More-detailed maps have now been prepared for the Y chromosome and chromosome 21, the two smallest chromosomes, and work continues on the others.

Mapping is useful in itself, but it is only an intermediate step on the way to the ultimate goal: the determination of the nucleotide sequence of every gene in the human genome. As of 1997, a gene register maintained by Dr. Victor McKusick and other researchers at Johns Hopkins University contained more than 6800 genes, with roughly 5000 of them assigned to specific chromosomes. These numbers are a small percentage of the total, but the pace of identification and localization is increasing rapidly. The sequencing process, which is expected to cost $3 billion and to be completed by the year 2005, will provide basic information about the location of genes on normal human chromosomes.

Gene Manipulation and Modification

Suppose that the location of a defective gene has been pinpointed. Before attempting to remedy the defect in an individual, a clinician would have to determine the nature of the genetic abnormality. For example, the gene could be inactive, overactive, or producing an abnormal protein. It could even be missing entirely. Finally, it would be necessary to decide how to remedy the defect. Can the gene be turned on, turned off, modified, or replaced?

What's the problem? This can be a particularly difficult question to answer. Many of the 2500 inheritable genetic disorders are classified according to general patterns of symptoms rather than any specific protein or enzyme deficiency. In some cases, the approximate location of the gene has been determined, but the identity of the protein responsible for the clinical symptoms remains a mystery. There may be several possible abnormalities in this protein that can result in different patterns of clinical disease.

What can be done? If the gene is present but overproducing or underproducing, its activity might be controlled by introducing chemical repressors or inducers. Another approach relies on **gene splicing** to produce a protein that is missing or present in inadequate quantities in the affected individual. Gene splicing (Figure A-10, p. 31) begins with the localization of the gene, followed by its isolation. A "healthy" copy of that gene is then spliced

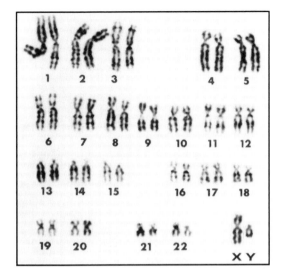

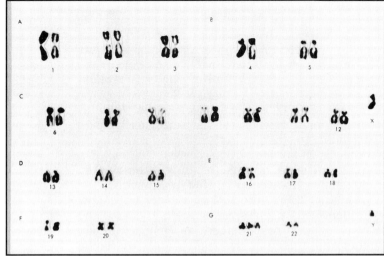

Figure A-9 Normal and Abnormal Karyotypes.
(a) A micrograph of the normal human chromosome set; the chromosomes have been arranged in this sequence for ease of comparison. (b) The chromosomes of an individual with Down Syndrome. Note the extra copy of chromosome 21. (c) Down syndrome is associated with mental retardation, cardiovascular problems, and a variety of physical abnormalities. The boy has the characteristic appearance of an individual with Down syndrome; compare his features with those of his sister, who does not have the disorder.

into the relatively simple DNA strand of a bacterium, creating *recombinant DNA*. Bacteria grow and reproduce rapidly under laboratory conditions, and before long there is a colony of identical bacteria. All the members of the colony will carry the introduced gene and manufacture the corresponding protein. The protein can be extracted, concentrated, and administered to individuals whose diseases represent deficiencies in the activity of that particular gene. *Hemophilia* (a deficiency of blood clotting factors) and a form of diabetes caused by insulin deficiency can be treated in this way.

Gene splicing is also used to obtain large quantities of proteins normally found in very small concentrations. *Interferon*, an antiviral protein, and human growth hormone are compounds now being produced commercially by means of gene-splicing technology.

The most revolutionary strategies involve "fixing" abnormal cells by giving them copies of normal genes. In general, this method poses significant targeting problems, for the gene must be introduced into the right kind of cell. For example, placing liver enzymes in fingernails would not

correct a metabolic disorder. But when the target cells can be removed and isolated, as in the case of bone marrow, the technique is promising. Actual removal of a defective gene does not appear to be a practical approach, and the focus has been on adding genes that can take over normal functions.

In September 1990, the first gene therapy trials were initiated. The procedure was used to treat a 4-year-old girl who had *adenosine deaminase deficiency (ADA)*. ADA is a rare condition that each year affects only about 20 children worldwide. Without this enzyme, toxic chemicals build up in cells of the immune system. As these cells die, the body's defenses break down.

ADA results in a complex of symptoms known as *severe combined immunodeficiency disease*, or *SCID*. SCID can also be caused by other enzyme disorders affecting cells of the immune system. Symptoms include chronic respiratory infections, diarrhea, and a low resistance to viral or bacterial infections. Most children with ADA die from infections that would pose no threat to normal children. A new drug called *PEG-ADA*, an altered form of the

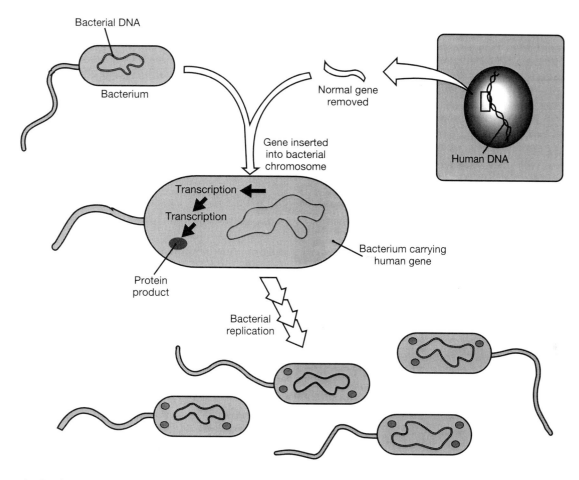

Figure A-10   Gene splicing.
A gene is removed from a human cell nucleus and attached to the DNA in a bacterium, where it directs production of a human protein. Bacterial replication creates a colony of bacteria that share the introduced gene and can yield large quantities of the protein product.

missing enzyme, can prolong life, but it does not cure the condition.

In the 1990 clinical trial, blood cells were collected, and *lymphocytes,* the primary cells of the immune system, were removed. Short segments of DNA containing the normal gene for adenosine deaminase production were then inserted into the nuclei of these cells, and the modified lymphocytes were returned to the body. Roughly a billion modified cells were reintroduced. Over time, these modified cells divided to produce a large population of normal immune cells. The experiment was a success, and this method has been used to treat several other children with ADA deficiency. Each has regained and retained immune function, although there is concern that the engineered lymphocytes may not recognize and attack as great a variety of antigens (foreign compounds, pathogens, or toxins) as can the lymphocytes of normal individuals. The treatment must be repeated periodically (in the original case, at 3–5 month intervals). In more-recent experiments on individuals with ADA, the genes for bone marrow stem cell production were modified; these patients

regained apparently normal immune function without the need for repetitive treatments.

The major limitation of gene therapy for disorders other than ADA, which involves blood cells, has been the lack of a reliable mechanism for the insertion of genes into specific cells inside the body. Viruses are an option, but there are concerns about using a potential pathogen to deliver genetic material. The use of viruses to deliver genes to defective cells has been approved for clinical trials for the treatment of about 20 disorders. Instead, small lipid packets called *liposomes* can be used to deliver genes to cells that line the respiratory tract; they are administered in a fine mist that can be inhaled. As is not unusual for a new method of treatment, there have been more failures than successes so far. For example, recent attempts to treat cystic fibrosis, an inherited disorder characterized by respiratory problems (see p. 156), and a common form of muscular dystrophy (p. 64) by liposome inhalation have been unsuccessful.

These procedures attempt to relieve the symptoms of disease by inserting genes into defective somatic cells. The new genes do not change the genetic structure of reproductive cells; because the eggs or sperm retain the original genetic pattern, the genetic defect will be passed to future generations. Researchers are much further away from practical methods of changing the genetic characteristics of reproductive cells. Mouse eggs fertilized outside the body have been treated and transplanted into the uterus of a second mouse for development. The gene added was one for a growth hormone obtained from a rat, and the large "supermouse" that resulted demonstrated that such manipulations can be performed. The possibilities for manipulating the characteristics of valuable animal stocks, such as cattle, sheep, or chickens, are quite exciting. The potential for altering the genetic characteristics of human beings is intimidating. Before any clinical variations on this theme are tested, our society will have to come to grips with some difficult ethical issues!

Just for a sense of the kinds of problems we might have to deal with, discuss the following questions with your friends and classmates.

Genetic Engineering—Questions to Think About

1. Your 51-year-old father has recently been diagnosed with a hereditary disorder affecting the brain. The prognosis is poor, and there is currently no cure for the disorder. The physician advises that you be tested for the presence of the faulty gene.

    • Would you be tested (would you want to know if you had the defective gene)?

    If you are tested, who should have access to the results:

    Your insurance company?

    Your children, spouse, or fiancé?

    Your employer?

2. *Eugenics* is the control of the hereditary characteristics of individuals to improve the species. The science of eugenics became distorted through the work of scientists under Adolph Hitler's control. Prenatal testing now permits the diagnosis of a variety of inherited disorders before birth. This information could be used to "improve the species" by selectively terminating pregnancies. Is this practice advisable or ethical?

3. Many scientists today, such as the Nobel laureate James Watson, consider the human genome to be the blueprint for a human being. If an individual's genetic code determines every characteristic of that person, can people accused of crimes be held legally responsible for their actions?

4. Studies in the past have shown that men with an extra Y chromosome (XYY) are more violent and predisposed to crime than are men with the XY genotype. More recent studies have revealed no greater tendency toward violence among XYY individuals than among XY males. A prejudice against XYY males still exists, even though the original studies are now known to be seriously flawed.

    • What type of controls are needed to ensure that new information concerning genetic abnormalities is not released before it is confirmed?

    • Not every person with a specific genotype will develop the same characteristics to the same degree. How can possible stereotypes be avoided when information is released?

5. You are an airline employer trying to offer your employees the least expensive health insurance available. The insurance company requires a blood test on each potential employee to determine genetic abnormalities. You learn that a candidate for a job as a pilot has a genetic predisposition for a heart attack. Would this information affect your decision to hire that individual?

☤ A CLOSER LOOK: CANCER

FAP pp. 103, 141

Twenty-five percent of all people in the United States develop cancer at some point in their lives. During 1996, roughly 560,000 people died of some form of cancer in the United States, making it Public Health Enemy Number 2, second only to heart disease.

Causes of Cancer

Relatively few types of cancer are inherited; 18 hereditary types have been identified to date, including two forms of leukemia. Most cancers develop through the interaction of genetic and environmental factors, and it is difficult to separate the two completely.

Genetic Factors. Two related genetic factors are involved in the development of cancer: (1) *hereditary predisposition* and (2) *oncogene activation.*

An individual born with genes that increase the likelihood of cancer is said to have a hereditary predisposition for the disease. Under these conditions a cancer is not guaranteed, but it is a lot more likely than on average. The inherited genes generally affect tissue abilities to metabolize toxins, control mitosis and growth, perform repairs after injury, or identify and destroy abnormal tissue cells. As a result, body cells become sensitive to local or environmental factors that would have little effect on normal tissues.

Cancers may also result from somatic mutations that modify genes involved with cell growth, differentiation, or mitosis. As a result, an ordinary cell is converted into a cancer cell. The modified genes are called **oncogenes** (ON-kō-jēnz); the normal genes are called **proto-oncogenes.** Oncogene activation occurs by alteration of normal somatic genes. Because these mutations do not affect reproductive cells, the cancers caused by active oncogenes are not inherited.

A proto-oncogene, like other genes, has a regulatory component that turns the gene ON and OFF and a structural component that contains the mRNA triplets that determine protein structure. Mutations in either portion of the gene may convert it to an active oncogene. A change of just one nucleotide out of a chain of 5000 can convert a normal proto-oncogene to an active oncogene. In some cases, a viral infection can trigger activation of an oncogene. For example, one of the human papilloma viruses appears to be responsible for many cases of cervical cancer.

More than 50 proto-oncogenes have been identified. In addition, a group of anticancer genes has been discovered. These genes, called *tumor-suppressing genes (TSG),* or *anti-oncogenes,* suppress division and growth in normal cells. Mutations that alter TSGs make oncogene activation more likely. TSG mutation has been suggested as important in promoting several cancers, including several blood cell cancers, breast cancer, and ovarian cancer. Examples of important suppressor genes include the genes *p53* and *p16.* Mutations affecting the p53 gene are responsible for the majority of cancers of the colon, breast, and liver. Abnormal p16 gene activity may be involved in as many as half of all cancer cases.

Environmental Factors. Many cancers can be directly or indirectly attributed to environmental factors called *carcinogens* (kar-SIN-ō-jenz). Carcinogens stimulate the conversion of a normal cell to a cancer cell. Some carcinogens are *mutagens* (MŪ-ta-jenz)—that is, they damage DNA strands and may cause chromosomal breakage. Radiation is a mutagen that has carcinogenic effects.

There are many different chemical carcinogens in the environment. Plants manufacture poisons that protect them from insects and other predators, and although their carcinogenic activities are often relatively weak, many common spices, vegetables, and beverages contain compounds that can be carcinogenic if consumed in large quantities. Animal tissues may also store or concentrate toxins, and hazardous compounds of many kinds can be swallowed in contaminated food. A variety of laboratory and industrial chemicals, such as coal tar derivatives and synthetic pesticides, have been shown to be carcinogenic. Cosmic radiation, X-rays, UV radiation, and other radiation sources can also cause cancer. It has been estimated that 70–80 percent of all cancers are the result of chemical or environmental factors, and 40 percent are due to a single stimulus: cigarette smoke.

Specific carcinogens will affect only those cells capable of responding to that particular physical or chemical stimulus. The responses vary because differentiation produces cell types with specific sensitivities. For example, benzene can produce a cancer of the blood; cigarette smoke, a lung cancer; and vinyl chloride, a liver cancer. Very few stimuli can produce cancers throughout the body; radiation exposure is a notable exception. In general, cells undergoing mitosis are most likely to be vulnerable to chemical or radiational carcinogens. As a result, cancer rates are highest in epithelial tissues, where stem cell divisions occur rapidly, and lowest in nervous and muscle tissues, where divisions do not normally occur.

Detection and Incidence of Cancer

Physicians who specialize in the identification and treatment of cancers are called **oncologists** (on-KOL-ō-jists; *onkos,* "mass"). Pathologists and oncologists classify cancers according to their cellular appearance and their sites of origin. Over a hundred kinds have been described, but broad categories are usually used to indicate the location of the primary tumor. Table A-5 summarizes information about benign and malignant tumors (cancers) associated with the major tissues of the body.

A statistical profile of cancer incidence and survival rates in the United States is shown in Table A-6 (p. 35). The numbers from other countries are different. For example, bladder cancer is common in Egypt, stomach cancer in Japan, and liver cancer in Africa. Differences in the combination of genetic factors and dietary, infectious, and other environmental factors are thought to be responsible for these differences.

Clinical Staging and Tumor Grading

Detection of a cancer often begins during a routine physical examination, when the physician notices an abnormal lump or growth. A tumor or neoplasm is defined as a "new growth" resulting from uncontrolled cell division. A tumor may be malignant or benign, metastasizing rapidly or spreading very slowly; only malignant tumors are called cancers.

Many laboratory and diagnostic tests are necessary for the correct diagnosis of cancer. Information is usually obtained through examination of a tissue sample, or *biopsy,* typically supplemented by medical imaging and blood studies. A biopsy is one of the most significant diagnostic procedures, because it permits a direct look at the tumor cells. Not only do malignant cells have an abnormally high mitotic rate, but they are structurally distinct from healthy body cells.

Table A-5 Benign and Malignant Tumors in the Major Tissue Types

Tissue	Description
Epithelia	
Carcinoma	Any cancer of epithelial origin
Adenocarcinoma	Cancers of glandular epithelia
Angiosarcomas	Cancers of endothelial cells
Mesotheliomas	Cancers of mesothelial cells
Connective tissues	
Fibromas	Benign tumors of fibroblast origin
Lipomas	Benign tumors of adipose tissue
Liposarcomas	Cancers of adipose tissue
Leukemias	Cancers of blood-forming tissues
Lymphomas	Cancers of lymphoid tissues
Chondromas	Benign tumors in cartilage
Chondrosarcomas	Cancers of cartilage
Osteomas	Benign tumors in bone
Osteosarcomas	Cancers of bone
Muscle tissues	
Myxomas	Benign muscle tumors
Myosarcomas	Cancers of skeletal muscle tissue
Cardiac sarcomas	Cancers of cardiac muscle tissue
Leiomyomas	Benign tumors of smooth muscle tissue
Leiomyosarcomas	Cancers of smooth muscle tissue
Neural tissues	
Gliomas	Cancers of neuroglial origin
Neuromas	Cancers of neuronal origin

If the tissue appears cancerous, other important questions must be answered, including the following:

- What is the measurable size of the primary tumor?
- Has the tumor invaded surrounding tissues?
- Has the cancer already metastasized to develop secondary tumors?
- Are any regional lymph nodes affected?

The answers to these questions are combined with observations from the physical exam, the biopsy results, and information from any imaging procedures to develop an accurate diagnosis and prognosis.

In an attempt to develop a standard system, national and international cancer organizations have developed the *TNM system* for staging (that is, identifying the stage of progression of) cancers. The letters stand for *tumor* (T) size and invasion, *lymph node* (N) involvement, and degree of metastasis (M):

- Tumor size is graded on a scale of 0 to 4. T0 indicates the absence of a primary tumor, and the largest dimensions and greatest amount of invasion are categorized as T4.

- Lymph nodes filter the tissue fluids from nearby capillary beds. The fluid, called *lymph,* then returns to the general circulation. Once cancer cells have entered the lymphatic system, they can spread very quickly throughout the body. Lymph node involvement is graded on a scale of 0 to 3. A designation of N0 indicates that no lymph nodes have been invaded by cancer cells. A classification of N1 to N3 indicates the involvement of increasing numbers of lymph nodes:

N1. Indicates involvement of a single lymph node less than 3 cm in diameter.

N2. Includes one medium-sized (3–6 cm) node or multiple nodes smaller than 6 cm.

N3. Indicates the presence of a single lymph node larger than 6 cm in diameter, whether or not other nodes are involved.

- Metastasis is graded on a scale of 0 to 1. M0 indicates that there is no evidence of metastasis, whereas M1 indicates that the cancer cells have produced secondary tumors in other portions of the body.

This grading system provides a general overview of the progression of the disease. For example, a tumor classified as T1N1M0 has a better prognosis than one classified as T4N2M1. The latter tumor will be much more difficult to treat. The grading system alone does not provide all the information needed to plan treatment, however, because different types of cancer progress in different ways, so therapies must vary accordingly. Thus *leukemia,* a cancer of the blood-forming tissues, will be treated differently than will colon cancer. We will consider specific treatments in discussions dealing with cancers that affect individual body systems. The next section provides a general overview of the strategies used to treat cancer.

Cancer Treatment

It is unfortunate that the media tend to describe cancer as though it were one disease rather than many. This simplistic perspective fosters the belief that some dietary change, air ionizer, or wonder drug will be found that can prevent or cure the affliction. There can be no single, universally effective cure for cancer, because there are too many separate causes, possible mechanisms, and individual differences.

Table A-6   Cancer Incidence and Survival Rates in the United States

Site	Estimated New Cases (1997)	Estimated Deaths (1997)	5-Year Survival Rates Diagnosis Date 1970–73	5-Year Survival Rates Diagnosis Date 1986–91
Digestive tract				
Esophagus	12,500	11,500	4%	11%
Stomach	22,400	14,000	13%	19%
Colon and rectum	131,200	54,900	47%	60%
Respiratory tract				
Lung and bronchus	178,100	160,400	10%	14%
Urinary tract				
Kidney and other urinary structures	30,900	11,820	46%	59%
Bladder	54,500	11,700	61%	82%
Reproductive system				
Breast	181,600	44,190	68%	84%
Ovary	26,800	14,200	36%	44%
Testis	7,200	350	72%	95%
Prostate gland	334,500	41,800	63%	87%
Nervous system	17,600	13,200	20%	28%
Circulatory system	103,200	57,490	22%	41%
Skin (melanoma only)	40,300	7,300	68%	87%

Data courtesy of the American Cancer Society.

The goal of cancer treatment is to achieve **remission.** A tumor in remission either ceases to grow or decreases in size. The treatment of malignant tumors must accomplish one of these two objectives to produce remission:

1. *Surgical removal or destruction of individual tumors.* Tumors containing malignant cells can be surgically removed or destroyed by radiation, heat, or freezing. These techniques are very effective if the treatment is undertaken before extensive metastasis has occurred. For this reason, early detection is important in improving survival rates for all forms of cancer.

2. *Killing metastatic cells throughout the body.* This is much more difficult and potentially dangerous, because healthy tissues are likely to be damaged at the same time. At present, the most widely approved treatments are *chemotherapy* and radiation.

Chemotherapy involves the administration of drugs that will either kill the cancerous tissues or prevent mitotic divisions. These drugs typically affect stem cells in normal tissues, and the side effects are usually unpleasant. For example, because chemotherapy slows the regeneration and maintenance of epithelia of the skin and digestive tract, patients lose their hair and experience nausea and vomiting. Several different drugs are often administered simultaneously or in sequence, because over time cancer cells can develop a resistance to a single drug. Chemotherapy is often used in the treatment of many kinds of metastatic cancer.

Massive doses of radiation are sometimes used to treat advanced cases of *lymphoma,* a cancer of the immune system. In this rather drastic procedure, enough radiation is administered to kill all the blood-forming cells in the body. After treatment, new blood cells must be provided by a bone marrow transplant. In later sections dealing with the lymphatic system, we will discuss marrow transplants, lymphomas, and other cancers of the blood.

An understanding of molecular mechanisms and cell biology is leading to new approaches that may revolutionize cancer treatment. One approach focuses on the fact that cancer cells are ignored by the immune system. In **immunotherapy** chemicals are administered that help the immune system recognize and attack cancer cells. More elaborate experimental procedures involve the creation of customized antibodies by the gene-splicing techniques discussed on p. 30. The resulting antibodies are specifically designed to attack the tumor cells in each particular patient. Although this technique shows promise, it remains difficult, costly, and very labor-intensive.

A second new procedure builds on the first. In **boron neutron capture therapy (BNCT),** antibodies made to attack cancer cells are labeled with an isotope of boron (B). After these antibodies are administered, the patient is irradiated with neutrons. These neutrons do not damage normal tissues. However, the boron atoms absorb neutrons and release alpha particles (2 neutrons + 2 protons). This radiation kills the cancer cells quite effectively. Because the cancer cells absorb the radiation, healthy tissues are unaffected.

Cancer and Survival

Advances in chemotherapy, radiation procedures, and molecular biology have produced significant improvements in the survival rates of several types of cancer. However, the improved survival rates indicated in Table A-6 reflect advances not only in therapy but also in early detection. Much of the credit goes to increased public awareness and con-

cern about cancer. In general, the odds of survival increase markedly if the cancer is detected early, especially before it undergoes metastasis. Despite the variety of possible cancers, the American Cancer Society has identified seven "warning signs" that mean it's time to consult a physician. These are presented in Table A-7.

Table A-7   Seven Warning Signs of Cancer

Change in bowel or bladder habits

A sore that does not heal

Unusual bleeding or discharge

Thickening or lump in breast or elsewhere

Indigestion or difficulty in swallowing

Obvious change in a wart or mole

Nagging cough or hoarseness

The Body Systems: Clinical and Applied Topics

This section relates aspects of the normal anatomy and physiology of each body system to specific clinical conditions, diagnostic procedures, and other relevant topics. We will briefly review each body system in order to identify patterns among the most common disorders that affect each system. Critical-Thinking Questions about each system and Clinical Problems involving multiple systems will give you the chance to apply the concepts you have learned.

The Integumentary System

The structures of the integumentary system include the skin, hair, nails, and several types of exocrine glands. The integumentary system has a variety of functions, including the protection of the underlying tissues, the maintenance of body temperature, the excretion of salts and water in sweat, cutaneous sensation, and the production of vitamin D_3.

The skin is the most visible organ of the body. As a result, abnormalities are easily recognized. A bruise, for example, typically creates a swollen and discolored area where the walls of blood vessels have been damaged. Changes in skin color, skin tone, and the overall condition of the skin commonly accompany illness or disease. These changes can assist in diagnosis. For example, extensive bruising without obvious cause may indicate a blood clotting disorder; a yellow color in the skin and mucous membranes may indicate *jaundice,* a sign that generally indicates some type of liver disorder. The general condition of the skin may also be significant. For example, color changes or changes in skin flexibility, elasticity, dryness, or sensitivity commonly follow the malfunctions of other organ systems.

EXAMINATION OF THE SKIN

FAP *p. 148*

When examining a patient, dermatologists use a combination of investigative interviews ("What has been in contact with your skin lately?" or "How does it feel?") and physical examination to arrive at a diagnosis. The condition of the skin is carefully observed. Notes are made concerning the presence of **lesions,** which are changes in skin structure caused by disease processes. Lesions are also called **skin signs,** because they are measurable, visible abnormalities of the skin surface. Figure A-11 diagrams the most common skin signs and related disorders.

The distribution of lesions may be an important clue to the source of the problem. For example, in *shingles* (herpes zoster) there are painful vesicular eruptions on the skin that follow the path of peripheral sensory nerves. A ring of slightly raised, scaly (papular) lesions is typical of fungal infections that may affect the trunk, scalp, and nails. Examples of skin disorders caused by infection or by allergic reactions are included in Table A-8, with descriptions of the related lesions. We consider skin lesions caused by trauma in the section titled "A Classification of Wounds" (p. 42).

Table A-8 (p. 39) considers signs on the skin surface, but signs involving the accessory organs of the skin can also be important. For example,

- Nails have a characteristic shape that can change due to an underlying disorder. An example is *clubbing* of the nails, commonly a sign of *emphysema* or *congestive heart failure.* In these conditions, the fingertips broaden, and the nails become distinctively curved.

- The condition of the hair can be an indicator of overall health. For example, depigmentation and coarseness of hair occur in the protein deficiency disease *kwashiorkor.*

Diagnosing Skin Conditions

Table A-9 (p. 39) introduces several major types of vascular lesions. A single vascular lesion, such as a hematoma, may have multiple causes. This is one of the challenges facing dermatologists; the signs may be apparent, but the underlying causes may not. Making matters more difficult, many different skin disorders produce the same signs of uncomfortable sensations. For example, **pruritis** (proo-RĪ-tus), an irritating itching sensation, is an extremely common symptom associated with a variety of skin conditions. Questions con-

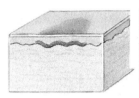

A flat **macule** is a localized change in skin color. Example: freckles

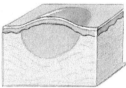

Accumulation of fluid in the papillary dermis may produce a **wheal**, a localized elevation of the overlying epidermis. Example: hives

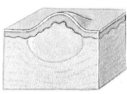

A **papule** is a solid elevated area containing epidermal and papillary dermal components. Example: mosquito or other insect bite

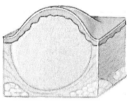

Nodules are large papules that may extend into the subcutaneous layer. Example: cyst

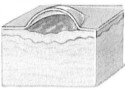

A **vesicle**, or blister, is a papule with a fluid core. A large vesicle may be called a bulla. Example: second-degree burn

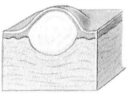

A **pustule** is a papule-sized lesion filled with pus. Example: acne pimple

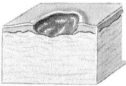

An **erosion**, or ulcer, may occur following the rupture of a vesicle or pustule. Eroded sites have lost part or all of the normal epidermis. Example: decubitis ulcer

A **crust** is an accumulation of dried sebum, blood, or interstitial fluid over the surface of the epidermis. Examples: seborrheic dermatitis, scabs, impetigo

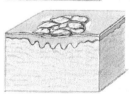

Scales form as a result of abnormal keratinization. They are thin plates of cornified cells. Example: psoriasis

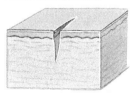

A **fissure** is a split in the integument that extends through the epidermis and into the dermis. Example: athlete's foot

Figure A-11   Skin Signs

cerning medical history, medications, possible sources of infection, environmental exposure, and other signs, such as bleeding at the gums, can be the key to making an accurate diagnosis.

Pain is another common symptom of many skin disorders. Although pain is unwelcome, cutaneous sensation is an important function of the integumentary system. Its importance is dramatically demonstrated in the condition of *Hansen's disease,* or *leprosy.* Hansen's disease is caused by a bacterium that has an affinity for cool regions of the body. The bacterium destroys cutaneous nerve endings that are sensitive to touch, pain, hot, and cold. Damage to the extremities then occurs and accumulates because the individual is no longer aware of painful stimuli. We will consider Hansen's disease in more detail in a later section.

Examples of diagnostic tests that may prove useful include the following:

- Scrapings of affected tissue, a process often performed to check for fungal infections
- Culturing of bacteria removed from a lesion to aid in identification and to determine drug sensitivity
- Biopsy of affected tissue to view cell structure
- Skin tests. Various types of disorders can be detected through use of a *skin test.* In a skin test, a localized area of the skin is exposed to an inactivated pathogen, a portion of a pathogen, or a substance capable of producing an allergic reaction in sensitive individuals. Exposure may be via injection or surface application. For example, in a tuberculosis skin test a small quantity of tuberculosis antigens is injected *intradermally* (*intra*, within). If the individual has been infected in the past or currently has tuberculosis, *erythema* (*erythros*, red; a change in skin color) and swelling will occur at the injection site 24–72 hours later *Patch testing* is used to check sensitivity to *allergens*, environmental agents that can cause allergic reactions. In a patch test, the allergen is applied to the skin surface. If erythema, swelling, and/or itching develop, the individual is sensitive to that allergen.

Disorders of Keratin Production
FAP *p. 151*

Not all skin signs are the result of infectious, traumatic, or allergic conditions. Excessive production of keratin if called **hyperkeratosis** (hī-per-ker-a-TŌ-sis). The most obvious effects are easily observed as calluses and corns. Calluses are thickened patches that appear on already thick-skinned areas, such as the palms of the hands or

Table A-8 Skin Signs of Various Disorders

Cause	Examples	Resulting Skin Lesions
Viral infections	Chickenpox	Lesions begin as macules and papules but develop into vesicles
	Measles (rubeola)	A maculopapular rash that begins at the face and neck and spreads to the trunk and extremities
	Erythema infectiousum (Fifth's disease)	A maculopapular rash that begins on the cheeks (slapped-cheek appearance) and spreads to the extremities
	Herpes simplex	Raised vesicles that heal with a crust
Bacterial infections	Impetigo	Vesiculopustular lesions with exudate and yellow crusting
Fungal infections	Ringworm	An annulus (ring) of scaly papular lesions with central clearing
Parasitic infections	Scabies	Linear burrows with a small papule at one end
	Lice (pediculosis)	Dermatitis: excoriation (scratches) due to pruritis (itching)
Allergies to medication	Penicillin	Wheals (urticaria or hives)
Food allergies	Eggs, certain fruits	Wheals
Environmental allergies	Poison ivy	Vesicles

the soles or heels of the foot, in response to chronic abrasion and distortion. Corns are more localized areas of excessive keratin production that form in areas of thin skin on or between the toes.

In **psoriasis** (sō-RĪ-a-sis), the stratum germinativum becomes unusually active, causing hyperkeratosis in specific areas, including the scalp, elbows, palms, soles, groin, and nails. Normally, an individual stem cell divides once every 20 days, but in psoriasis it may divide every day and a half. Keratinization is abnormal and typically incomplete by the time the outer layers are shed. The affected areas appear to be covered with small, silvery scales that continuously flake off. Psoriasis develops in 20–30 percent of the individuals with an inherited tendency for the condition. Roughly 5 percent of the U.S. population has psoriasis to some degree, often triggered by stress and anxiety. Most cases are painless and treatable.

Xerosis (ze-RŌ-sis), or dry skin, is a common complaint of the elderly and people who live in arid climates. Under these conditions, cell membranes in the outer layers of skin gradually deteriorate, and the stratum corneum becomes more a collection of scales than a single sheet. The scaly surface is much more permeable than an intact layer of keratin, and the rate of insensible perspiration increases. In persons afflicted with severe xerosis, the rate of insensible perspiration may increase by up to 75 times.

 Skin Cancers   FAP *p. 152*

Almost everyone has several benign tumors of the skin; freckles and moles are examples. Skin cancers are the most common form of cancer. The most common skin cancers are caused by prolonged exposure to the ultraviolet radiation in sunlight.

Table A-9 Examples of Vascular Lesions

Lesion	Features	Some Possible Causes
Ecchymosis	Reddish purple, blue, or yellow bruising related to trauma	Trauma; blood clotting disorder if bruising is abnormal; some vitamin deficiencies; thrombocytopenia; increased tendency to bruise can be aging or sun-damaged skin
Hematoma	Pooling of blood from a broken vessel forming a mass; associated with pain and swelling	
Petechiae	Small red to purple pinpoint dots appearing in clusters	Leukemia; septicemia (toxins in blood); thrombocytopenia
Erythema	Red flushed color of skin due to dilation of blood vessels in the skin	Extensive: drug reactions; Localized: burns, dermatitis

A **basal cell carcinoma** (Figure A-12a) is a malignant cancer that originates in the germinativum (basal) layer. This is the most common skin cancer. Roughly two-thirds of these cancers appear in body areas subjected to chronic UV exposure. Genetic factors have been identified that predispose people to this condition. **Squamous cell carcinomas** are less common but almost totally restricted to areas of sun-exposed skin. Metastasis seldom occurs in squamous cell carcinomas and virtually never in basal cell carcinomas, and most people survive these cancers. The usual treatment involves surgical removal of the tumor, and 95 percent of patients survive 5 years or more after treatment. (This statistic, the 5-year survival rate, is a common method of reporting long-term prognoses.)

Compared with these common and seldom life-threatening cancers, **malignant melanomas** (mel-a-NO-mas) (Figure A-12b) are extremely dangerous. In this condition, cancerous melanocytes grow rapidly and metastasize through the lymphatic system. The outlook for long-term survival changes dramatically, depending on when the condition is diagnosed. If the cancer is localized, the 5-year survival rate is 99 percent; if it is widespread, the survival rate drops to 14 percent.

To detect melanoma at an early stage, you must know what to look for when you examine your skin. The mnemonic ABCD makes it easy to remember the key points of detection:

A is for *asymmetry.* Melanomas tend to be irregular in shape. Typically they are raised; they may ooze or bleed.

B is for *border.* Generally unclear, irregular, and in some cases notched.

C is for *color.* Generally mottled, with many different colors (tan, brown, black, red, pink, white, and/or blue).

D is for *diameter.* A growth more than about 5 mm (0.2 in.) in diameter, or roughly the area covered by the eraser on a pencil, is dangerous.

A new experimental treatment for melanoma uses genetic engineering technology to manufacture antibodies that target the MSH (melanocyte-stimulating hormone) receptors on the surfaces of melanocytes. Melanocytes coated with these antibodies are then recognized and attacked by cells of the immune system.

Fair-skinned individuals who live in the tropics are most susceptible to all forms of skin cancer, because their melanocytes are unable to shield them from the UV radiation. Sun damage can be prevented by avoiding exposure to the sun during the middle of the day and by using a sunblock (not a tanning oil)—a practice that also delays the cosmetic problems of aging and wrinkling. *Everyone* who spends any time out in the sun should choose

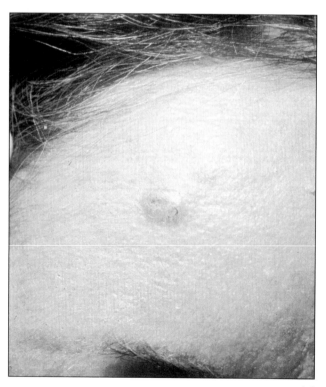

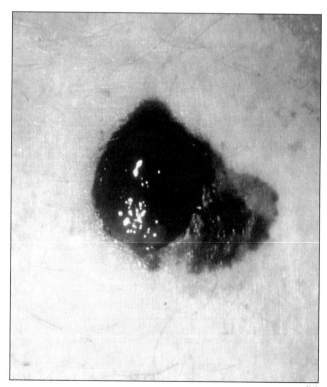

Figure A-12   Skin Cancers.
(a) Basal cell carcinoma. (b) Melanoma.

(a)

(b)

a broad-spectrum sunblock with a sun protection factor (SPF) of at least 15; blonds, redheads, and people with very fair skin are better off with an SPF of 20 to 30. (The risks are the same for those who spend time in a tanning salon or tanning bed.) Wearing a hat with a brim and panels to shield the neck and face provides added protection.

The use of sunscreens will be even more important as the ozone gas in the upper atmosphere is further destroyed by our industrial emissions. Ozone absorbs UV before it reaches Earth's surface, and in doing so, is assists the melanocytes in preventing skin cancer. Australia, the continent that is most affected by the depletion of ozone above the South Pole (the "ozone hole"), is already reporting an increased incidence of skin cancers.

DISORDERS OF THE ACCESSORY ORGANS OF THE SKIN

There are many disorders that affect accessory organs of the skin, especially the hair and the exocrine glands. We will consider only three relatively common examples: *baldness, hirsutism,* and *acne.*

Baldness and Hirsutism FAP *p. 159*

Two factors interact to determine baldness. A bald individual has a genetic susceptibility triggered by large quantities of male sex hormones. Many women carry the genetic background for baldness, but unless major hormonal abnormalities develop, as in certain endocrine tumors, nothing happens. (In some women, however, hair thins after menopause.)

Male pattern baldness affects the top of the head and forehead first, only later reducing the hair density along the sides. Thus hair follicles can be removed from the sides and implanted on the top or front of the head, temporarily delaying a receding hairline. This procedure is expensive, and not every hair transplant is successful.

Alopecia areata (al-ō-PĒ-shē-uh ar-ē-AH-ta) is a localized hair loss that can affect either gender. The cause is not known, and the severity of hair loss varies from case to case. This condition is associated with several disorders of the immune system. It has also been suggested that periods of stress may promote alopecia areata in individuals who are genetically prone to baldness.

Hairs are dead, keratinized structures, and no amount of oiling, shampooing, or dousing with kelp extracts, vitamins, or nutrients will influence the follicle buried in the dermis. Skin conditions that affect follicles can contribute to hair loss; temporary baldness can also result from exposure to radiation or to many of the toxic (poisonous) drugs used in cancer therapy.

Untested treatments for baldness were banned by the Food and Drug Administration (FDA) in 1984. When rubbed onto the scalp, *Minoxidil,* a drug originally marketed for the control of blood pressure, appears to stimulate inactive follicles. It is now available on a nonprescription basis (*Rogaine*™). Treatment involves applying a 2 percent solution to the scalp twice daily. In one study, after 4 months, more than one-third of patients reported satisfactory results. It is most effective in preventing the progression of early hair loss, and it must be continued indefinitely.

Hirsutism (HER-soot-izm; *hirsutus,* "bristly") refers to the growth of hair on women in patterns generally characteristic of men. Because considerable overlap exists between the two genders in terms of normal hair distribution, and because there are significant racial and genetic differences, the precise definition is more often a matter of personal taste than of objective analysis. Age and sex hormones may play a role, for hairiness increases late in pregnancy and menopause produces a change in body hair patterns.

Severe hirsutism is commonly associated with abnormal androgen (male sex hormone) production, either in the ovaries or in anal glands. Unwanted follicles can be permanently "turned off" by plucking a growing hair and removing the papilla. *Electrocautery,* which destroys the follicle with a jolt of electricity, requires the services of a professional, but the results are more reliable. Patients may also be treated with drugs that reduce or prevent androgen stimulation of the follicles.

Acne FAP *p. 160*

Individuals with a genetic tendency toward acne have larger-than-average sebaceous glands. When the ducts become blocked, the secretions accumulate, inflammation develops, and bacterial infection may occur. The condition generally surfaces at puberty, as sex hormone production accelerates and stimulates the sebaceous glands. Their secretory output may be further encouraged by anxiety, stress, physical exertion, certain foods, and drugs.

The visible signs of acne are called **comedos** (ko-MĒ-dōz). "Whiteheads" contain accumulated, stagnant secretions. "Blackheads" contain more solid material that has been invaded by bacteria. Although neither condition indicates the presence of dirt in the pores, washing may help keep superficial oiliness down.

Acne generally fades after sex hormone concentrations stabilize. Topical antibiotics, vitamin A derivatives, such as *Retin-A*™, or peeling agents may help reduce inflammation and minimize scarring. In cases of severe acne, the most effective usually involves the discouragement of bacteria by the administration of antibiotic drugs. Because oral antibiotic therapy has risks, including the development of antibiotic-resistant bacteria, however, this therapy is not used unless other treatment methods have failed. Truly dramatic improvements in severe cases have been obtained with the prescription drug *Accutane*™. This compound is structurally similar to vitamin A, and it reduces oil gland activity on a long-term basis. A number of minor side effects, including dry skin rashes, have been reported; these apparently disappear when the treatment ends. However, the use of Accutane during the first month of pregnancy carries a 25-times-normal risk of inducing birth defects, so women must avoid pregnancy while using this drug.

A CLASSIFICATION OF WOUNDS
FAP *p. 162*

Injuries, or **trauma,** involving the integument are very common, and a number of terms are used to describe them. Traumatic injuries generally affect all components of the integument, and each type of wound presents a different series of problems to clinicians attempting to limit damage and promote healing.

An **open wound** is an injury producing a break in the epithelium. The major categories of open wounds are illustrated in Figure A-13. **Abrasions** (Figure A-13a) are the result of scraping against a solid object. Bleeding may be slight, but a considerable area may be open to invasion by microorganisms. **Incisions** (Figure A-13b) are linear cuts produced by sharp objects. Bleeding may be severe if deep vessels are damaged. The bleeding may help flush the wound, and closing the incision with bandages or stitches can limit the area open to infection while healing is underway. A **laceration** (Figure A-13c) is a jagged, irregular tear in the surface produced by solid impact or by an irregular object. Tissue damage is extensive, and repositioning the opposing sides of the injury may be difficult. Despite the bleeding that generally occurs, lacerations are prone to infection. **Punctures** (Figure A-13d) result when slender, pointed objects pierce the epithelium. Little bleeding results, and any microbes delivered under the epithelium in the process are likely to find conditions to their liking. In an **avulsion** (Figure A-13e), chunks of tissue are torn away by brute force in, for example, an auto accident, explosion, or other such incident. Bleeding may be considerable, and even more serious internal damage may be present.

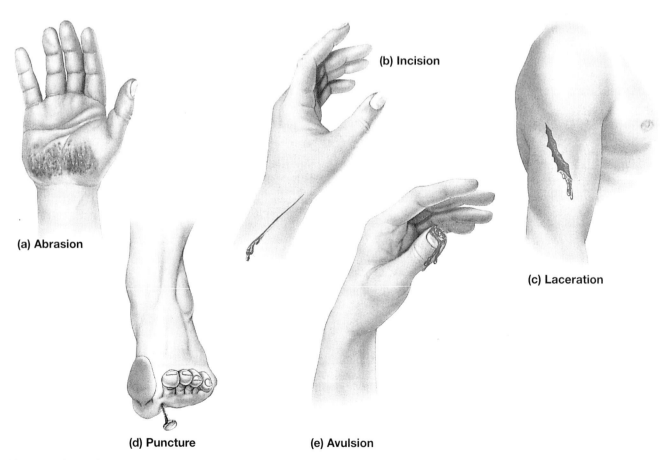

(a) Abrasion

(b) Incision

(c) Laceration

(d) Puncture

(e) Avulsion

Figure A-13   Major Types of Open Wounds

Closed wounds may affect any internal tissue, but because the epithelium is intact, the likelihood of infection is reduced. A **contusion** is a bruise causing bleeding in the dermis. "Black and blue" marks are familiar examples of contusions; most are not dangerous, but contusions of the head, such as "black eyes," may be accompanied by life-threatening intracranial bleeding. Closed wounds affecting internal organs and organ systems are almost always serious threats to life.

TREATMENT OF SKIN CONDITIONS

Topical **anti-inflammatory drugs,** such as the steroid **hydrocortisone,** can be used to reduce the redness and itching that accompany a variety of skin conditions. Some systemic (injected or swallowed) drugs may also be helpful; aspirin is a familiar systemic drug with anti-inflammatory properties. Isolated growths, such as skin tumors or warts, may be surgically removed. Alternatively, abnormal cells may be destroyed by electrical currents (**electrosurgery**) or by freezing (**cryosurgery**). Ultraviolet radiation may help conditions such as acne or psoriasis, whereas use of a sunscreen or a sunblock is important in controlling outbreaks caused by sensitivity to the sun.

℞ Synthetic Skin   FAP *p. 168*

Traditional skin grafts involve covering areas where skin has been completely destroyed or lost. Pieces of undamaged skin from other areas of the body are used. If the damaged area is large, there may not be enough normal skin available for grafting. *Epidermal culturing* can produce a new epithelial layer to cover a burn site. The cells in a 3-cm^2 epidermal sample from a burn patient can be cultured in a controlled environment that contains epidermal growth factors, fibroblast growth factors, and other stimulatory chemicals. The number of cells doubles in the first 18 hours. The cells are then separated, and culturing continues. After 3 days in culture, the number of cells has increased by more than 16 times. After a week, the number has increased by over 300 times. The cells are now in small clusters, and they have begun forming lay-

ers. Germinative cells are on the bottom, attached to the glass of the culturing vessel. The more superficial layers of cells roughly resemble those of normal epidermis, although keratin production does not occur. This artificially produced epidermis can now be transplanted to cover an injury site. The larger the area that must be covered, the longer the culturing process continues. After 3 or 4 weeks of culturing, the cells obtained in the original sample can provide enough epidermis to cover the entire body surface of a typical adult. The absence of a dermis makes this grafter epidermis less flexible and more fragile—a particular problem around joints and areas of friction. Also, these skin grafts do not contain accessory organs, such as hairs or sweat glands.

The success of grafting is limited by the contamination of the wound site by bacteria while the epidermis is being cultured. The contamination is prevented by the use of a skin *allograft.* In this procedure, skin from a frozen cadaver is removed and placed over the wound as a temporary method of sealing the surface. Before the immune system of the patient attacks it, the graft will be partially or completely removed to provide a binding site for the epidermal transplant. After grafting, the complete reorganization and repair of the dermis and epidermis at the injury site takes approximately 5 years.

A second new procedure provides a model for dermal repairs that takes the place of normal tissue. A special synthetic skin is used. The imitation has a plastic (silastic) "epidermis" and a dermis composed of collagen fibers and ground cartilage. The collagen fibers are taken from cow skin, and the cartilage is taken from sharks. Over time, fibroblasts migrate among the collagen fibers and gradually replace the model framework with their own. The silastic epidermis is intended only as a temporary cover that will be replaced by either skin grafts or a cultured epidermal layer.

Several other procedures have either been approved or are awaiting FDA approval. All involve the application of a temporary dermal layer that can guide the repair process. The most interesting approach, awaiting FDA approval, involves the application of a dermal layer created by culturing fibroblasts obtained from the foreskins of circumcised newborn male infants.

CRITICAL-THINKING QUESTIONS

1-1. Charlie is badly burned in a fireworks accident on the Fourth of July. When he reaches the emergency room, the examining physician determines the severity of the incident as a third-degree burn. The physician would likely order

    a. IV (intravenous) fluids and electrolytes

    b. antibiotics

    c. a high-nutrient diet

    d. all of the above

1-2. John has sailed boats professionally for years. John's wife, Jane, has noticed a mole on the back of his neck. The mole has suspicious characteristics, so John visits his dermatologist. The dermatologist makes note of the irregular border of the nevus (mole), the unusual red and white coloration, and the surrounding inflammation. Jane reports that the mole has undergone these changes in the last month. The dermatologist recommends immediate biopsy of the nevus, suspecting

    a. basal cell carcinoma

    b. melanoma

    c. squamous cell carcinoma

    d. hemangioma

1-3. Carrie, a vegetarian, tells her physician about a yellow discoloration of the palms of her hands and the soles of her feet. Further inspection reveals that Carrie's forehead and the area around her nose are also yellow. The physician determines that Carrie has not been exposed to hepatitis and that the results of her liver function tests, including serum bilirubin levels, are normal. What might be the problem?

1-4. Sam likes to work on his motorcycle on weekends, and he spends a great deal of time cleaning engine parts with organic solvents. He doesn't wear gloves while he works. After he is finished, Sam notices that his hands feel painfully dry for several hours. What is the probable cause?

1-5. During pregnancy, it is not uncommon for skin and hair color to become darker, especially around the genitals, nipples, and face. What is the probable cause of these changes?

NOTES:

The Skeletal System

The skeletal framework of the body is composed of at least 206 bones and the associated tendons, ligaments, and cartilages. The skeletal system has a variety of important functions, including the support of soft tissues, blood cell production, mineral and lipid storage, and, through its relationships with the muscular system, the support and movement of the body as a whole. Skeletal system disorders can thus affect many other systems. The skeletal system is in turn influenced by the activities of other systems. For example, weakness or paralysis of skeletal muscles will lead to a weakening of the associated bones.

Although the bones you study in the lab may seem rigid and permanent structures, the living skeleton is dynamic and undergoes continuous remodeling. The remodeling process involves bone deposition by osteoblasts and osteolysis, or dissolution of bone matrix, by osteoclasts. As indicated in Figure A-14, the net result of the remodeling varies with the following five factors:

1. *The age of the individual.* During development, bone deposition occurs faster than bone resorption; as the amount of bone increases, the skeleton grows. At maturity, bone deposition and resorption are in balance. As the aging process continues, the rate of bone deposition declines, and the bones become less massive. This gradual weakening, called *osteopenia,* begins at age 30–40 and may ultimately progress to osteoporosis (p. 52).

2. *The applied physical stresses.* Heavily stressed bones become thicker and stronger, and lightly stressed bones become thinner and weaker. Skeletal weakness can therefore result from muscular disorders, such as *myasthenia gravis* (p. 65) and the *muscular dystrophies* (p. 64), and from conditions that affect CNS motor neurons, such as spinal cord injuries (p. 76), *demyelination disorders* (p. 73), and *multiple sclerosis* (p. 75).

3. *Circulating hormone levels.* Changing levels of growth hormone, androgens and estrogens, thyroid hormones, parathyroid hormone, and calcitonin increase or decrease the rate of mineral deposition in bone. As a result, many disorders of the endocrine system affect the skeletal system. For example,

   • Conditions affecting the skin, liver, or kidneys can interfere with calcitriol production.

   • Thyroid and parathyroid disorders can alter thyroid hormone, parathyroid hormone, or calcitonin levels.

   • Pituitary gland disorders and liver disorders can affect GH or somatomedin production.

   • Reproductive system disorders can alter circulating levels of androgens or estrogens.

   We will describe many of these conditions in the section dealing with the endocrine system (pp. 96–105).

4. *Rates of calcium and phosphate absorption and excretion.* For bone mass to remain constant, the rate of calcium and phosphate excretion, primarily at the kidneys, must be balanced by the rate of calcium and phosphate absorption at the digestive tract. Dietary calcium deficiencies or problems that reduce calcium and phosphate absorption at the digestive tract will thus have a direct effect on the skeletal system.

5. *Genetic or environmental factors.* Genetic or environmental factors may affect the structure of bone or the remodeling process. There are a number of inherited abnormalities of skeletal development, such as *Marfan's syndrome* and *achondroplasia* (p. 50). When bone fails to form embryonically in certain areas, underlying tissues can be exposed and associated function can be altered. This abnormality occurs in a *cleft palate* (FAP *p. 217*) and in *spina bifida* (FAP *p. 219*). Environmental stresses can alter the shape and contours of developing bones. For example, some cultures use lashed boards to form an infant's or child's skull to a shape considered fashionable. Environmental forces can also result in the formation of bone in unusual locations. These *heterotopic bones* (p. 48) may develop in a variety of connective tissues exposed to chronic friction, pressure, or mechanical stress. For example, cowboys in the nineteenth century sometimes developed heterotopic bones in the dermis of the thigh, from friction with the saddle.

Figure A-15 (p. 47) diagrams the major classes of skeletal disorders that affect the structure and function of bones (Figure A-15a) and joints (Figure A-15b). Some of these disorders are the result of conditions that affect primarily the skeletal system (*osteosarcoma, osteomyelitis*); others result from problems originating in other systems (*acromegaly, rickets*).

Traumatic injuries, such as fractures or dislocations, and infections also damage cartilages, tendons, and ligaments. A somewhat different array of conditions affect the soft tissues of the bone marrow. Areas of red bone marrow contain the stem cells for red blood cells, white blood cells, and platelets. Blood diseases characterized by blood cell overproduction (*leukemia, polycythemia,* pp. 118, 110) or underproduction (several *anemias,* pp. 116, 117) result in bone marrow abnormalities.

6

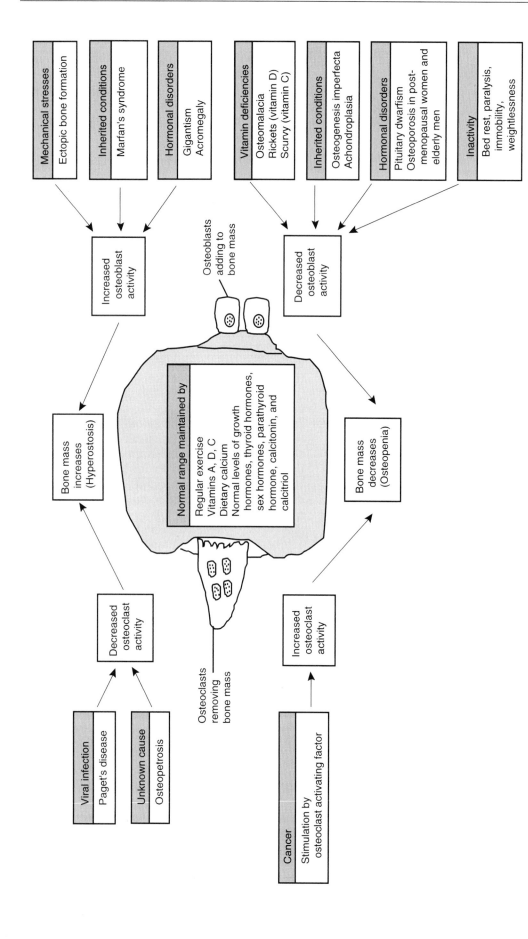

Figure A-14   Factors Affecting Bone Mass

6

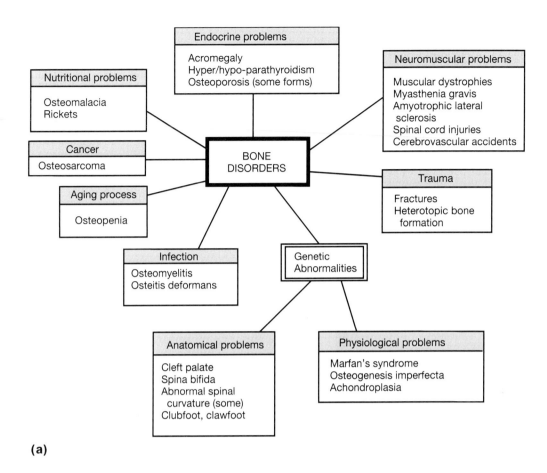

(a)

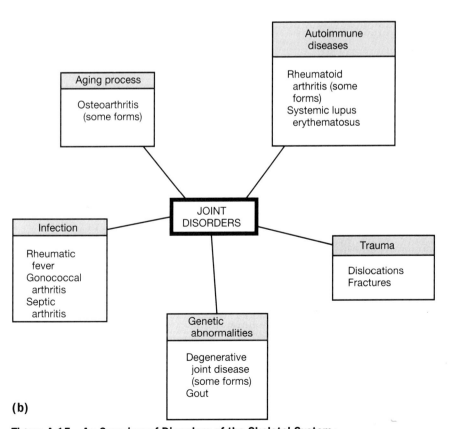

(b)

Figure A-15   An Overview of Disorders of the Skeletal System.
(a) Bone disorders. (b) Joint disorders.

THE SYMPTOMS OF BONE AND JOINT DISORDERS

A common symptom of a skeletal system disorder is pain. Bone pain and joint pain are common symptoms associated with many bone disorders. As a result, the presence of pain does not provide much help in identifying a specific bone or joint disorder. A person may be able to tolerate chronic, aching bone or joint pain and therefore not seek medical assistance until more-definitive symptoms appear. By then the condition is relatively advanced. For example, a symptom that may require immediate attention is a *pathologic fracture.* Pathologic fractures are the result of weakening of the skeleton by disease processes, such as *osteosarcoma* (a bone cancer). These fractures can be caused by physical stresses that are easily tolerated by normal bones.

EXAMINATION OF THE SKELETAL SYSTEM   FAP *p. 179*

The bones of the skeleton cannot be seen without relatively sophisticated equipment. However, there are a number of physical signs that can assist in the diagnosis of a bone or joint disorder. Important factors noted in the physical examination include the following:

- *Limitation of movement or stiffness.* Many joint disorders, such as the various forms of arthritis, will restrict movement or produce stiffness at one or more joints.

- *The distribution of joint involvement and inflammation.* In a *monoarthritic* condition, only one joint is affected. In a *polyarthritic* condition, several joints are affected simultaneously.

- *Sounds associated with joint movement. Bony crepitus* (KREP-i-tus) is a crackling or grating sound generated during movement of an abnormal joint. The sound may result from the movement and collision of bone fragments following an articular fracture or from friction at an arthritic joint.

- *The presence of abnormal bone deposits.* Thickened, raised areas of bone develop around fracture sites during the repair process. Abnormal bone deposits may also develop around the joints in the fingers. These deposits are called *nodules* or *nodes.* When palpated, nodules are solid and painless. Nodules, which can restrict movement, commonly form at the interphalangeal joints of the fingers in osteoarthritis.

- *Abnormal posture.* Bone disorders that affect the spinal column can result in abnormal posture. This result is most apparent when the condition alters the normal spinal curvature.

Examples include *kyphosis, lordosis,* and *scoliosis* (p. 53). A condition involving an intervertebral joint, such as a herniated disc, will also produce abnormal posture and movement.

Table A-10 summarizes descriptions of the most important diagnostic procedures and laboratory tests that can be used to obtain information about the status of the skeletal system.

☤ Heterotopic Bone Formation

FAP *p. 180*

Heterotopic (*hetero,* different + *topos,* place), or **ectopic** (*ektos,* outside), bones are those that develop in unusual places. Such bones dramatically demonstrate the adaptability of connective tissues. Physical or chemical events can stimulate the development of osteoblasts in normal connective tissues. For example, sesamoid bones develop within tendons near points of friction and pressure. Bone may also appear within a large blood clot at an injury site or within portions of the dermis subjected to chronic abuse. Other triggers include foreign chemicals and problems that affect calcium excretion and storage.

Almost any connective tissues may be affected. Ossification within a tendon or around joints can painfully interfere with movement. Bone formation may also occur within the kidneys, between skeletal muscles, within the pericardium, in the walls of arteries, or around the eyes.

Myositis ossificans (mī-ō-SĪ-tis os-SIF-i-kans) involves the deposition of bone around skeletal muscles. A muscle injury can trigger a minor case. Severe cases have no known cause, but they certainly provide the most dramatic demonstrations of heterotopic bone formation. If the process does not reverse itself, the muscles of the back, neck, and upper limbs will gradually be replaced by bone. The extent of the conversion can be seen in Figure A-16 (p. 50). Figure A-16a shows the skeleton of a healthy adult male; Figure A-16b shows the skeleton of a 39-year-old man with advanced myositis ossificans. Several of the vertebrae have fused into a sold mass, and major muscles of the back, shoulders, and hips have undergone extensive ossification. Compare this figure with Figures 11-3b and 11-14a in the text to identify the specific muscles.

☤ Abnormalities of Skeletal Development   FAP *p. 186*

There are several inherited conditions that result in abnormal bone formation. Three examples are *osteogenesis imperfecta, Marfan's syndrome,* and *achondroplasia.*

Osteogenesis imperfecta (im-per-FEK-ta) is an inherited condition, appearing in 1 individual in about 20,000, that affects the organization of collagen fibers. Osteoblast function is impaired, growth is abnormal, and the bones are very fragile,

Table A-10   Examples of Tests Used in the Diagnosis of Bone and Joint Disorders

Diagnostic Procedure	Method and Result	Representative Uses	Notes
X-ray of bone and joint	Standard X-ray; film sheet with radiodense tissues in white on a black background	Detection of fracture, tumor, dislocation, reduction in bone (mass, and bone infection osteomyelitis)	
Bone scans	Injected radiolabeled phosphate accumulates in bones, and radiation emitted is converted into an image	Especially useful in the diagnosis of metastatic bone cancer; can also be used to detect fractures, early infection and degenerative bone diseases	
Arthrocentesis	Insertion of a needle into joint for aspiration of synovial fluid	See section on analysis of synovial fluid (below)	
Arthroscopy	Insertion of fiber-optic tubing into a joint cavity; attached camera displays joint interior	Detection of abnormalities of the menisci and ligamental tears; useful in differential diagnosis of joint disorders	
MRI	Standard MRI (FAP p. 26), produces computer-generated image	Observation of bone and soft tissue abnormalities	

Laboratory Test	Normal Values in Blood Plasma or Serum	Significance of Abnormal Values	Notes
Alkaline phosphatase	Adults: 30–85 mIU/ml Children: 60–300 mIU/ml (higher values occur during adolescent bone growth)	Elevated levels in adults due to abnormal osteoblast activity; elevated levels occur in bone cancer, osteitis deformans, multiple myeloma	Isoenzymes are used to differentiate between liver and bone disorders; ALP_2 is the isoenzyme studied in bone diseases
Calcium	Adults: 8.5–10.5 mg/dl Children 8.5–11.5 mg/dl	Elevated levels occur in bone cancers and multiple fractures and in patients who are immobilized for a prolonged period	
Phosphorus	Adults: 2.3–4.7 mg/dl Children: 4.0–7.0 mg/dl	Typically elevated when calcium levels are low; elevated levels occur in acromegaly, parathyroid disorders, and bone tumors	
Uric acid	Adult Males: 3.5–8.0 mg/dl Adult Females: 2.8–6.8 mg/dl	Elevated levels occur with gout, a form of arthritis that develops when uric acid crystals build up in the joint	Uric acid is a product of purine metabolism; levels vary during certain times of day
Rheumatoid factor	Adults: A negative result is normal; 1:20–1:80 titer or higher is positive for rheumatoid arthritis	On average, about 75% of people diagnosed with rheumatoid arthritis have a positive test for this factor	Collagen diseases and systemic lupus erythematosus may give positive results
Synovial fluid analysis	WBC:<200/mm^3 RBC: none Glucose: < 10 mg/dl below serum glucose levels Protein: 1.8 g/dl No uric acid crystals	Elevation of white blood cell count indicates bacterial infection; mild elevation indicates inflammatory process; decreased glucose levels and decreased viscosity of fluid indicates inflammation of joint; uric acid crystals indicate gout	Sample can be cultured for infecting microorganism

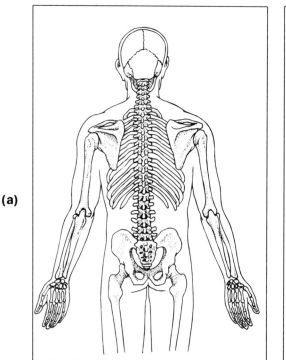
(a)

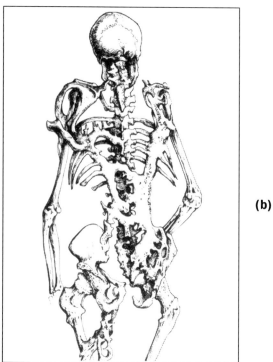
(b)

Figure A-16 Heterotopic Bone Formation.
(a) Skeleton of a healthy adult male, posterior view. (b) Skeleton of an adult male with advanced myositis ossificans. [(a) 1974, L. B. Halstead Wykerman Publications (London) Ltd.]

leading to progressive skeletal deformation and repeated fractures. Fibroblast activity is also affected, and the ligaments and tendons are very "loose," permitting excessive movement at the joints.

Marfan's syndrome (Figure A-17a) is also linked to defective connective tissue structure. Extremely long and slender limbs, the most obvious physical indication of this disorder, result from excessive cartilage formation at the epiphyseal plates. (Marfan's syndrome is discussed further on p. 126.)

Achondroplasia (ā-kon-drō-PLĀ-sē-uh) (Figure A-17b) also results from abnormal epiphyseal activity. In this case the epiphyseal plates grow unusually slowly, and the individual develops short, stocky limbs. Although other skeletal abnormalities occur, the trunk is normal in size, and sexual and mental development remain unaffected. The adult will be an *achondroplastic dwarf.*

In **osteomalacia** (os-tē-ō-ma-LĀ-shē-uh; *malakia,* softness) the size of the skeletal elements remains the same, but their mineral content decreases, softening the bones. The osteoblasts are working hard, but the matrix isn't accumulating enough calcium salts. This condition can occur in adults or children whose diet contains inadequate levels of calcium or vitamin D_3. The excessive formation of bone is termed **hyperostosis** (hī-per-os-TŌ-sis). In **osteopetrosis** (os-tē-ō-pe-TRŌ-sis; *petros,* stone), the total mass of the skeleton gradually increases because of a decrease in osteoclast activity. Remodeling stops, and the

shapes of the bones gradually change. Osteopetrosis in children produces a variety of skeletal deformities. The primary cause of this relatively rare condition is unknown.

In **acromegaly** (*akron,* extremity + *megale,* great) (Figure A-17c) an excessive amount of growth hormone is released after puberty, when most of the epiphyseal plates have already closed. Cartilages and small bones respond to the hormone, however, resulting in abnormal growth at the hands, feet, lower jaw, skull, and clavicle.

℞ Stimulation of Bone Growth and Repair FAP *p. 189*

Despite the considerable capacity for bone repair, every fracture does not heal as expected. A *delayed union* is one that proceeds more slowly than anticipated. *Nonunion* may occur as a result of complicating infection, continued movement, or other factors preventing complete callus formation.

There are several techniques for inducing bone repair. Surgical bone grafting is the most common treatment for nonunion. This method immobilizes the bone fragments and provides a bony model for the repair process. Dead bone or bone fragments can be used; alternatively, living bone from another site, such as the iliac crest or part of a rib, can be inserted. Surgeons can insert a shaped patch, made by mixing crushed bone and water, as an

6

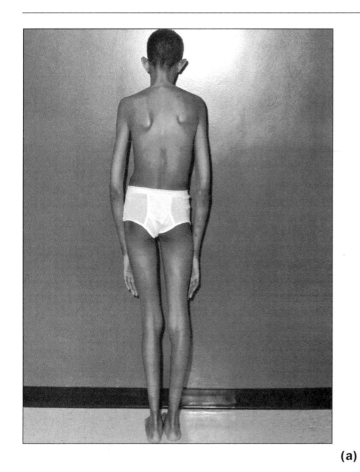

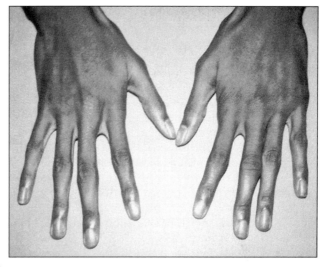

(a)

(b)

(c)

Figure A-17   Disorders of Bone Formation.
(a) Marfan's syndrome. (b) Achondroplasia. (c) Acromegaly.

alternative to bone grafting. Bone transplants from cadavers have been used, but unless these bones are thoroughly sterilized, blood-borne diseases, including AIDS, may be transmitted. The calcium carbonate skeleton of corals has been sterilized and used as another alternative.

Another approach involves the stimulation of osteoblast activity by strong electrical fields at the injury site. This procedure has been used to promote bone growth after fractures have refused to heal normally. Wires may be inserted into the skin, implanted in the adjacent bone, or wrapped around a cast. The overall success rate of about 80 percent is truly impressive.

One experimental method of inducing bone repair involves the mixing of bone marrow cells into a soft matrix of bone collagen and ceramic. This combination is used like a putty at the fracture site. Mesenchymal cells in the marrow divide, producing chondrocytes that create a cartilaginous patch; the patch is later converted to bone by periosteal cells. A second experimental procedure uses a genetically engineered protein to stimulate the conversion of osteoprogenitor cells into active osteoblasts. Although results in animal experimentation have been promising, neither technique has yet been approved for human trials.

Osteoporosis and Age-Related Skeletal Abnormalities   FAP *p. 191*

Osteoporosis (os-tē-ō-por-ō-sis; *porosus*, porous) is a condition that produces a reduction in bone mass sufficient to compromise normal function. The distinction between the "normal" osteopenia of aging and the clinical condition of osteoporosis is a matter of degree. Current estimates indicate that 29 percent of women between the ages of 45 and 79 can be considered osteoporotic. The increase in incidence after menopause has been linked to decreases in the production of estrogens (female sex hormones). The incidence of osteoporosis in men of the same age is estimated at 18 percent.

The excessive fragility of the bones frequently leads to breakage, and subsequent healing is impaired. Vertebrae may collapse, distorting the vertebral articulations and putting pressure on spinal nerves. Therapies that boost estrogen levels, dietary changes to elevate calcium levels in the blood, exercise that stresses bones and stimulates osteoblast activity, and calcitonin administration by injection or nasal spray appear to slow but not completely prevent the development of osteoporosis.

Osteoporosis can also develop as a secondary effect of many cancers. Cancers of the bone marrow, breast, or other tissues release a chemical known as *osteoclast-activating factor*. This compound increases both the number and activity of osteoclasts and produces a severe osteoporosis.

Infectious diseases that affect the skeletal system become more common as individuals age. In part, this fact reflects the higher incidence of fractures, combined with slower healing and the reduction of immune defenses.

Osteomyelitis (os-tē-ō-mī-e-LĪ-tis; *myelos*, marrow) is a painful bone infection generally caused by bacteria. This condition, most common in people over 50 years of age, can lead to dangerous systemic infections. A virus appears to be responsible for **Paget's disease**, also known as **osteitis deformans** (os-tē-Ī-tis de-FOR-mans). This condition may affect up to 10 percent of the population over 70. Osteoclast activity accelerates, producing areas of acute osteoporosis, and osteoblasts produce abnormal matrix proteins. The result is a gradual deformation of the skeleton.

Phrenology Then and Now   FAP *p. 215*

Phrenology (fre-NOL-o-jē was a popular "science" of the late eighteenth century. It was argued that because the brain is responsible for intellectual abilities and the skull houses the brain, any bumps and lumps on the skull should reflect features of the brain responsible for each individual's unique, specific traits and abilities. Although this theory fell into disfavor in the early 1800s, it eventually resurfaced a century later as investigators analyzed the size, weight, and superficial features of the brain. Prestigious institutions devoted considerable time and effort to acquiring brain collections in the hopes of establishing correlations between brain characteristics and intellectual abilities, personality traits, and suitable professions. These attempts failed, as no such correlation exists. However, the collections continue to make interesting paperweights and educational displays.

Brain size and shape appear to be quite variable factors. As long as growth is permitted, the brain's functional capabilities will not be affected. In some cultures, an infant's head is immobilized and compressed to alter the shape of the skull. For example, head boards may be used to flatten the forehead or reshape the back of the skull. Premature infants with muscles too weak to move their heads may initially have elongate skulls as a result of lying with the head to one side. Bracing the head on the occipital bone can eventually restore normal shape.

Although phrenology is seldom mentioned in scientific journals, the basic concepts continue to appear with new labels in the popular literature. Since the mid-1980s, attempts have been made to link skull shape to various behavioral problems, ranging from learning disabilities to Alzheimer's disease. The treatments proposed have involved painful compression or manipulation of the roofing bones of the skull, supposedly to remove pressure on the tissues of the brain. There is no recognized scientific basis for this practice, and the treatments may be both dangerous and painful if administered to young children.

Kyphosis, Lordosis, and Scoliosis
FAP *p. 218*

In **kyphosis** (kī-FŌ-sis), the normal thoracic curvature becomes exaggerated posteriorly, producing a "roundback" appearance (Figure A-18a). This condition can be caused by (1) osteoporosis with compression fractures affecting the anterior portions of vertebral bodies, (2) chronic contractions in muscles that insert on the vertebrae, or (3) abnormal vertebral growth. In **lordosis** (lor-DŌ-sis), or "swayback," the abdomen and buttocks protrude due to an anterior exaggeration of the lumbar curvature (Figure A-18b).

Scoliosis (skō-lē-Ō-sis) is an abnormal lateral curvature of the spine (Figure A-18c). This lateral deviation may occur in one or more of the movable vertebrae. Scoliosis is the most common distortion of the spinal curvature. This condition may result from developmental problems, such as incomplete vertebral formation, or from muscular paralysis affecting one side of the back. In four out of five cases, the structural or functional cause of the abnormal spinal curvature is impossible to determine. Scoliosis generally appears in girls during adolescence, when periods of growth are most rapid. Treatment consists of a combination of exercises, braces, and in some cases surgical modifications of the affected vertebrae.

Problems with the Ankle and Foot
FAP *p. 249*

The ankle and foot are subjected to a variety of stresses during normal daily activities. In a *sprain*, a ligament is stretched to the point at which some of the collagen fibers are torn. The ligament remains functional, and the structure of the joint is

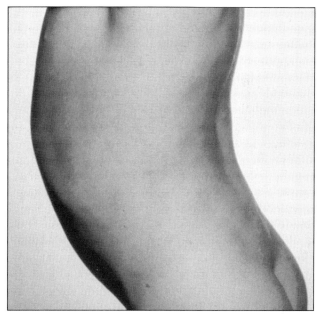

(b)

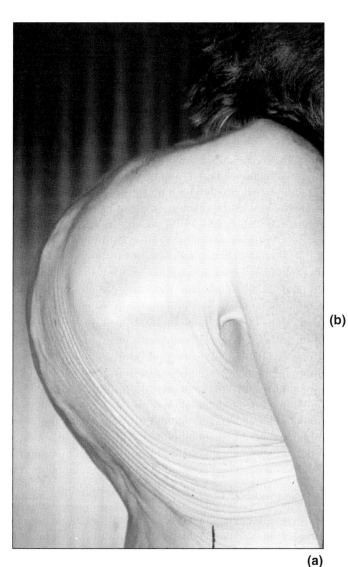

(a)

Figure A-18   Spinal Deformities.
(a) Kyphosis. (b) Lordosis. (c) Scoliosis.

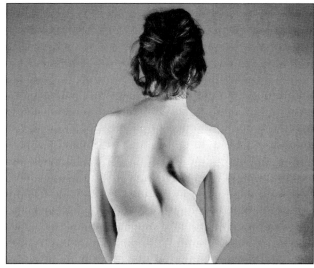

(c)

not affected. The most common cause of a **sprained ankle** is a forceful inversion of the foot that stretches the lateral ligament. An ice pack is generally required to reduce swelling. With rest and support, the ankle should heal in about 3 weeks.

In more serious incidents, the entire ligament may be torn apart, or the connection between the ligament and the lateral malleolus may be so strong that the bone breaks instead of the ligament. In general, a broken bone heals more quickly and effectively than does a completely torn ligament. A dislocation typically accompanies such injuries.

In a **dancer's fracture,** the proximal portion of the fifth metatarsal is broken. Most such cases occur while the body weight is being supported by the longitudinal arch. A sudden shift in weight from the medial portion of the arch to the lateral, less elastic border breaks the fifth metatarsal close to its distal articulation.

Individuals with abnormal arch development are most likely to suffer metatarsal injuries. Someone with flat feet loses or never develops the longitudinal arch. "Fallen arches" may develop as tendons and ligaments stretch and become less elastic. Obese individuals or those who must constantly stand or walk on the job are likely candidates. Children have very mobile articulations and elastic ligaments, so they commonly have flexible flat feet. Their feet look flat only while they are standing, and the arch appears when they stand on their toes or sit down. In most cases, the condition disappears as growth continues.

Clawfeet are also produced by muscular abnormalities. In this case, the median longitudinal arch becomes exaggerated because the plantar flexors overpower the dorsiflexors. Muscle cramps or nerve paralysis may be responsible. The condition tends to develop in adults and gets progressively worse with age.

Congenital talipes equinus, or *clubfoot,* is an inherited developmental abnormality detailed in the text (FAP *p. 249*).

A Matter of Perspective FAP *p. 249*

Several professions focus on the structure and function of the human skeleton. Each specialty has a different perspective, with its own techniques, traditions, and biases. People seeking medical advice may find it difficult to choose the most appropriate specialist. Unfortunately, misinformation and misconceptions have polarized opinions regarding the relative merits of the various specialists.

An **orthopedist,** or **orthopedic surgeon,** is a physician (M.D.) who specializes in the structure and function of bones and articulations. Orthopedic surgery attempts to restore lost function or to repair damage. An orthopedist might typically be involved in the reconstruction of a limb after a severe fracture, the replacement of a hip, knee

surgery, and so forth. Subspecialties focus on particular regions such as the hands. An orthopedist is trained to recognize the underlying condition when an injury is not the primary cause of a particular problem. As a physician, an orthopedist must complete (1) a 4-year college program, (2) an osteopathic medical school, (3) an internship, (4) a residency program, and (5) national boards (examinations) and then be licensed by the state in which he or she practices medicine. It could take up to 12 or 13 years of training after high school to complete all the requirements.

Osteopaths (D.O.) and **osteopathic surgeons** have a slightly different perspective, but like orthopedists, osteopaths must complete undergraduate college and an osteopathic medical school, serve an internship, and pass national and state examinations. Additional residency training may also be required. A century ago, when osteopathy was founded, drug therapies for the treatment of diseases were unreliable, and osteopaths put their trust in rest, cleanliness, and proper nutrition, in some cases accompanied by manipulation of the skeleton. Osteopaths today in general place a greater emphasis than do orthopedists on preventive medicine, such as the maintenance of normal structural relationships, proper nutrition, exercise, and a healthy environment. In terms of evaluating and treating bone or joint conditions, there is no substantive difference today between osteopaths and orthopedists other than the amount of skeletal manipulation prescribed by osteopaths.

The viewpoint of a **chiropractor** (D.C.) differs considerably from that of either orthopedists or osteopaths. In its original form, chiropractic proposed that all forms of disease result from abnormal functioning of the nervous system. This rather sweeping premise was modified as physiological principles became better understood. *Straight chiropractors* continue to place primary emphasis on vertebral alignment as the source of spinal nerve compression that causes peripheral symptoms. These practitioners use modern radiological techniques to evaluate vertebral position but generally avoid the use of drugs. Treatment involves physical manipulation of the spinal column to bring the vertebral elements into proper alignment.

Chiropractors not classified as "straight" may have widely varying approaches and techniques. Many claim that spinal manipulation alone can cure conditions totally unrelated to the skeletal or nervous system. There is no known mechanism that would lend plausibility to those claims, nor is there any valid supporting evidence. (Valid evidence must be testable, unbiased, and repeatable—criteria discussed in the section on the scientific method, pp. 1–4.)

Because there is so much variability among the practitioners, generalizations are difficult if not impossible to make. However, chiropractors in every state are licensed. For example, to be licensed in Hawaii an applicant needs more than 2 years of undergraduate credits (60 semester hours), a diploma from 1 of 16 accredited 4-year chiropractic colleges, and completion of a national examination.

Podiatrists (D.P.M.), or *chiropodists,* are specialists in foot structure. They commonly treat sports injuries, design supports for feet affected by trauma or developmental problems, and perform limited surgical procedures. Podiatrists must be graduates of an accredited podiatry college and may be required to pass national and state examinations.

Where you take your aching bones is your choice, but the way the practitioner views your problem will obviously affect your treatment. Be wary of any individual who claims to have the answer to every question. A competent medical practitioner knows his or her personal and professional limitations and is willing to refer a patient to another specialist when appropriate.

⚕ Rheumatism, Arthritis, and Synovial Function FAP *p. 256*

Rheumatism (ROO-ma-tizm) is a general term that indicates pain and stiffness affecting the skeletal system, the muscular system, or both. There are several major forms of rheumatism. **Arthritis** (ar-THRĪ-tis) includes all the rheumatic diseases that affect synovial joints. Arthritis always involves damage to the articular cartilages, but the specific cause may vary. For example, arthritis can result from bacterial or viral infection, injury to the joint, metabolic problems, or autoimmune disorders.

Proper synovial function requires healthy articular cartilages. When an articular cartilage has been damaged, the matrix begins to break down, and the exposed cartilage changes from a slick, smooth gliding surface to a rough feltwork of bristly collagen fibers. This feltwork drastically increases friction, damaging the cartilage further and producing pain. Eventually the central area of the articular cartilage may completely disappear, exposing the underlying bone.

Fibroblasts are attracted to areas of friction, and they begin tying the opposing bones together with a network of collagen fibers. This network may later be converted to bone, locking the articulating elements into position. Such a bony fusion, called **ankylosis** (an-kē-LŌ-sis), eliminates the friction, but only by the drastic remedy of making movement impossible.

The diseases of arthritis are usually considered as either **degenerative** or **inflammatory** in nature. Degenerative diseases begin at the articular cartilages, and modification of the underlying bone and inflammation of the joint occur secondarily. Inflammatory diseases start with the inflammation of synovial tissues, and damage later spreads to the articular surfaces. We will consider a single example of each type.

Osteoarthritis (os-tē-ō-ar-THRĪ-tis), also known as **degenerative arthritis** or **degenerative joint disease** (DJD), generally affects older individuals. In the U.S. population, 25 percent of women and 15 percent of men over 60 years of age show signs of this disease. The condition seems to result from cumulative wear and tear on the joint surfaces. Some individuals, however, may have a genetic predisposition to develop osteoarthritis, for researchers have recently isolated a gene linked to the disease. This gene codes for an abnormal form of collagen that differs from the normal protein in only 1 of its 1000 amino acids.

Rheumatoid arthritis is an inflammatory condition that affects roughly 2.5 percent of the adult U.S. population. The cause is uncertain, although allergies, bacteria, viruses, and genetic factors have all been proposed. The synovial membrane becomes swollen and inflamed, a condition known as **synovitis** (sī-nō-VĪ-tis). The cartilaginous matrix begins to break down, and the process accelerates as dying cartilage cells release lysosomal enzymes.

Advanced stages of inflammatory and degenerative forms of arthritis produce an inflammation that spreads into the surrounding area. Ankylosis, common in the past when complete rest was routinely prescribed for arthritis patients, is rarely seen today. Regular exercise, physical therapy, drugs that reduce inflammation (such as aspirin), and even immune suppressants (such as *methotrexate*) may slow the progression of the disease. Nutritional supplements such as chondroitin sulfate and other compounds sold to "help repair cartilage" are broken down by digestive enzymes and have no effect on chondrocyte activity. Surgical procedures can realign or redesign the affected joint. In extreme cases involving the hip, knee, elbow, or shoulder, the defective joint can be replaced by an artificial one. Prosthetic (artificial) joints, such as those shown in Figure A-19, are weaker than natural ones, but elderly people seldom stress them to their limits.

🏃 Hip Fractures, Aging, and Professional Athletes FAP *p. 269*

Today, two very different groups of people suffer hip fractures: (1) individuals over age 60, whose bones have been weakened by osteoporosis, and (2) young, healthy professional athletes, who subject their hips to extreme forces. When the injury is severe, the vascular supply to the joint is damaged. As a result, two problems gradually develop:

1. *Avascular necrosis.* The mineral deposits in the bone of the pelvis and femur are turned over very rapidly, and osteocytes have high energy demands. A reduction in blood flow first injures and then kills the osteocytes. When bone maintenance stops in the affected region, the matrix begins to break down. This process is called *avascular necrosis.*

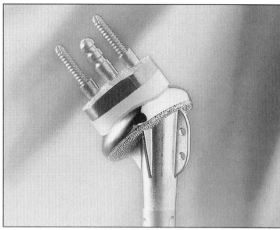

(a) Shoulder

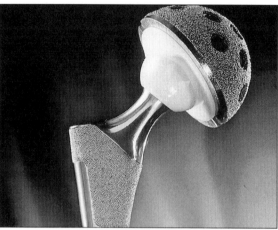

(b) Hip

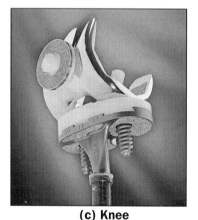

(c) Knee

Figure A-19   Artificial Joints

2. *Degeneration of articular cartilages.* The chondrocytes in the articular cartilages absorb nutrients from the synovial fluid, which circulates around the joint cavity as the bones change position. A fracture of the femoral neck is generally followed by joint immobility and poor circulation to the synovial membrane. The combination results in a gradual deterioration of the articular cartilages of the femur and acetabulum.

In recent years the frequency of hip fractures among young, healthy professional athletes has increased dramatically; the best-known example is Bo Jackson. At age 28, he was playing professional football with the LA Raiders and professional baseball with the Kansas City Royals; and he was starring in both sports. But on January 13, 1991, things changed dramatically when Bo was tackled in an NFL playoff game. The combination of pressure and twisting applied by the tackler and the tremendous power of Bo's thigh muscles produced a fracture-dislocation of the hip. Roughly 15 percent of the inferior acetabular fossa was broken away. The femur was not broken, but it was dislocated. Although Jackson experienced severe pain, the immediate damage to the femur and hip appeared to be sufficiently limited.

The initial optimism began to fade when it became apparent that the complications of the injury were more damaging than the injury itself. The dislocation tore blood vessels in the capsule and along the femoral neck, where the capsular fibers attach. The result was avascular necrosis and the degeneration of the articular cartilages at the hip.

After more than a year of rehabilitation, Jackson had surgery to replace the damaged joint with an artificial hip. In this "total hip" procedure, the damaged portion of the femur is removed, and an artificial femoral head and neck is attached by a spike that extends into the marrow cavity of the shaft. Special cement may be used to anchor it in place and to attach a new articular surface to the acetabulum. After this procedure, Jackson reentered professional baseball for a short time as a designated hitter and occasional first baseman, but his speed was diminished.

Joint replacement eliminates the pain and restores full range of motion. However, prosthetic (artificial) joints, such as those shown in Figure A-19, are weaker than natural ones. They are generally implanted in older people who seldom stress them to their limits. Whether an artificial hip will be able to tolerate the stresses of professional sports is an open question. It is also an important question, because the frequency of such injuries is increasing. By early 1994, hip fractures and the associated potential for avascular necrosis had sidelined Marvin Jones of the Jets and Mike Sherrard of the Giants.

® Arthroscopic Surgery   FAP *p. 272*

An **arthroscope** uses fiber optics to permit exploration of a joint without major surgery. Optical fibers are thin threads of glass or plastic that conduct light. The fibers can be bent around corners, so they can be introduced into a knee or other joint and moved around, enabling the physician to see what is going on inside the joint. If necessary, the apparatus can be modified to perform surgical modification of the joint at the same time. This procedure, called **arthroscopic surgery,** has greatly simplified the treatment of knee and other joint injuries. Figure A-20a is an arthroscopic view of the interior of an injured knee, showing a damaged meniscus. Small pieces of cartilage may be removed and the meniscus surgically trimmed. A total **meniscectomy,** the removal of the affected cartilage, is generally avoided, as it leaves the joint prone to develop DJD. New tissue-culturing techniques may someday permit the replacement of the meniscus or even the articular cartilage.

An arthroscope cannot show the physician soft tissue details outside the joint cavity, and repeated arthroscopy eventually leads to the formation of scar tissue and other joint problems. Magnetic resonance imaging (MRI) is a cost-effective and non-invasive method of viewing, without injury, and examining soft tissues around the joint. Figures A-20b and A-20c are MRI views of the knee joint. Note the image clarity and the soft tissue details visible in these scans.

Figure A-20   Arthroscopy and the Knee.
(a) Arthroscopic view of a damaged knee, showing the torn edge of an injured meniscus. (b) A transverse and (c) a frontal MRI scan of the knee joint. For identification of the structures visualized in these images, see scans on *p. S-5.*

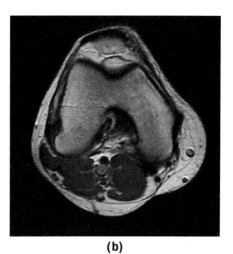

(b)

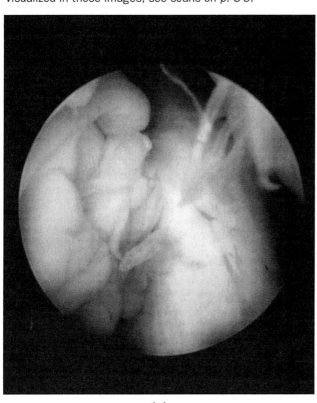

(a)

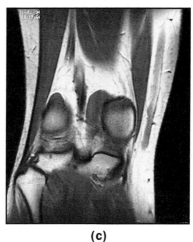

(c)

CRITICAL-THINKING QUESTIONS

2-1. Diane, a 65-year-old woman, is brought into the emergency room by her daughter, Mary. Diane has had severe back pain since she accidentally fell yesterday. Her daughter insists it was a minor fall and cannot explain the severity of the pain and the tenderness over the thoracic spine. X-rays of the thoracic vertebrae reveal fractures of T_{10}, T_{11}, and T_{12}. The X-rays also reveal decreased bone density in all the vertebrae. Laboratory tests are within normal limits. The physician diagnoses the fracture as a vertebral fracture most likely due to

a. osteomalacia   b. osteomyelitis
c. osteoporosis   d. Paget's disease

2-2. Ted, a high school football player, complains of increasing right knee pain and limited motion since he played in the football game 3 days ago. Upon examination, swelling was noted around the joint. The range of motion of Ted's right knee was considerably limited compared with the left knee. Routine lab tests were within normal limits. Other test results are as follows:

X-ray of tibia, fibula, patella, and femur: no fracture

Arthrocentesis:

Appearance of fluid: blood tinged

WBC count: <200 WBC/mm^3

Glucose: within normal limits

Predict the possible results of an arthroscopic examination of the affected knee.

2-3. Beth, a dentist, has a patient who is complaining of a toothache on the lower right side of his mouth. Examination shows severe tooth decay in the first mandibular molar, with facial swelling along the mandibular angle. Beth suspects an infection and takes an X-ray that shows a reduction in bone density around the root of the decayed tooth. What is the likely diagnosis?

a. osteomalacia
b. osteomyelitis
c. fracture of the mandible
d. tumor of the mandible

2-4. While studying a 2000-year-old campsite, an archaeologist discovers several adult male and female skeletons. Chemical analysis of the bones of both genders indicates that osteopenia had been occurring for several years before their deaths. All the male skeletons showed heavy thickenings and prominences in the bones of the upper limbs. The bones of the lower limbs were of normal size, and the markings indicated that only minimal muscle stress had been applied during life. How old do you think these individuals were when they died, and what does the skeletal information tell you about the lifestyle of the males?

2-5. Two patients sustain hip fractures. In one patient, a pin is inserted into the joint and the injury heals well. In the other, the fracture fails to heal. Identify the types of fractures that are probably involved. Why did the second patient's fracture not heal, and what steps could be taken to restore normal function?

NOTES:

The Muscular System

The muscular system includes more than 700 skeletal muscles that are directly or indirectly attached to the skeleton by tendons or aponeuroses. The muscular system produces movement as the contractions of skeletal muscles pull on the attached bones. Muscular activity does not always result in movement, however; it can also be important in stabilizing skeletal elements and preventing movement. Skeletal muscles are also important in guarding entrances or exits of internal passageways, such as those of the digestive, respiratory, urinary, and reproductive systems, and in generating heat to maintain our stable body temperatures.

Skeletal muscles contract only under the command of the nervous system. For this reason, clinical observation of muscular activity may provide indirect information about the nervous system as well as direct information about the muscular system. Assessment of facial expressions, posture, speech, and gait can be an important part of a physical examination. Classical signs of muscle disorders include the following:

- *Gower's sign* is a distinctive method of standing from a sitting or lying position on the floor. This method is used by children with *muscular dystrophy* (p. 64). Young children with this disorder move from a sitting position to a standing position by pushing the trunk off the floor with the hands and then moving the hands to the knees. The hands are then used as braces to force the body into the standing position. This extra support is necessary because the pelvic muscles are too weak to swing the weight of the trunk over the legs.

- *Ptosis* is a drooping of the upper eyelid. It may be seen in *myasthenia gravis* (p. 65), *botulism* (p. 65), and *myotonic dystrophy* (p. 64), or following damage to the cranial nerve that innervates the *levator palpebrae superioris*, a muscle of the eyelid.

- A *muscle mass*, or an abnormal dense region, may sometimes be seen or felt within a skeletal muscle. A muscle mass may result from torn muscle tissue; a hematoma; a parasitic infection, such as *trichinosis* (p. 63); or bone deposition, as in *myositis ossificans* (p. 48).

- Abnormal contractions may indicate problems with the muscle tissue or its innervation. *Muscle spasticity* exists when a muscle has excessive muscle tone. A *muscle spasm* is a sudden, strong, and painful involuntary contraction.

- *Muscle flaccidity* exists when the relaxed skeletal muscle appears soft and loose and its contractions are very weak.

- *Muscle atrophy* is skeletal muscle deterioration, or *wasting*, due to disuse, immobility, or interference with the normal muscle innervation.

- Abnormal patterns of muscle movement, such as *tics, choreiform movements,* or *tremors,* and muscular paralysis are generally caused by nervous system disorders. We will describe these movements further in sections dealing with abnormal nervous system function.

SIGNS AND SYMPTOMS OF MUSCULAR SYSTEM DISORDERS

Two common symptoms of muscular disorders are *pain* and *weakness* in the affected skeletal muscles. The potential causes of muscle pain include the following:

- *Muscle trauma.* Examples of traumatic injuries to a skeletal muscle include a laceration, a deep bruise or crushing injury, a muscle tear, and a damaged tendon.

- *Muscle infection.* Skeletal muscles may be infected by viruses, as in some forms of myositis, or colonized by parasitic worms, such as those responsible for *trichinosis* (p. 63). These infections generally produce pain that is restricted to the involved muscles. Diffuse muscle pain may develop in the course of other infectious diseases, such as influenza or measles.

- *Related problems with the skeletal system.* Muscle pain may result from skeletal problems, such as arthritis (p. 55) or a sprained ligament near the point of muscle origin or insertion.

- *Problems with the nervous system.* Muscle pain may be related to inflammation of sensory neurons or stimulation of pain pathways in the CNS.

Muscle strength can be evaluated by applying an opposite force against a specific action. For example, an examiner might exert a gentle extending force on a patient's forearm while asking the patient to flex the elbow. Because the muscular and nervous systems are so closely interrelated, a single symptom, such as muscle weakness, can have a variety of causes (Figure A-21). Muscle weakness may also develop as a consequence of a condition that affects the entire body, such as anemia or starvation.

Diagnostic Tests

Table A-11 (p. 61) provides information about representative diagnostic procedures for muscular system abnormalities. Figure A-22 (p. 62) presents a simple diagnostic chart of representative muscle disorders.

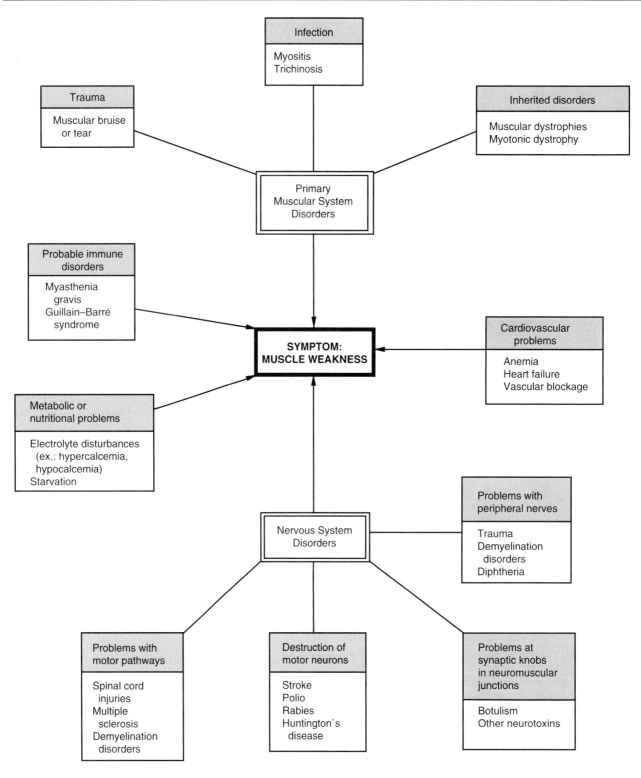

Figure A-21   Some Possible Causes of Muscle Weakness

⚕ **Disruption of Normal Muscle Organization**   FAP *p. 279*

A variety of disorders are characterized by a disruption in the structural organization of skeletal muscles. We will consider only a few representative examples: necrotizing fasciitis, characterized by a breakdown in the connective tissues of skeletal muscles; trichinosis, characterized by the colonization of muscles by parasites; fibromyalgia and chronic fatigue syndrome, two muscle disorders of uncertain origin.

Table A-11 Examples of Tests Used in the Diagnosis of Muscle Disorders

Diagnostic Procedure	Method and Result	Representative Uses	Notes
Muscle biopsy	Removal of a small amount of affected muscle tissue	Determination of muscle disease; also used to detect cyst formation or larvae in trichinosis	Certain forms of muscular dystrophy have degenerating muscle fibers and an increase in fibrous tissue and fat
Electromyography	Insertion of a probe that transmits information to an instrument that measures the electrical activity in contracting muscles	Abnormal EMG readings occur in disorders such as myasthenia gravis, amyotrophic lateral sclerosis (ALS), and muscular dystrophy	
MRI	Standard MRI	Useful in the detection of muscle diseases and associated soft tissue damage	

Laboratory Test	Normal Values in Blood Plasma or Serum	Significance of Abnormal Values	Notes
Aldolase	Adults: less than 8 U/l 22–59 mU/l (SI units)	Elevated levels occur in muscular dystropy but not in myasthenia gravis or multiple sclerosis	This enzyme is important in glycolysis; found in many tissues; particularly high in skeletal muscle tissue
Aspartate aminotransferase	Adults: 7–40 U/l Children: 15–55 U/l	Elevated levels occur in some muscle diseases	This enzyme is found in heart, liver (AST or SGOT), and skeletal muscle. Exercise can cause an increase in values
Creatine phosophokinase (CKP, or CK)	Adults: 30–180 IU/L	Elevated levels occur in muscular dystrophy, myositis	An isoenzyme of this enzyme is found in cardiac muscle; CPK/CK is the best enzyme indicator of muscle disease
Lactate dehydrogenase (LDH)	Adults: 100–190 U/l	Elevated levels occur in some muscle diseases and in lactic acidosis	
Electrolytes			
Potassium	Adults: 3.5–5.0 mEq/l	Decreased levels of potassium can cause muscle weakness	
Calcium	Adults: 8.5–10.5 mg/dl	Decreased calcium levels can cause muscle tremors and tetany; increased levels can cause muscle flaccidity	

10

Necrotizing Fasciitis

Several bacteria produce enzymes such as *hyaluronidase* or *cysteine protease.* Hyaluronidase breaks down hyaluronic acid and disassembles the associated proteoglycans. Cysteine protease breaks down connective tissue proteins. The bacteria that produce these enzymes are dangerous because they can spread rapidly by liquefying the matrix and dissolving the intercellular cement that holds epithelial cells together. The *streptococci* are one group of bacteria that secrete both of these enzymes. *Streptococcus A* bacteria are involved in many human diseases, most notably "strep

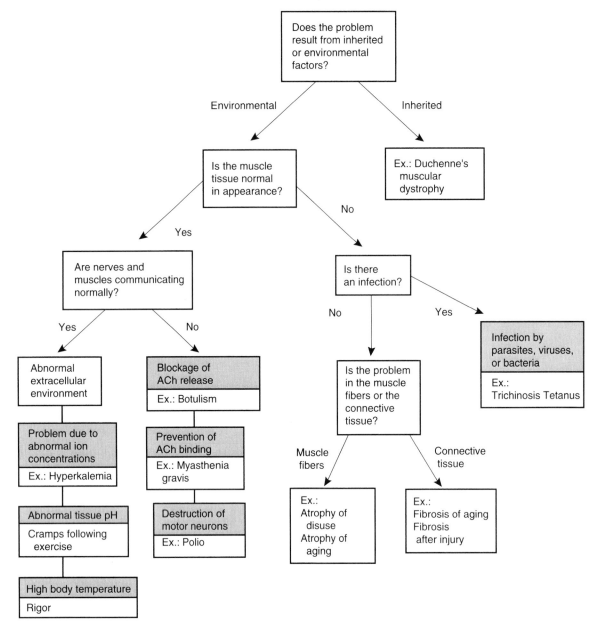

Figure A-22   A Simple Diagnostic Flow Chart for Disorders of the Muscular System

throat," a pharyngeal infection. In most cases, the immune response is sufficient to contain and ultimately defeat these bacteria before extensive tissue damage has occurred.

However, in 1994 tabloid newspapers had a field day recounting stories of "killer bugs" and "flesh-eating bacteria" that terrorized residents of Gloucester, England. The details were horrific: Minor cuts became major open wounds, and interior connective tissues dissolved. Although in fact there were only seven reported cases, five of the victims died. The pathogen responsible was a strain of *Streptococcus A* that overpowered immune defenses and swiftly invaded and

destroyed soft tissues. Moreover, the pathogens eroded their way along the fascial wrapping that covers skeletal muscles and other organs. The term for this condition is **necrotizing fasciitis.** In some cases, the muscle tissue was also destroyed, a condition called *myositis.*

The problem is not restricted to the United Kingdom. Some form of soft-tissue invasion occurs 75–150 times annually in the United States; this number is 5–10 percent of the invasive *Streptococcus A* wound infections reported each year. Early in 1997, five people in New York died of *Streptococcus A* infections; two had developed symptoms of necrotizing fasciitis. At present it is uncertain whether the recent surge in myositis and necrotizing fasciitis reflects increased awareness of the condition or the appearance of a new strain of strep bacteria.

Trichinosis

Trichinosis (trik-i-NŌ-sis; *trichos,* hair + *nosos,* disease) results from infection by the parasitic nematode worm *Trichinella spiralis,* whose life cycle is diagrammed in Figure A-23. Symptoms include diarrhea, weakness, and muscle pain. The muscular symptoms are caused by the invasion of skeletal muscle tissue by larval worms, which create small pockets within the perimysium and endomysium. Muscles of the tongue, eyes, diaphragm, chest, and legs are most often affected.

Larvae are common in the flesh of pigs, horses, dogs, and other mammals. The larvae are killed by cooking, and people are most often exposed by eating undercooked infected pork. Once eaten, the lar-

vae mature within the human intestinal tract, where they mate and produce eggs. The new generation of larvae then migrates through the body tissues to reach the muscles, where they complete their early development. The migration and subsequent settling produce a generalized achiness, muscle and joint pain, and swelling in infected tissues. An estimated 1.5 million Americans carry *Trichinella* in their muscles, and up to 300,000 new infections occur each year. The mortality rate for people who have symptoms severe enough to require treatment is approximately 1 percent.

Fibromyalgia and Chronic Fatigue Syndrome

Fibromyalgia (-*algia,* pain) is an inflammatory disorder that has formally been recognized only since the mid-1980s. Although first described in the early 1800s, the condition is still somewhat

10

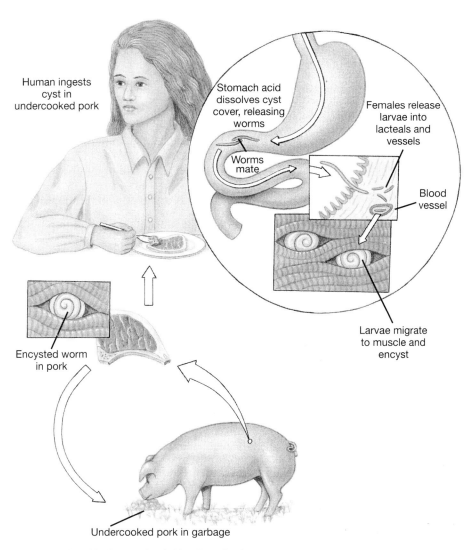

Figure A-23   Life Cycle of *Trichinella spiralis*

controversial because the reported symptoms cannot be linked to any anatomical or physiological abnormalities. However, physicians now recognize a distinctive pattern of symptoms that warrant consideration as a clinical entity.

Fibromyalgia may be the most common musculoskeletal disorder affecting women under 40 years of age. There may be as many as 6 million cases in the United States today. Symptoms include chronic aches, pain, and stiffness, and multiple tender points at specific, characteristic locations. The four most common tender points are (1) the medial aspect of the knee, (2) distal to the lateral epicondyle of the humerus, (3) near the external occipital crest of the skull, and (4) at the junction between the second rib and its costal cartilage. An additional clinical criterion is that the pains and stiffness cannot be explained by other mechanisms. Patients suffering from this condition invariably report chronic fatigue; they feel tired on awakening and often complain of awakening repeatedly during the night. Fibromyalgia may also be associated with other conditions, such as irritable bowel syndrome (p. 174) or chronic depression.

Most of these symptoms could be attributed to other problems. For example, chronic depression can by itself lead to fatigue and poor-quality sleep. As a result, the pattern of tender points is the diagnostic key to fibromyalgia. This symptom clearly distinguishes fibromyalgia from **chronic fatigue syndrome** (CFS). The current symptoms accepted as a definition of CFS include (1) sudden onset, generally following a viral infection, (2) disabling fatigue, (3) muscle weakness and pain, (4) sleep disturbance, (5) fever, and (6) enlargement of cervical lymph nodes. Roughly twice as many women as men are diagnosed with CFS.

Attempts to link either fibromyalgia or CFS to a viral infection or to some physical or psychological trauma have not been successful, and the cause remains mysterious. Treatment at present is limited to relieving symptoms when possible. For example, anti-inflammatory medications may help relieve pain, drugs can be used to promote sleep, and exercise programs may help maintain normal range of motion.

☤ The Muscular Dystrophies

FAP *p. 283*

The **muscular dystrophies** (DIS-trō-fēz) are inherited diseases that produce progressive muscle weakness and deterioration. One of the most common and best understood conditions is **Duchenne's muscular dystrophy (DMD).** This form of muscular dystrophy appears in childhood, commonly between the ages of 3 and 7. The condition generally affects only males; the incidence is roughly 30 per 100,000 male births. A progressive muscular weakness develops, and the child typi-

cally requires a wheelchair by age 12. Most individuals die before age 20 due to respiratory paralysis or cardiac problems. Skeletal muscles are primarily affected, although for some reason the facial muscles continue to function normally. In later stages of the disease, the facial muscles and cardiac muscle tissue may also become involved.

The skeletal muscle fibers in a DMD patient are structurally different from those of other individuals. Abnormal membrane permeability, cholesterol content, rates of protein synthesis, and enzyme composition have been reported. Individuals with DMD also lack the protein *dystrophin*, found in normal muscle fibers. Dystrophin is a large protein attached to the inner surface of the sarcolemma near the triads. Although the functions of this protein remain uncertain, dystrophin is suspected to play a role in the regulation of calcium ion channels in the sarcolemma. In children with DMD, calcium channels remain open for an extended period, and calcium levels rise in the sarcoplasm to the point at which key proteins denature, and the muscle fiber degenerates. Researchers have recently identified and cloned the gene for dystrophin production; that gene is located on the X chromosome. Rats with DMD have been cured by insertion of this gene into their muscle fibers, a technique that may eventually be used to treat human patients.

The inheritance of DMD is sex-linked: Women carrying the defective genes are unaffected, but each of their male children will have a 50 percent chance of developing DMD. Now that the specific location of the gene has been identified, it is possible to determine whether a woman is carrying the defective gene. It is also possible to use an innovative prenatal test to determine if a fetus has this condition. In this procedure, a small sample of fluid is collected from the membranous sac that surrounds the fetus. This fluid contains fetal cells called *amniocytes* that are collected and cultivated in the laboratory. Researchers then insert a gene called *MyoD*, which triggers their differentiation into skeletal muscle fibers. These cells can then be tested not only for the signs of muscular dystrophy but also for indications of other inherited muscular disorders.

Attempts have been made to give children with DMD injections that contain donated normal myoblasts, in the hope that these myoblasts will fuse with developing muscle fibers and provide normal dystrophin-production genes. Results have been inconclusive. Animal studies indicate that symptoms of DMD develop when fewer than 20 percent of the muscle fibers contain normal dystrophin. In one recent study that used muscle biopsies, 1–10 percent of the muscle fibers of treated children contained normal dystrophin. Efforts to improve the delivery method and increase the treatment's effects are underway.

Myotonic Dystrophy. Myotonic dystrophy is a form of muscular dystrophy that occurs in the

United States at an incidence of 13.5 per 100,000 population. Symptoms may develop in infancy, but more commonly they develop after puberty. As with other forms of muscular dystrophy, adults developing myotonic dystrophy experience a gradual reduction of muscle strength and control. Problems with other systems, especially the cardiovascular and digestive systems, typically develop. There is no effective treatment.

The inheritance of myotonic dystrophy is unusual because the children of an individual with myotonic dystrophy commonly develop more severe symptoms. The increased severity of the condition appears to be related to the presence of multiple copies of a specific gene on chromosome 19. For some reason, the nucleotide sequence of that gene gets repeated several times, and the number can increase from generation to generation. This phenomenon has been called a "genetic stutter." The greater the number of copies, the more severe the symptoms. It is not known why the stutter develops or how the genetic duplication affects the severity of the condition. There is evidence that the extra nucleotides in some way interfere with transcription of an adjacent gene involved with the control of muscle tone.

Problems with the Control of Muscle Activity
FAP pp. 287, 303

Another group of disorders interfere with normal neuromuscular communication by affecting either the nerve's ability to issue commands or the muscle's ability to respond.

BOTULISM

Botulinus (bot-ū-LĪ-nus) **toxin** prevents the release of ACh at synaptic terminals. It thus produces a severe and potentially fatal paralysis of skeletal muscles. A case of botulinus poisoning is called **botulism.**[1] The toxin is produced by the bacterium *Clostridium botulinum,* which does not need oxygen to grow and reproduce. Because the organism can live quite well in a sealed can or jar, most cases of botulism are linked to improper canning or storing procedures, followed by failure to cook the food adequately before it is eaten. Canned tuna or beets, smoked fish, and cold soups are the foods most commonly involved with cases of botulism. Boiling for a half-hour destroys both the toxin and the bacteria.

Symptoms generally begin 12–36 hours after a contaminated meal is eaten. The initial symptoms are typically disturbances in vision, such as double vision or a painful sensitivity to bright light. These symptoms are followed by other sensory and motor problems, including blurred speech and an inability to stand or walk. Roughly half of botulism

[1]This disorder was described more than 200 years ago by German physicians who treated patients poisoned by dining on contaminated sausages. *Botulus* is the Latin word for sausage.

patients experience intense nausea and vomiting. These symptoms persist for days to weeks, followed by a gradual recovery; some patients are still in recovery after a year.

The major risk of botulinus poisoning is respiratory paralysis and death by suffocation. Treatment is supportive: bed rest, observation, and, if necessary, use of a mechanical respirator. In severe cases drugs that promote the release of ACh, such as *guanidine hydrochloride,* may be administered. The overall mortality rate in the United States is about 10 percent; an antitoxin is now available.

MYASTHENIA GRAVIS

Myasthenia gravis (mī-as-THĒ-nē-uh GRA-vis) is characterized by a general muscular weakness that tends to be most pronounced in the muscles of the arms, head, and chest. The first symptom is generally a weakness of the eye muscles and drooping eyelids. Facial muscles are commonly weak as well, and the individual develops a peculiar smile known as the "myasthenic snarl." As the disease progresses, pharyngeal weakness leads to problems with chewing and swallowing, and holding the head upright becomes difficult.

The muscles of the upper chest and upper extremities are next to be affected. All the voluntary muscles of the body may ultimately be involved. Severe myasthenia gravis produces respiratory paralysis, with a mortality rate of 5–10 percent. However, the disease does not always progress to such a life-threatening stage. For example, roughly 20 percent of people with the disease experience no symptoms other than eye problems.

The condition results from a decrease in the number of ACh receptors on the motor end plate. Before the remaining receptors can be stimulated enough to trigger a strong contraction, the ACh molecules are destroyed by cholinesterase. As a result, muscular weakness develops.

The primary cause of myasthenia gravis appears to be a malfunction of the immune system. Roughly 70 percent of individuals with myasthenia gravis have abnormal thymus glands, organs involved with the maintenance of normal immune function. In myasthenia gravis, the immune response attacks the ACh receptors of the motor end plate as if they were foreign proteins. For unknown reasons, 1.5 times as many women as men are affected. The typical age at onset is 20–30 for women, versus over 60 for men. Estimates of the incidence of this disease in the United States range from 2 to 10 cases per 100,000 population.

One approach to therapy involves the administration of drugs, such as *Neostigmine,* that are termed **cholinesterase inhibitors.** As their name implies, these compounds are enzyme inhibitors; they tie up the active sites at which cholinesterase normally binds ACh. With cholinesterase activity reduced, the concentration of ACh at the synapse can rise enough to stimulate the surviving receptors and produce muscle contraction. Corticosteroid therapy is typically beneficial, as is the surgical removal of the thymus gland.

POLIO

Because skeletal muscles depend on their motor neurons for stimulation, disorders that affect the nervous system can have an indirect affect on the muscular system. The *poliovirus* is a virus that does not produce clinical symptoms in roughly 95 percent of infected individuals. The virus may produce variable symptoms in the remaining 5 percent. Some individuals develop a nonspecific illness resembling the flu. Other individuals develop a brief *meningitis* (p. 74), an inflammation of the protective membranes that surround the CNS. In still another group of people, the virus attacks somatic motor neurons in the CNS.

In this third form of the disease, the individual develops a fever 7–14 days after infection. The fever subsides but recurs roughly a week later, accompanied by muscle pain, cramping, and flaccid paralysis of one or more limbs. Respiratory paralysis may also occur, and the mortality rate of this form of polio is 2–5 percent of children and 15–30 percent of adults. If the individual survives, some degree of recovery generally occurs over a period of up to 6 months.

For as-yet unknown reasons, the survivors of paralytic polio may develop progressive muscular weakness 20–30 years after the initial infection. This *postpolio syndrome* is characterized by fatigue, muscle pain and weakness, and, in some cases, muscular atrophy. There is no treatment for this condition, although rest seems to help.

Polio has been almost completely eliminated from the U.S. population due to a successful immunization program. In 1954, there were 18,000 new cases in the United States; there were 8 in 1976, and there was one in 1994. The World Health Organization now reports that polio has been eradicated from the entire Western Hemisphere, and the virus may be eliminated worldwide by the year 2000. Unfortunately, many parents today refuse to immunize their children against the poliovirus on the assumption that the disease has been "conquered." Failure to immunize is a mistake, because (1) there is still *no cure* for polio, (2) the virus remains in the environment in many areas of the world, and (3) approximately 38 percent of children ages 1–4 have not been immunized. A major epidemic could therefore develop very quickly if the virus were brought into the United States from another part of the world.

For years, the vaccine that has been used is the oral Sabin vaccine, preferred for its ease of administration and better immune stimulation than the injectable vaccine. However, because there is an approximately 1 in 1 million risk of developing polio from the oral vaccine, in 1996 the Centers for Disease Control and Prevention (CDC) recommended the use of a combination of injected and oral vaccines.

✝ Hernias   FAP *p. 337*

When the abdominal muscles contract forcefully, pressure in the abdominopelvic cavity can increase dramatically. Those pressures are applied to internal organs. If the individual exhales at the same time, the pressure is relieved, because the diaphragm can move upward as the lungs collapse. But during vigorous isometric exercises or when lifting a weight while holding one's breath, pressure in the abdominopelvic cavity can rise to 106 kg/cm^2 (1500 lb/in.^2), roughly 100 times normal pressures. Pressures that high can cause a variety of problems, including the development of a *hernia.*

A **hernia** develops when a visceral organ protrudes abnormally through an opening. There are many types of hernias; we will consider only *inguinal* (groin) *hernias* and *diaphragmatic hernias* here. Late in the development of males, the testes descend into the scrotum by passing through the abdominal wall at the **inguinal canals.** In adult males, the sperm ducts and associated blood vessels penetrate the abdominal musculature at the **inguinal canals** as the *spermatic cords*, on their way to the abdominal reproductive organs. In an inguinal hernia (Figure A-24), the inguinal canal enlarges, and the abdominal contents such as a portion of the greater omentum, small intestine, or (more rarely) the bladder are forced into the inguinal canal. If the herniated structures become trapped or twisted within the inguinal sac, surgery may be required to prevent serious complications. Inguinal hernias are not always caused by unusually high abdominal pressures. Injuries to the abdomen, or inherited weakness or distensibility of the canal, may have the same effect.

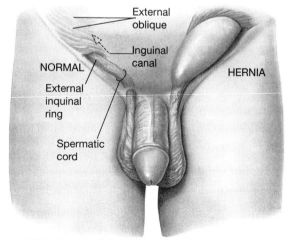

Figure A-24   Inguinal Hernia

The esophagus and major blood vessels pass through an opening in the diaphragm, the muscle that separates the thoracic and abdominopelvic cavities. In a **diaphragmatic hernia,** also called a *hiatal hernia* (hī-A-tal; *hiatus,* a gap or opening), abdominal organs slide into the thoracic cavity, typically through the *esophageal hiatus,* the opening used by the esophagus. The severity of the condition will depend on the location and size of the herniated organ or organs. Hiatal hernias are very common, and most go unnoticed. Radiologists see them in about 30 percent of patients whose upper gastrointestinal tracts are examined with barium contrast techniques. When clinical complications develop, they generally occur because abdominal organs that have pushed into the thoracic cavity are exerting pressure on the structure or organs there. As is the case with inguinal hernias, a diaphragmatic hernia may result from congenital factors or from an injury that weakens or tears the diaphragmatic muscle.

🏃 Sports Injuries   FAP *p. 345*

Sports injuries affect amateurs and professionals alike. A 5-year study of college football players indicated that 73.5 percent experienced mild injuries, 21.5 percent moderate injuries, and 11.6 percent severe injuries during their playing careers. Contact sports are not the only activities that show a significant injury rate; a study of 1650 joggers running at least 27 miles a week reported 1819 injuries in a single year.

Muscles and bones respond to increased use by enlarging and strengthening. Poorly conditioned individuals are therefore more likely than people in good condition to subject their bones and muscles to intolerable stresses. Training is also important in minimizing the use of antagonistic muscle groups and keeping joint movements within the intended ranges of motion. Planned warm-up exercises before athletic events stimulate circulation, improve muscular performance and control, and help prevent injuries to muscles, joints, and ligaments. Stretching exercises stimulate muscle circulation and help keep ligaments and joint capsules supple. Such conditioning extends the range of motion and prevents sprains and strains when sudden loads are applied.

Dietary planning can also be important in preventing injuries to muscles during endurance events, such as marathon running. Emphasis has commonly been placed on the importance of carbohydrates, leading to the practice of "carbohydrate loading" before a marathon. But while operating within aerobic limits, muscles also utilize amino acids extensively, and an adequate diet must include both carbohydrates and proteins.

Improved playing conditions, equipment, and regulations also play a role in reducing the incidence of sports injuries. Jogging shoes, ankle or knee braces, helmets, mouth guards, and body padding are examples of equipment that can be effective. The substantial penalties now earned for "personal fouls" in contact sports have reduced the numbers of neck and knee injuries.

Several injuries common to those engaged in active sports may also affect nonathletes, although the primary causes may differ considerably. A partial listing of activity-related conditions would include the following:

- *Bone bruise.* Bleeding within the periosteum of a bone

- *Bursitis.* Inflammation of the bursae around one or more joints

- *Muscle cramps.* Prolonged, involuntary, and painful muscular contractions

- *Sprains.* Tears or breaks in ligaments or tendons

- *Strains.* Tears in muscles

- *Stress fractures.* Cracks or breaks in bones subjected to repeated stresses or trauma

- *Tendinitis.* Inflammation of the connective tissue surrounding a tendon

We have discussed many of these conditions in related sections of the text.

Finally, many sports injuries would be prevented if people who engage in regular exercise would use common sense and recognize their personal limitations. It can be argued that some athletic events, such as the ultramarathon, place such excessive stresses on the cardiovascular, muscular, respiratory, and urinary systems that these events cannot be recommended, even for athletes in peak condition.

11

CRITICAL-THINKING QUESTIONS

3-1. A patient experiencing severe hyperkalemia could have the following related problems:

   a. a below-normal potassium ion concentration of the interstitial fluid
   b. a more-negative membrane potential of nerves and muscles
   c. unresponsive skeletal muscles and cardiac arrest
   d. muscle weakness and increased strength of twitch contractions
   e. all of the above

3-2. Making hospital rounds, Dr. R., an anesthesiologist, meets with CeCe, a first-semester anatomy and physiology student who is scheduled for surgery the next day. Having just finished the unit on skeletal muscles and the nervous system, CeCe is eager to learn about the anesthesia that will be used during the surgery. Dr. R. explains that she will be using the drug *tubocurarine chloride,* which competes with the neurotransmitter acetylcholine and blocks its action at the neuromuscular junction. What effect will this drug have on CeCe's skeletal muscles?

   a. produce paralysis of all the skeletal muscles
   b. cause tetany of the skeletal muscles
   c. increase the force and strength of muscle contractions

CeCe answers that question correctly but immediately becomes concerned about this effect on a select group of skeletal muscles. Which muscle group is CeCe concerned about?

3-3. Tom broke his right leg in a football game. After he had worn a cast for 6 weeks, the cast was finally removed. When he took his first few steps, Tom lost his balance and fell. What is the most likely explanation?

   a. the bone fracture was not completely healed
   b. the right leg muscles had atrophied from disuse
   c. Tom had an undiagnosed neuromuscular disorder

3-4. Samples of muscle tissue are taken from a champion tennis player and from a nonathlete of the same age and gender. The two samples are subjected to enzyme analysis. How would you expect them to differ?

3-5. Calvin steps into a pothole and twists his ankle. He is in a great deal of pain and cannot stand on the ankle. In the hospital, the examining physician notes that Calvin can plantar flex and dorsiflex the foot but cannot perform inversion without extreme pain. Which structures may Calvin have injured?

Clinical Problems

1. Ann, who is 35 years old, complains of progressively worsening pain in several joints. Pain is worst in the early morning. Physical examination reveals limited movement of the jaw and inflammation of several other joints. Slight ulnar deviation of the fingers is noted. There are no skin lesions present. Many different disorders can cause joint pain. Study the following information to make a preliminary diagnosis:

- Synovial fluid analysis: cloudy, reduced viscosity, absence of bacteria, no uric acid crystals

- X-ray studies of the hand: detectable deterioration of articular cartilages in metacarpophalangeal joints

- Serum rheumatoid factor: 1:60 titer

2. The mother of a 5-year-old boy reports that her son cannot rise from a sitting to a standing position without support. He also falls frequently. Physical examination reveals proximal muscle weakness, especially in the lower limbs. The physician orders bone X-rays, a serum aldolase test, and nerve conduction tests. Two of the test results are abnormal, and the physician makes a preliminary diagnosis of muscular dystrophy. This diagnosis is later confirmed by muscle biopsy. Which of the three results was (were) abnormal?

The Nervous System

The nervous system is a highly complex and interconnected network of neurons and supporting neuroglia. Neural tissue is extremely delicate, and the characteristics of the extracellular environment must be kept within narrow homeostatic limits. When homeostatic regulatory mechanisms break down under the stress of genetic or environmental factors, infection, or trauma, symptoms of neurological disorders appear.

There are hundreds of disorders of the nervous system. A *neurological examination* attempts to trace the source of the problem through evaluation of the sensory, motor, behavioral, and cognitive functions of the nervous system. Figure A-25 introduces several major categories of nervous system disorders. We will discuss many of these examples in the sections that follow.

THE SYMPTOMS OF NEUROLOGICAL DISORDERS

The nervous system has varied and complex functions, and the symptoms of neurological disorders are equally diverse. However, there are a few symptoms that accompany many different disorders:

- *Headache.* Roughly 90 percent of headaches are *tension headaches,* which are thought to be due to muscle tension, or *migraine headaches,* which have both neurological and circulatory origins (see p. 70). Neither of these conditions is life-threatening.

- *Muscle weakness.* Muscle weakness can have an underlying neurological basis, as we noted in the section on muscle disorders (see Figure A-21, p. 60). The examiner must determine the origin of the symptom. Myopathies (muscle disease) must be differentiated from neurological diseases such as demyelinating disorders, neuromuscular junction dysfunction, and peripheral nerve damage.

- *Paresthesias.* Loss of feeling, numbness, or tingling sensations may develop after damage to (1) a sensory nerve (cranial or spinal nerve) or (2) sensory pathways inside the CNS. The effects may be temporary or permanent. For example, a *pressure palsy* (p. 77) may last a few

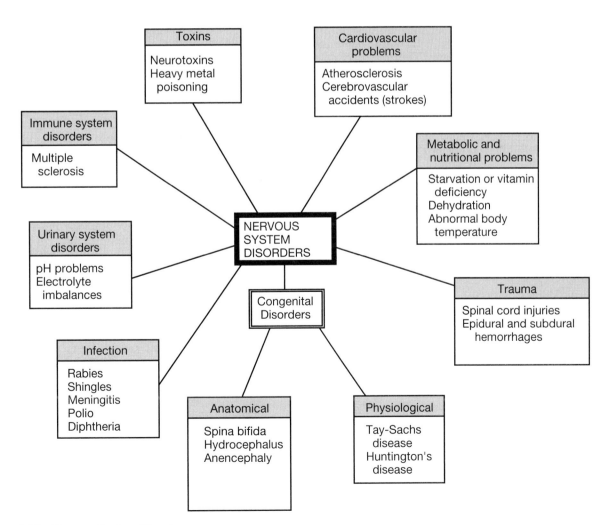

Figure A-25   Nervous System Disorders

minutes, whereas the paresthesia that develops distal to an area of severe spinal cord damage (p. 77) will probably be permanent.

THE NEUROLOGICAL EXAMINATION   FAP *p. 369*

During a physical examination, information about the nervous system is obtained indirectly, by assessing sensory, motor, and intellectual functions. Examples of factors noted in the physical examination include:

- *State of consciousness.* There are many different levels of consciousness, ranging from unconscious and incapable of being aroused, to fully alert and attentive, to hyperexcitable. We introduce the names assigned to the various levels of consciousness in a later section (p. 84).

- *Reflex activity.* The general state of the nervous system, and especially the state of peripheral sensory and motor innervation, can be checked by testing specific reflexes (p. 78). For example, the *knee-jerk reflex* will not be normal if there has been damage to associated segments of the lumbar spinal cord, their spinal nerve roots, or the peripheral nerves involved in the reflex.

- *Abnormal speech patterns.* Normal speech involves intellectual processing, motor coordination at the speech centers of the brain, precise respiratory control, regulation of tension in the vocal cords, and adjustment of the musculature of the palate and face. Problems with the selection, production, or use of words commonly follows damage to the cerebral hemispheres, as in a stroke (p. 127).

- *Abnormal motor patterns.* An individual's posture, balance, and mode of walking, or *gait*, are useful indicators of the level of motor coordination. Clinicians also ask about abnormal involuntary movements that may indicate a *seizure*, a temporary disorder of cerebral function (p. 83).

A number of diagnostic procedures and laboratory tests can be used to obtain additional information about the status of the nervous system. Table A-12 summarizes information about these procedures.

Headaches

Almost everyone has experienced a **headache** at one time or another. Diagnosis and treatment pose a number of problems, primarily because, as we noted earlier, headaches can be produced by a wide variety of underlying conditions. The most common causes of headache are either vascular or muscular problems.

Most headaches do not merit a visit to a neurologist. The vast majority of headaches are associated with muscle tension, such as tight neck muscles, but a variety of other factors may be responsible. For example, headaches may develop due to one of the following problems:

1. *CNS problems*, such as infections *(meningitis, encephalitis, rabies)* or *brain tumors*
2. *Trauma*, such as a blow to the head (p. 80)
3. *Cardiovascular disorders*, such as a *stroke* (p. 127)
4. *Metabolic disturbances*, such as low blood sugar
5. *Related muscle tension*, such as stiff neck or temporomandibular joint (TMJ) syndrome

Migraine headaches affect roughly 5 percent of the population. An individual with a *classic migraine* experiences visual or other sensory signals that an attack is imminent. The headache pain may then be accompanied by disturbances in vision or somatic sensation, extreme anxiety, nausea, or disorientation. The symptoms generally persist for several hours. A *common migraine* attack typically lacks any warning signs.

There is now good evidence that migraine headaches begin at a portion of the mesencephalon known as the *dorsal raphe.* Electrical stimulation of the dorsal raphe can produce changes in cerebral blood flow; several drugs with anti-migraine action inhibit neurons at this location. The most effective drugs stimulate a class of serotonin receptors that are found in high concentration within the dorsal raphe.

The trigger for **tension headaches** probably involves a combination of factors, but sustained contractions of the neck and facial muscles are most commonly implicated. Tension headaches last for days or even months, typically without the throbbing, pulsing sensations characteristic of migraine headaches. Instead, the victim may complain of a feeling of pressure or viselike compression. Some tension headaches do not involve the muscles but accompany severe depression or anxiety.

Axoplasmic Transport and Disease   FAP *p. 374*

With a soft flutter of wings, dark shapes drop from the sky onto the backs of grazing cattle. Each shape is a small bat whose scientific name, *Desmodus rotundus*, is less familiar than the popular term *vampire bat.* Vampire bats inhabit tropical and semitropical areas of North, Central, and South America. They range from the Texas coast to Chile and southern Brazil. These rather aggressive animals are true vampires, subsisting on a diet of fresh blood. Over the next hour, every bat in the flight—which may number in the hundreds—will consume about 65 ml of blood through small slashes in the skin of their prey.

Table A-12 Representative Diagnostic and Laboratory Tests for Nervous System Disorders and Their Common Uses

Diagnostic Procedure	Method and Result	Representative Uses	Notes
Lumbar puncture (spinal tap)	Needle aspiration of CSF from the subarachnoid space in the lumbar area of the spinal cord	See CSF analysis for diagnostic uses	Attaching manometer to needle allows for CSF pressure determination
Skull X-ray	Standard X-ray	Detection of fracture and possible sinus involvement	Malignant processes and infections of the bone can result in erosion of bony structures normally present on X-ray
Electroencephalography (EEG)	Electrodes placed on the scalp detect electrical activity of the brain, and EEG produces graphic record	Detection of abnormalities in frequency and amplitude of brain waves, due to cranial trauma or neurological disorders such as seizures	Abnormally low-amplitude waves in specific areas may indicate cranial trauma or a neurological disorder
Computerized tomography (CT) scan of the brain	Standard CT; contrast media are commonly used	Detection of tumors, cerebrovascular abnormalities, such as aneurysms (weakened areas in vessel walls), scars, strokes, or areas of edema	
Cerebral angiography and digital subtraction angiography	Dye is injected into an artery of the neck, and the movement of the dye is observed via serial X-rays; digital subtraction angiography transfers dye location information to a computer that can store images or have part of the image (such as that of a bony area) blocked out to clarify the image of the blood vessel	Detection of abnormalities in the cerebral vessels, such as aneurysms or blockages	
Positron emission tomography (PET) scan	Radiolabeled compounds injected into the circulation accumulate at specific areas of the brain; the radiation emitted is monitored by a computer that generates a reconstructed image	Determination of blood flow to the brain; detection of focal points of brain activity; also useful in the diagnosis of Parkinson's disease and Alzheimer's disease	
Magnetic resonance imaging (MRI)	Standard MRI; contrast media are commonly used to enhance visualization	Detection of brain tumors, hemorrhaging, edema, spinal cord injury, and other structural abnormalities	

As unpleasant as this blood collection may sound, it is not the blood loss that is the primary cause for concern. The major problem is that these bats can be carriers for the rabies virus. **Rabies** is an acute disease of the central nervous system. The rabies virus can infect any mammal, wild or domestic. With few exceptions, the result is death within 3 weeks. For unknown reasons, bats can survive rabies infection for an indefinite period. As a result, an infected bat can serve as a carrier for the disease. Because many bat species, including vampire bats, form dense colonies, a single infected individual can spread the disease through the entire colony.

Table A-12 *(continued)*

Laboratory Test	Normal Values	Significance of Abnormal Values	Notes
Analysis of CSF			
Pressure of CSF	<200 cm H_2O	Pressure higher than 200 cm H_2O is considered abnormal, possibly indicating hemorrhaging, tumor formation, or infection	
Color of CSF	Clear and colorless	Increased turbidity suggests hemorrhage or faulty puncture technique	Increased WBC count indicates infection; presence of RBCs indicates hemorrhaging
Glucose in CSF	50–75 mg/dl	Decreased levels are found when CSF tumors or infections are present	
Protein in CSF	15–45 mg/dl	Elevated levels occur in some infectious processes, such as meningitis and encephalitis; may be elevated during inflammatory processes or following tumor formation	Plasma proteins generally do not cross the blood–brain barrier; increased permeability occurs with infection or tumor formation
Cells present in CSF	No RBCs present; WBC count should be less than 5 per mm^3	RBCs appear with subarachnoid hemorrhage; increased number of neutrophils occurs in bacterial infections, such as bacterial meningitis; increase in lymphocyte numbers occurs in viral meningitis	The CSF is also examined for cancerous cells from a brain tumor
IgG in CSF	Should be in a normal ratio to other proteins	Elevated CSF IgG levels occur with autoimmune disorders, such as multiple sclerosis	
Culture of CSF	Organism causing infectious process in brain or spinal cord can be cultured for identification and determination of sensitivity to antibiotics	Determination of causative agent in meningitis or brain abscess	

The deaths of cattle and pigs are economic hardships; human infection is a disaster. Rabies is generally transmitted to people through the bite of a rabid animal. There are an estimated 15,000 cases of rabies each year worldwide, the majority of them the result of dog bites. Only about five of those cases, however, are diagnosed in the United States. Because most dogs and cats in the United States are vaccinated against rabies, most cases there are caused by the bites of raccoons, foxes, skunks, or bats.

Although these bites generally involve peripheral sites, such as the hand or foot, the symptoms are caused by CNS damage. The virus present at the injury site is absorbed by the synaptic knobs of peripheral nerves in the region. It then gets a free ride to the CNS, courtesy of retrograde flow. During the first few days after exposure, the individual may experience headache, fever, muscle pain, nausea, and vomiting. The victim then enters a phase marked by extreme excitability, hallucinations, muscle spasms, and disorientation. There is difficulty in swallowing, and the accumulation of saliva makes the individual appear to be "foaming at the mouth." Coma and death soon follow.

Preventive treatment, which must begin almost immediately after exposure, consists of injections that contain antibodies against the rabies virus followed by a series of vaccinations against rabies. This postexposure treatment may not be sufficient after a massive infection, which can lead to death in as little as 4 days. Individuals such as veterinarians or field biologists who are at high risk of exposure commonly take a preexposure series of vaccinations. These injections bolster the immune defenses and improve the effectiveness of the postexposure treatment. Without preventive treatment, rabies infection in humans is always fatal.

Rabies is perhaps the most dramatic example of a clinical condition directly related to axoplasmic flow. However, many toxins, including heavy metals, some pathogenic bacteria, and other viruses use this mechanism to enter the CNS.

☤ Demyelination Disorders   FAP *p. 380*

Demyelination disorders are linked by a common symptom: the destruction of myelinated axons in the CNS and PNS. The mechanism responsible for this loss differs in each of these disorders. We will consider only the major categories of demyelination disorders:

* *Heavy metal poisoning.* Chronic exposure to heavy metal ions, such as arsenic, lead, or mercury, can lead to glial cell damage and demyelination. As demyelination occurs, the affected axons deteriorate, and the condition becomes irreversible. Historians note several interesting examples of heavy metal poisoning with widespread impact. For example, lead contamination of drinking water has been cited as one factor in the decline of the Roman Empire. In the seventeenth century, the great physicist Sir Isaac Newton is thought to have suffered several episodes of physical illness and mental instability brought on by his use of mercury in chemical experiments. Well into the nineteenth century, mercury used in the preparation of felt presented a serious occupational hazard for those employed in the manufacture of stylish hats. Over time, mercury absorbed through the skin and across the lungs accumulated in the CNS, producing neurological damage that affected both physical and mental function. (This effect is the source of the expression "mad as a hatter.") More recently, Japanese fishermen working in Minamata Bay, Japan, collected and consumed seafood contaminated with mercury discharged from a nearby chemical plant. Levels of mercury in their systems gradually rose to the point at which clinical symptoms appeared in hundreds of people. Making matters worse, mercury contamination of developing embryos and fetuses caused severe, crippling birth defects.

* *Diphtheria.* **Diphtheria** (dif-THĒ-rē-uh; *diphtheria,* membrane + *-ia,* disease) is a disease that results from a bacterial infection of the respiratory tract. In addition to restricting airflow and sometimes damaging the respiratory surfaces, the bacteria produce a powerful toxin that injures the kidneys and adrenal glands, among other tissues. In the nervous system, diphtheria toxin damages Schwann cells and destroys myelin sheaths in the PNS. This demyelination leads to sensory and motor problems that may ultimately produce a fatal paralysis. The toxin also affects cardiac muscle cells, and heart enlargement and heart failure may occur. The fatality rate for untreated cases ranges from 35 to 90 percent, depending on the site of infection and the subspecies of bacterium. Because an effective vaccine exists, cases are relatively rare in countries with adequate health care.

* *Multiple sclerosis.* **Multiple sclerosis** (skler-Ō-sis; *sklerosis,* hardness), or **MS,** is a disease characterized by recurrent incidents of demyelination that affects axons in the optic nerve, brain, and spinal cord. Common symptoms include partial loss of vision and problems with speech, balance, and general motor coordination. The time between incidents and the degree of recovery vary from case to case. In about one-third of all cases, the disorder is progressive, and each incident leaves a greater degree of functional impairment. The average age at the first attack is 30–40; the incidence among women is 1.5 times that among men. There is no effective treatment at present, although corticosteroid injections and interferon may slow the progression of the disease. We discuss MS in more detail in a later section (p. 75).

* *Guillain–Barré syndrome.* **Guillain–Barré syndrome** is characterized by a progressive but reversible demyelination. Symptoms initially involve weakness of the legs, which spreads rapidly to muscles of the trunk and arms. These symptoms generally increase in intensity for 1–2 weeks before subsiding. The mortality rate is low (under 5 percent), but there may be some permanent loss of motor function. The cause is unknown, but because roughly two-thirds of Guillain–Barré patients develop symptoms within 2 months after they have experienced a viral infection, it is suspected that the condition may result from a malfunction of the immune system. (We consider the mechanism involved in Chapter 22 of the text; see FAP *p. 804.*)

12

✠ Neurotoxins in Seafood   FAP *p. 389*

Several forms of human poisoning result from the ingestion of seafood that contains neurotoxins, poisons that primarily affect neurons. **Tetrodotoxin** (te-TRŌ-dō-tok-sin), or **TTX,** is found in the liver, gonads, and blood of certain Pacific puffer fish species, and a related compound is found in the skin glands of some salamanders. Tetrodotoxin selectively blocks voltage-regulated sodium ion channels, effectively preventing nerve cell activity. The usual result is death from paralysis of the respiratory muscles. Despite the risks, the Japanese consider the puffer fish a delicacy, served under the name *fugu.* Specially licensed chefs prepare these meals, carefully removing the potentially toxic organs. Nevertheless, a mild tingling and sense of intoxication are considered desirable, and several people die each year as a result of improper preparation of the dish.

Saxitoxin (sak-si-TOK-sin), or **STX,** can have a similarly lethal effect. Saxitoxin and related poisons are produced by several species of marine microorganisms. When these organisms undergo a population explosion, they color the surface waters, producing a "red tide." Eating seafood that has become contaminated by feeding on the toxic microbes can result in symptoms of **paralytic shellfish poisoning (PSP;** from clams, mussels, or oysters) or **ciguatera** (sēg-wa-TER-uh; *cigua,* a sea snail) **(CTX;** from fish). Mild cases produce symptoms of paresthesias and a curious reversal of hot-versus-cold sensations. Severe cases, which are relatively rare, result in coma and death due to respiratory paralysis.

℞ Neuroactive Drugs   FAP *p. 402*

The text discussion on FAP *p. 402* introduced the concept that many drugs and toxins work by interfering with normal synaptic function. Table A-13 provides additional information on specific chemical compounds, their uses, and their sites of action.

✠ Tay-Sachs Disease   FAP *p. 408*

Tay-Sachs disease is a genetic abnormality that involves the metabolism of *gangliosides,* important components of neuron cell membranes. It is a *lysosomal storage disease* caused by abnormal lysosome activity. Individuals with this condition lack the enzyme needed to break down one particular ganglioside, which accumulates within the lysosomes of CNS neurons and causes them to deteriorate. Affected infants seem normal at birth, but within 6 months, neurological problems begin to appear. The progression of symptoms typically includes muscular weakness, blindness, seizures, and death, generally before age 4. No effective treatment exists, but prospective parents can be tested to determine whether they are carrying the gene

responsible for this condition. The disorder is most prevalent in Ashkenazi Jews of Eastern Europe. A prenatal test is available to detect this condition in a fetus.

✠ Meningitis   FAP *p. 420*

The warm, dark, nutrient-rich environment of the meninges provides ideal conditions for a variety of bacteria and viruses. Microorganisms that cause meningitis include bacteria associated with middle ear and sinus infections; with pneumonia, streptococcal ("strep"), staphylococcal ("staph"), or meningococcal infections; and with tuberculosis. These pathogens may gain access to the meninges by traveling within blood vessels or by entering at sites of vertebral or cranial injury. Headache, chills, high fever, disorientation, and rapid heart and respiratory rates appear as higher centers are affected. Without treatment, delirium, coma, convulsions, and death may follow within hours.

The most common clinical assessment involves checking for a "stiff neck" by asking the patient to touch the chin to the chest. Meningitis affecting the cervical portion of the spinal cord results in a marked increase in the muscle tone of the extensor muscles of the neck. So many motor units become activated that voluntary or involuntary flexion of the neck becomes painfully difficult, if not impossible.

The mortality rate for viral and bacterial forms of meningitis ranges from 1 to more than 50 percent, depending on the type of virus or bacteria, the age and health of the patient, and other factors. There is no effective treatment for viral meningitis, but bacterial meningitis can be combated with antibiotics and the maintenance of proper fluid and electrolyte balance. The incidence of one form of bacterial meningitis, caused by *Haemophilus influenzae,* can be reduced through childhood immunization.

℞ Spinal Anesthesia   FAP *p. 420*

Injecting a local anesthetic around a nerve produces a temporary blockage of nerve function. This can be done peripherally, as when skin lacerations are sewn up, or at sites around the spinal cord to obtain more widespread anesthetic effects. Although an *epidural block,* the injection of an anesthetic into the epidural space, has the advantage of affecting only the spinal nerves in the immediate area of the injection, epidural anesthesia may be difficult to achieve in the upper cervical, midthoracic, and lumbar regions, where the epidural space is extremely narrow. **Caudal anesthesia** involves the introduction of anesthetics into the epidural space of the sacrum. Injection at this site paralyzes lower abdominal and perineal structures. Caudal anesthesia may be used instead of epidural blocks in the lower lumbar or sacral regions to control pain during childbirth.

Table A-13   Drugs Affecting Acetylcholine Activity at Synapses

Drug	Mechanism	Effects	Remarks
Hemicholinium	Blocks ACh synthesis	Produces symptoms of synaptic fatigue	
Botulinus toxin	Blocks ACh release directly	Paralyzes voluntary muscles	Produced by bacteria; responsible for a deadly type of food poisoning
Barbiturates	Decrease rate of ACh release	Muscular weakness; depression of CNS activity	Administered as sedatives and anesthetics
Procaine (Novocain)	Reduces membrane permeability to sodium	Prevents stimulation of sensory neurons	Used as a local anesthetic
Tetrodotoxin (TTX) Saxitoxin (STX) Cigautoxin (CTX)	Blocks sodium ion channels	Eliminates production of action potentials	Produced by some marine organisms during normal metabolic activity
Neostigmine	Prevents ACh inactivation by cholinesterase	Sustained contraction of skeletal muscles; other effects on cardiac muscle, smooth muscle, and glands	Used clinically to treat myasthenia gravis and to counteract overdoses of tubocurarine; related compound produced by Calabar bean
Insecticides (malathion, parathion, etc.), and nerve gases	As for Neostigmine	As for Neostigmine	Related compounds used in military nerve gases
d-tubocurarine	Prevents ACh binding to postsynaptic receptor sites	Paralyzes voluntary muscles	Curare produced by South American plant
Nicotine	Binds to ACh receptor sites	Low doses facilitate voluntary muscles; high doses cause paralysis	An active ingredient in cigarette smoke, addictive for most people
Succinylcholine	Reduces sensitivity to ACh	Paralyzes voluntary muscles	Used to produce temporary muscular relaxation during surgery
Atropine	Competes with ACh for binding sites on postsynaptic membrane	Reduced heart rate, smooth muscle activity; decreased salivation; dilation of pupils; skeletal muscle weakness develops at high doses	Produced by deadly nightshade plant

13

Local anesthetics may also be introduced into the subarachnoid space of the spinal cord. However, the effects spread as CSF circulation and diffusion distribute the anesthetic along the spinal cord. As a result, precise control of the regional effects can be difficult to achieve. Problems with overdosing are seldom serious, because the diaphragmatic breathing muscles are controlled by upper cervical spinal nerves. Thus respiration continues even when the thoracic and abdominal segments have been paralyzed.

☤ Multiple Sclerosis   FAP *p. 423*

Multiple sclerosis (MS), introduced in the discussion of demyelination disorders, is a disease that produces muscular paralysis and sensory losses through demyelination. The initial symptoms appear as the result of myelin degeneration within the white matter of the lateral and posterior columns of the spinal cord or along tracts within the brain. For example, spinal cord involvement may produce weakness, tingling sensations, and a loss of "position sense" for the limbs. During subsequent attacks, the effects become more widespread. The cumulative sensory and motor losses may eventually lead to a generalized muscular paralysis.

Recent evidence suggests that this condition may be linked to a defect in the immune system that causes it to attack myelin sheaths. MS patients have lymphocytes that do not respond normally to foreign proteins. Because several viral proteins have amino acid sequences similar to those of normal myelin, it has been proposed that MS results from a case of mistaken identity. For unknown reasons, MS appears to be associated with cold and temperate climates. It has been suggested that individuals who develop MS may have an inherited susceptibility to a virus and that this susceptibility is exaggerated by environmental conditions. The yearly incidence within the United States averages around 50 cases for every 100,000 in the population. Improvement has been noted in some patients treated with *interferon*, a peptide secreted by cells of the immune system, and recently corticosteroid treatment has been linked to a slowdown in the progression of MS.

℞ Spinal Cord Injuries and Experimental Treatments FAP *p. 423*

At the outset, any severe injury to the spinal cord produces a period of sensory and motor paralysis termed **spinal shock.** The skeletal muscles become flaccid; neither somatic nor visceral reflexes function; and the brain no longer receives sensations of touch, pain, heat, or cold. The location and severity of the injury determine how long these symptoms persist and how completely the individual recovers.

Violent jolts, such as those associated with blows or gunshot wounds near the spinal cord, may cause **spinal concussion** without visibly damaging the spinal cord. Spinal concussion produces a period of spinal shock, but the symptoms are only temporary and recovery may be complete in a matter of hours. More serious injuries, such as vertebral fractures, generally involve physical damage to the spinal cord. In a **spinal contusion,** hemorrhages occur in the meninges, pressure rises in the cerebrospinal fluid, and the white matter of the spinal cord may degenerate at the site of injury. Gradual recovery over a period of weeks may leave some functional losses. Recovery from a **spinal laceration** by vertebral fragments or other foreign bodies tends to be far slower and less complete. **Spinal compression** occurs when the spinal cord becomes squeezed or distorted within the vertebral canal. In a **spinal transection,** the spinal cord is completely severed. At present, surgical procedures cannot repair a severed spinal cord.

Many spinal cord injuries involve some combination of compression, laceration, contusion, and partial transection. Relieving pressure and stabilizing the affected area through surgery (such as *spinal fusion*, the immobilization of adjacent vertebrae) may prevent further damage and allow the injured spinal cord to recover as much as possible.

Two avenues of research are being pursued, one biological and the other electronic. A major biological line of investigation involves the biochemical control of nerve growth and regeneration. Neurons are influenced by a combination of growth promoters and growth inhibitors. Damaged myelin sheaths apparently release an inhibitory factor that slows the repair process. Researchers have made an antibody, *IN-1*, that will inactivate the inhibitory factor released in the damaged spinal cords of rats. The treatment stimulates repairs, even in severed spinal cords. Because there is no myelin in gray matter, the chances of regrowth of severed axons are enhanced if the proximal cut end is guided into the gray matter rather than being left in the white matter. Researchers are now attempting to devise a suitable means of accomplishing this rerouting.

Nerve growth and regeneration are stimulated by a variety of recently identified chemicals. A partial listing includes **nerve growth factor** (NGF), **brain-derived neutrophic factor** (BDNF), **neurotrophin-3** (NT-3), **neurotrophin-4** (NT-4), **glial growth factor, glial maturation factor, ciliary neurotrophic factor,** and **growth-associated protein 43** (GAP-43). Many of these factors have now been synthesized by using gene-splicing techniques, and sufficient quantities are available to permit their use in experiments on humans and other mammals. Initial results are promising, and these factors in various combination are being evaluated for treatment of CNS injuries and the chronic degeneration seen in Alzheimer's disease and Parkinson's disease.

Other research teams are experimenting with the use of computers to stimulate specific muscles and muscle groups electrically. The technique is called *functional electrical stimulation*, or FES. This approach commonly involves the implantation of a network of wires beneath the skin with their tips in skeletal muscle tissue. The wires are connected to a small computer worn at the waist. The wires deliver minute electrical stimuli to the muscles, depolarizing their membranes and causing contractions. With this equipment and lightweight braces, quadriplegics have walked several hundred yards and paraplegics several thousand. The Parastep™ system, which uses a microcomputer controller, is now undergoing clinical trials.

Equally impressive results have been obtained by using a network of wires woven into the fabric of close-fitting garments. This network provides the necessary stimulation without the complications and maintenance problems that accompany implanted wires. A paraplegic woman in a set of electronic "hot pants" completed several miles of the 1985 Honolulu Marathon, and more recently a paraplegic woman walked down the aisle at her wedding.

Such technological solutions can provide only a degree of motor control without accompanying sensation. Everyone would prefer a biological procedure that would restore the functional integrity of the nervous system. For now, however, computer-assisted programs such as FES can improve the quality of life for thousands of paralyzed individuals. The 1995 horseback-riding injury of actor Christopher Reeve has brought publicity and research funds to this area.

✝ Shingles and Hansen's Disease

FAP *p. 425*

In **shingles,** or *herpes zoster,* the *herpes varicella-zoster* virus attacks neurons within the dorsal roots of spinal nerves and sensory ganglia of cranial nerves. This disorder produces a painful rash whose distribution corresponds to that of the affected sensory nerves (Figure A-26a). Shingles develops in adults who were first exposed to the virus as children. The initial infection produces symptoms known as chickenpox. After this encounter, the virus remains dormant within neurons of the anterior gray horns of the spinal cord. It is not known what triggers reactivation of this pathogen. Fortunately for those affected, attacks of shingles generally heal and leave behind only unpleasant memories.

Most people suffer only a single episode of shingles in their adult lives. However, the problem may recur in people with weakened immune systems, including those with AIDS or some forms of cancer. Treatment for shingles typically involves large doses of the antiviral drug *acyclovir (Zovirax®).*

The condition traditionally called **leprosy,** now more commonly known as **Hansen's disease,** is an infectious disease caused by a bacterium, *Mycobacterium leprae.* It is a disease that progresses slowly, and symptoms may not appear for up to 30 years after infection. The bacterium invades peripheral nerves, especially those in the skin, producing initial sensory losses. Over time, motor paralysis develops, and the combination of sensory and motor loss can lead to recurring injuries and infections. The eyes, nose, hands, and feet may develop deformities as a result of neglected injuries (Figure A-26b). There are several forms of this disease; peripheral nerves are always affected, but some forms also involve extensive lesions of the skin and mucous membranes.

Only about 5 percent of those exposed to *Mycobacterium leprae* develop symptoms; people living in the tropics are at greatest risk. There are about 2000 cases in the United States, and an estimated 12–20 million cases worldwide. If detected before deformities occur, the disease can generally be treated successfully with drugs such as rifampin and dapsone. Treated individuals are not infectious, and the practice of confining "lepers" in isolated compounds has been discontinued.

✝ Peripheral Neuropathies   FAP *p. 426*

Peripheral nerve palsies, or peripheral neuropathies, are characterized by regional losses of sensory and motor function as the result of nerve trauma or compression. **Brachial palsies** result from injuries to the brachial plexus or its branches. **Crural palsies** involve the nerves of the lumbosacral plexus.

Palsies may appear for several reasons. The *pressure palsies* are especially interesting; a familiar but mild example is the experience of having an arm or leg "fall asleep." The limb becomes numb, and afterward an uncomfortable "pins-and-needles" sensation, or **paresthesia,** accompanies the return to normal function.

These incidents are seldom of clinical significance, but they provide graphic examples of the effects of more serious palsies that can last for days to months. In **radial nerve palsy,** pressure on the back of the arm interrupts the function of the radial nerve, so the extensors of the wrist and fingers

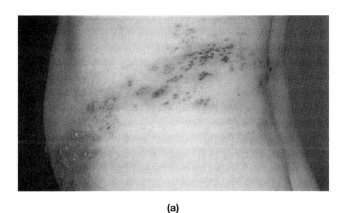

(a)

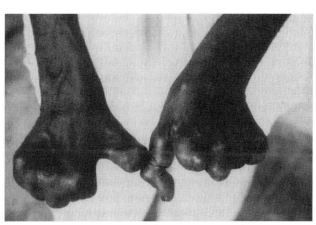

(b)

Figure A-26 Shingles and Hansen's Disease.
(a) The left side of a person with shingles. The skin eruptions follow the distribution of dermatomal innervation. (b) The distal extremities are gradually deformed as untreated Hansen's disease progresses.

are paralyzed. This condition is also known as "Saturday night palsy," for falling asleep on a couch with your arm over the seat back (or beneath someone's head) can produce the right combination of pressures. Students may also be familiar with **ulnar palsy,** which can result from prolonged contact between elbow and desk. The ring and little fingers lose sensation, and the fingers cannot be adducted. We considered *carpal tunnel syndrome,* a neuropathy resulting from compression of the *median nerve* at the wrist, in FAP *(pp. 239, 350).*

Men with large wallets in their hip pockets may develop symptoms of **sciatic compression** after they drive or sit in one position for extended periods. As nerve function declines, the individuals notice some lumbar or gluteal pain, a numbness along the back of the leg, and a weakness in the leg muscles. Similar symptoms result from the compression of nerve roots that form the sciatic nerve by a distorted lumbar intervertebral disc. This condition is termed **sciatica,** and one or both legs may be affected, depending on the site of compression. Finally, sitting with your legs crossed may produce symptoms of a **peroneal palsy.** Sensory losses from the top of the foot and side of the leg are accompanied by a decreased ability to dorsiflex or evert the foot.

® Damage and Repair of Peripheral Nerves
FAP *p. 427*

If a peripheral axon is damaged but not displaced, normal function may eventually return as the cut stump grows across the injury site away from the soma and along its former path. The mechanics of this process were detailed at the close of Chapter 12. For normal function to be restored, several things must happen. The severed ends must be relatively close together (1–2 mm); they must remain in proper alignment; and there must not be any physical obstacles between them, such as the collagen fibers of scar tissue. These conditions can be created in the laboratory, using experimental animals and individual axons or small fascicles. But in accidental injuries to peripheral nerves, the edges are likely to be jagged; intervening segments may be lost entirely, and elastic contraction in the surrounding connective tissues may pull the cut ends apart and misalign them.

Until recently, the surgical response would involve trimming the injured nerve ends, neatly sewing them together, and hoping for the best. This procedure was typically unsuccessful, in part because scalpels do not produce a smoothly cut surface and because the thousands of broken axons would never be perfectly aligned. Moreover, axons are not highly elastic, so if a large segment of the nerve was removed, crushed, or otherwise destroyed, there would be no way to bring the intact ends close enough to permit any regeneration. In such instances, a **nerve graft** could be inserted,

using a section from some other, less important peripheral nerve. The functional results were even less likely to be wholly satisfactory, for the growing axonal tips had to find their way across not one but two gaps, and the chances for successful alignment were proportionately smaller. Nevertheless, any return of function was better than none at all!

Research now focuses on the physical and biochemical control of nerve regeneration, using the growth factors introduced earlier (p. 76). Another promising strategy is the use of a synthetic sleeve to guide nerve growth. The sleeve is a tube with an outer layer of silicone around an inner layer of cowhide collagen bound to the proteoglycans from shark cartilage. Using this sleeve as a guide, axons can grow across gaps as large as 20 mm (0.75 in.). The procedure has yet to be tried on humans, and functional restoration in other mammals is generally incomplete because proper alignment does not always occur.

Reflexes and Diagnostic Testing
FAP *p. 440*

Many reflexes can be assessed through careful observation and the use of simple tools. The procedures are easy to perform, and the results can provide valuable information about damage to the spinal cord or spinal nerves. By testing a series of spinal and cranial reflexes, a physician can assess the function of sensory pathways and motor centers throughout the spinal cord and brain.

Neurologists test many different reflexes; only a few are so generally useful that physicians make them part of a standard physical examination. These reflexes are shown in Figure A-27.

The *patellar reflex* (Figure 13-16, FAP *p. 435*), *ankle jerk* (Figure A-27a), *biceps reflex* (Figure A-27b), and *triceps reflex* (Figure A-27c) are stretch reflexes controlled by specific segments of the spinal cord. Testing these reflexes provides information about the corresponding spinal segments. For example, a normal patellar reflex, or knee jerk, indicates that spinal nerves and spinal segments L_2–L_4 are undamaged. The Babinski sign (Figure 13-18, FAP *p. 440*) is normally absent in the adult because of descending spinal inhibition. Spinal cord injury or strokes may reduce this inhibition, resulting in reemergence of the Babinski sign.

The **abdominal reflex** (Figure A-27d), present in the normal adult, results from descending spinal facilitation. In this reflex, a light stroking of the skin produces a reflexive twitch in the abdominal muscles that moves the navel toward the stimulus. This reflex disappears after descending tracts have been damaged.

ABNORMAL REFLEX ACTIVITY

In **hyporeflexia,** normal reflexes are weak but apparent, especially with reinforcement. In **areflexia** (Ā-rē-FLEK-sē-uh; *a-*, without), normal re-

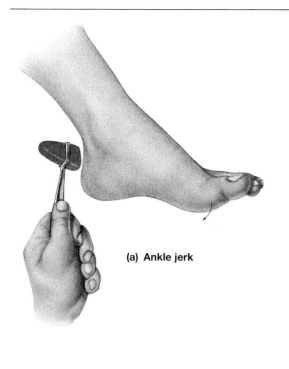

(a) Ankle jerk

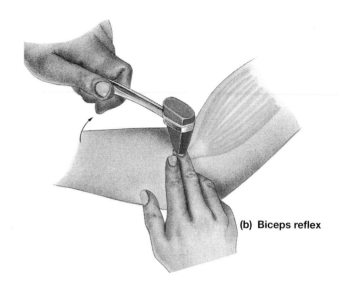

(b) Biceps reflex

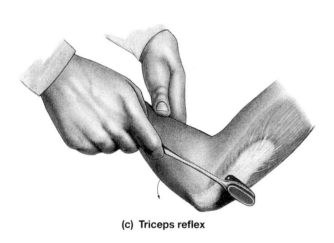

(c) Triceps reflex

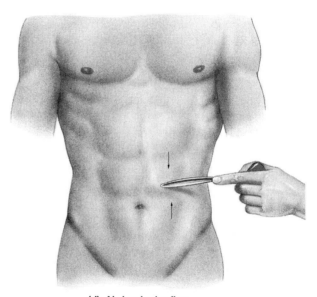

(d) Abdominal reflex

Figure A-27 Reflexes and Diagnostic Testing

flexes fail to appear, even with reinforcement. Hyporeflexia or areflexia may indicate temporary or permanent damage to skeletal muscles, dorsal or ventral nerve roots, spinal nerves, the spinal cord, or the brain.

Hyperreflexia occurs when higher centers maintain a high degree of facilitation along the spinal cord. Under these conditions, reflexes are easily triggered, and the responses may be grossly exaggerated. This effect can also result from spinal cord compression or diseases that target higher centers or descending tracts. One potential result of hyperreflexia is the appearance of alternating contractions in opposing muscles. When one muscle contracts, it stimulates the stretch receptors in the other. The stretch reflex

then triggers a contraction in that muscle, and this contraction stretches receptors in the original muscle. This self-perpetuating sequence, which can be repeated indefinitely, is called **clonus** (KLŌ-nus). In a hyperreflexive person, a tap on the patellar tendon will set up a cycle of kicks rather than just one or two.

A more extreme hyperreflexia develops if the motor neurons of the spinal cord lose contact with higher centers. In many cases, after a severe spinal injury, the individual first experiences a temporary period of areflexia known as spinal shock, discussed on p. 76. When the reflexes return, they respond in an exaggerated fashion, even to mild stimuli. For example, the lightest touch on the skin surface may produce a massive withdrawal reflex.

The reflex contractions may occur in a series of intense muscle spasms potentially strong enough to break bones. In the **mass reflex,** the entire spinal cord becomes hyperactive for several minutes, issuing exaggerated skeletal muscle and visceral motor commands.

♈ Cranial Trauma   FAP *p. 451*

Cranial trauma is a head injury resulting from harsh contact with another object. Head injuries account for over half the deaths attributed to trauma. There are roughly 8 million cases of cranial trauma each year in the United States, and over a million of these involve intracranial hemorrhaging, concussion, contusion, or laceration of the brain. We presented the characteristics of spinal concussion, contusion, and laceration on p. 76; comparable descriptions are applied to injuries of the brain.

Concussions typically accompany even minor head injuries. A concussion involves a temporary loss of consciousness and some degree of amnesia (p. 83). Physicians examine any concussed individual quite closely and may X-ray the skull to check for skull fractures or cranial bleeding. Mild concussions produce a brief interruption of consciousness and little memory loss. Severe concussions produce extended periods of unconsciousness and abnormal neurological functions. Severe concussions are typically associated with **contusions** (bruises) or **lacerations** (tears) of the brain tissue; the possibilities for recovery vary with the areas affected. Extensive damage to the reticular formation may produce a permanent state of unconsciousness, whereas damage to the lower brain stem generally proves fatal.

♈ Aphasia   FAP *p. 460*

Aphasia is a disorder affecting the ability to speak or read. *Global aphasia,* discussed in the text, results from extensive damage to the general interpretive area or to the associated sensory tracts. There are several other forms of aphasia. **Major motor aphasia** may develop in an individual after a brief period of global aphasia. This condition is extremely frustrating for the individual, who can understand language and knows how to respond but lacks the motor control necessary to produce the right combinations of sounds. It is also known as *nonfluent,* or *expressive, aphasia.* In *fluent,* or *receptive, aphasia,* the person does not understand what is heard or make sense while speaking. The individual words and sounds come easily, but they convey no meaning.

Lesser degrees of aphasia commonly follow a minor stroke. There is no initial period of global aphasia, and the individual can understand spoken and written words. The problems encountered with speaking or writing gradually fade. Many individuals with minor aphasia recover completely.

♈ The Cerebral Nuclei and Parkinson's Disease   FAP *p. 471*

The cerebral nuclei contain two discrete populations of neurons. One group stimulates motor neurons by releasing ACh, and the other inhibits motor neurons by releasing GABA. Under normal conditions, the excitatory neurons remain inactive, and the descending tracts are primarily responsible for inhibiting motor neuron activity. If the descending tracts are severed in an accident, the loss of inhibitory control leads to a generalized state of muscular contraction known as **decerebrate rigidity.**

The excitatory neurons are quiet because they are continuously exposed to the inhibitory effects of the neurotransmitter dopamine. This compound is manufactured by neurons in the substantia nigra and is carried by axoplasmic flow to synapses in the cerebral nuclei. If the ascending tract or the dopamine-producing neurons are damaged, this inhibition is lost, and the excitatory neurons become increasingly active. This increased activity produces the motor symptoms of **Parkinson's disease,** or *paralysis agitans.*

Parkinson's disease is characterized by a pronounced increase in muscle tone. Voluntary movements become hesitant and jerky, a condition called **spasticity,** for a movement cannot occur until one muscle group manages to overpower its antagonists. Individuals with Parkinson's disease show spasticity during voluntary movement and a continual **tremor** when at rest. A tremor represents a tug of war between antagonistic muscle groups that produces a background shaking of the limbs, in this case at a frequency of 4–6 cycles per second. Individuals with Parkinson's disease also have difficulty starting voluntary movements. Even changing one's facial expression requires intense concentration, and the individual acquires a blank, static expression. Finally, the positioning and preparatory adjustments normally performed automatically no longer occur. Every aspect of each movement must be voluntarily controlled, and the extra effort requires intense concentration that can prove tiring and extremely frustrating. In the late stages of this condition, other CNS effects, such as depression, hallucinations, and dementia commonly appear.

Providing the cerebral nuclei with dopamine can significantly reduce the symptoms for two-thirds of Parkinson's patients. Intravenous dopamine injection is not effective, because the molecule cannot cross the blood–brain barrier. The most common procedure involves the oral administration of the drug L-DOPA (levodopa), a compound that crosses the capillaries and is then converted to dopamine. Unfortunately, it appears that with repeated treatment, the capillaries become less permeable to L-DOPA, so the required dosage must increase. The effectiveness of L-DOPA can be increased by giving it in combination with other drugs, such as *Amantadine* or *Bromocriptine*. Amantadine accelerates dopamine release at synaptic terminals, and Bromocriptine, a dopamine agonist, stimulates dopamine receptors on postsynaptic membranes. The drug *deprenyl* appears to slow the progression of the disease in some patients by affecting the biochemical pathways of dopamine removal.

Surgery to control Parkinson's symptoms focuses on the destruction of large areas within the cerebral nuclei or thalamus to control the motor symptoms of tremor and rigidity. The high success rate of drug therapy has greatly reduced the number of surgical procedures. Recent attempts to transplant tissues that produce dopamine or related compounds into the cerebral nuclei have met with limited success. Variable results have been obtained with the transplantation of tissue from the adrenal gland. The transplantation of fetal tissue into adult brains has been more successful, with several research groups reporting relatively long-term (up to 3 years after surgery) improvement in motor skills.

Most individuals with Parkinson's disease are elderly. The champion boxer Muhammad Ali developed symptoms relatively early in life (in his forties), possibly as a result of repeated cranial trauma during his career. Since 1983, an increasing number of young people have developed this condition. In that year, a drug appeared on the streets rumored to be "synthetic heroin." In addition to the compound that produced the "high" sought by users, the drug contained several contaminants, including a complex molecule with the abbreviated name **MPTP**. This accidental byproduct of the synthetic process destroys neurons of the substantia nigra, eliminating the manufacture and transport of dopamine to the cerebral nuclei. As a result of exposure to this drug, approximately 200 young, healthy adults have developed symptoms of severe Parkinson's disease. Why MPTP targets these particular neurons, and not all the CNS neurons that produce dopamine, remains a mystery.

☤ Cerebellar Dysfunction   FAP *p. 473*

Cerebellar function may be altered permanently by trauma or a stroke or temporarily by drugs such as alcohol. Such alterations can produce disturbances in motor control. In severe ataxia, balance problems are so great that the individual cannot sit or stand upright. Less-severe conditions cause an obvious unsteadiness and irregular patterns of movement. The individual typically watches his or her feet to see where they are going and controls ongoing movements by intense concentration and voluntary effort. Reaching for something becomes a major exertion, for the only information available must be gathered by sight or touch while the movement is taking place. Without the cerebellar ability to adjust movements while they are occurring, the individual becomes unable to anticipate the time course of a movement. Most commonly, a reaching movement ends with the hand overshooting the target. This inability to anticipate and stop a movement precisely is called **dysmetria** (dis-MET-rē-uh; *dys-*, bad + *metron*, measure). In attempting to correct the situation, the person usually overshoots again in the opposite direction, and then again. The hand oscillates back and forth until either the object can be grasped or the attempt is abandoned. This oscillatory movement is known as an **intention tremor.**

Clinicians check for ataxia by watching an individual walk in a straight line; the usual test for dysmetria involves touching the tip of the index finger to the tip of the nose. Because many drugs impair cerebellar performance, the same tests are used by police officers to check drivers suspected of alcohol or other drug abuse.

ANALYSIS OF GAIT AND BALANCE

To check gait and balance, the examiner typically asks the person to walk a line, first with the eyes open and then with the eyes closed. This procedure checks how much the individual is relying on visual information to fine-tune motor functions. If the gait is normal with the eyes open but abnormal and unsteady with the eyes closed, there are problems with the balance pathways and perhaps the cerebellum as well. Heel-to-toe walking on a straight line will magnify any gait abnormalities or loss of balance sensations.

While the subject is walking, the examiner also watches how the heels and toes are placed and how the arms swing back and forth. The pattern of limb movement during walking is normally regulated by the cerebral nuclei. Problems with these nuclei, as in Parkinson's disease, will upset the pace and rhythm of these movements.

The *Romberg test* is used to check balance and equilibrium sensations. A *positive Romberg sign* exists if the individual cannot maintain balance with the feet together and eyes closed. A positive Romberg sign may be found in persons with multiple sclerosis (p. 75), peripheral neuropathies (p. 77), or several vestibular disorders.

☤ Tic Douloureux   FAP *p. 478*

Tic douloureux, or **trigeminal neuralgia,** affects 1 individual out of every 25,000. Sufferers complain of severe, almost totally debilitating pain that arrives with a sudden, shocking intensity and then disappears. In most cases, only one side of the face is involved, and the pain is along the sensory path of the maxillary and mandibular branches of the trigeminal nerve. This condition generally affects adults over 40 years of age; the cause is unknown. The pain can often be temporarily controlled by drug therapy, but surgical procedures may eventually be required. The goal of the surgery is to

14

destroy the afferents that carry the pain sensations. This goal can be attempted by actually cutting the nerve, a procedure called a **rhizotomy** (*rhiza*, root), or by injecting chemicals such as alcohol or phenol into the nerve at the foramina ovale and rotundum. The sensory fibers may also be destroyed by inserting an electrode and cauterizing the sensory nerve trunks as they leave the semilunar ganglion.

Cranial Nerve Tests FAP *p. 485*

A variety of tests are used to monitor the condition of specific cranial nerves. For example,

- The olfactory nerve (N I) is assessed by using different types of odoriferous chemicals. In this test, the subject is asked to distinguish among the different odors.

- Cranial nerves II, III, IV, and VI are assessed while the vision and movement of the eyes are checked. First, the person is asked to hold the head still and track the movement of the examiner's finger with the eyes. For the eyes to track the finger through the visual field, the oculomotor muscles and cranial nerves must be functioning normally. For example, if the person cannot track with the right eye a finger that is moving from left to right, there may be damage to the right lateral rectus muscle or to N VI on the right side.

- Cranial nerve V, which provides motor innervation to the muscles of mastication, can be checked by asking the person to clench the teeth. The jaw muscles are then palpated; if motor components of N V on one side are damaged, the muscles on that side will be weak or flaccid. Sensory components of N V can be tested by lightly touching areas of the forehead and side of the face.

- The facial nerve (N VII) is checked by watching the muscles of facial expression or asking the person to perform particular facial movements. Wrinkling the forehead, raising the eyebrows, pursing the lips, and smiling are controlled by the facial nerve. If a branch of N VII has been damaged, there will be muscle weakness or drooping on the affected side. For example, the corner of the mouth may sag and fail to curve upward when the person smiles. Special sensory components of N VII can be checked by placing solutions known to stimulate taste receptors on the anterior third of the tongue.

- The glossopharyngeal and vagus nerves (N IX and N X) can be evaluated by watching the person swallow something. Examination of the soft palate arches and uvula for normal movement is also important.

- The accessory nerve (N XI) can be checked by asking the person to shrug the shoulders. Atrophy of the sternocleidomastoid or trapezius muscles may also indicate problems with the accessory nerve.

- The hypoglossal nerve (N XII) can be checked by having the person extend the tongue and move it from side to side.

Amyotrophic Lateral Sclerosis
FAP *p. 502*

Demyelinating disorders affect both sensory and motor neurons, producing losses in sensation and motor control. **Amyotrophic lateral sclerosis** (ALS) is a progressive disease that affects specifically motor neurons, leaving sensory neurons intact. As a result, individuals with ALS experience a loss of motor control but have no loss of sensation in the affected regions. Motor neurons throughout the CNS are destroyed. Neurons involved with the innervation of skeletal muscles are the primary targets. Neurons involved with sensory information and intellectual functions are unaffected.

ALS is a disease of adulthood; symptoms generally do not appear until the individual is over age 40. It occurs at an incidence of 3–5 cases per 100,000 population worldwide. The disorder is somewhat more common among males than among females. The pattern of symptoms varies with the specific motor neurons involved. When motor neurons in the cerebral hemispheres of the brain are the first to be affected, the individual experiences difficulty in performing voluntary movements and has exaggerated stretch reflexes. If motor neurons in other portions of the brain and the spinal cord are targeted, the individual experiences weakness, initially in one limb but gradually spreading to other limbs and ultimately the trunk. When the motor neurons innervating skeletal muscles degenerate, there is a loss of muscle tone. Over time, the skeletal muscles atrophy. The disease progresses rapidly, and the average survival after diagnosis is just 3–5 years. Because intellectual functions remain unimpaired, a person with ALS remains alert and aware throughout the course of the disease. This is one of the most disturbing aspects of the condition.

The primary cause of ALS is uncertain, and only 5–10 percent of ALS cases appear to have a genetic basis. At the cellular level, it appears that the underlying problem lies at the postsynaptic membranes of motor neurons. It has been suggested that an abnormal receptor complex for the neurotransmitter *glutamate* in some way leads to the buildup of free radicals, such as NO (nitric oxide), that ultimately kill the neuron. In a recent clinical study, treatment of ALS patients with *riluzole*, an experimental drug that blocks glutamate receptors, has shown promise in extending the life of ALS patients. The Food and Drug Administration (FDA) has approved this drug for clinical use.

 Seizures and Epilepsies   FAP *p. 503*

A *seizure* is a temporary disorder of cerebral function, accompanied by abnormal, involuntary movements, unusual sensations, or inappropriate behavior. The individual may or may not lose consciousness for the duration of the attack. There are many different types of seizures. Clinical conditions characterized by seizures are known as *seizure disorders,* or *epilepsies.* The term **epilepsy** refers to more than 40 different conditions characterized by a recurring pattern of seizures over extended periods. In roughly 75 percent of patients, no obvious cause can be determined.

Seizures of all kinds are accompanied by a marked change in the pattern of electrical activity that is monitored in an electroencephalogram. The alteration begins in one portion of the cerebral cortex but may thereafter spread to adjacent regions, potentially involving the entire cortical surface. The neurons at the site of origin of the alteration are abnormally sensitive. When they become active, they may facilitate and subsequently stimulate adjacent neurons. As a result, the abnormal electrical activity can spread across the entire cerebral cortex.

The extent of the cortical involvement determines the nature of the observed symptoms. A **focal seizure** affects a relatively restricted cortical area, producing sensory or motor symptoms or both. The individual generally remains conscious throughout the attack. If the seizure occurs within a portion of the primary motor cortex, the activation of pyramidal cells will produce uncontrollable movements. The muscles affected or the specific sensations experienced provide an indication of the precise region involved. In a **temporal lobe seizure,** the disturbance spreads to the sensory cortex and association areas, so the individual may also experience unusual memories, sights, smells, or sounds. Involvement of the limbic system may also produce sudden emotional changes. The individual will typically lose consciousness at some point during the incident.

Convulsive seizures are associated with uncontrolled muscle contractions. In a **generalized seizure,** the entire cortical surface is involved. Generalized seizures may range from prolonged, major events to brief, almost unnoticed incidents. We will consider only two examples here, *grand mal* and *petit mal* seizures.

Some epileptic attacks involve powerful, uncoordinated muscular contractions that affect the face, eyes, and limbs. These are symptoms of a **grand mal seizure.** During a grand mal attack, the cortical activation begins at a single focus and then spreads across the entire surface. There may be no warning, but some individuals experience a vague apprehension or awareness that a seizure is about to begin. There follows a sudden loss of consciousness, and the individual drops to the floor as major muscle groups go into tonic contraction. The body remains rigid for several seconds before a rhythmic series of contractions occurs in the limb muscles. Incontinence may occur. After the attack subsides, the individual may appear disoriented or sleep for several hours. Muscles or bones subjected to extreme stresses may be damaged, and the person will probably be rather sore for days after the incident.

Petit mal epileptic attacks are very brief (under 10 seconds in duration) and involve few motor abnormalities. Typically, the individual loses consciousness suddenly, with no warning. It is as if an internal switch were thrown and the conscious mind turned off. Because the individual is "not there" for brief periods during petit mal attacks, the incidents are known as *absence seizures.* During the seizure, there may be small motor activities, such as fluttering of the eyelids or trembling of the hands.

Petit mal attacks generally begin between ages 6 and 14. They can occur hundreds of times a day, so the child lives each day in small segments separated by blank periods. The individual is aware of brief losses of consciousness that occur without warning but, due to embarrassment, does not seek help. He or she becomes extremely anxious about the timing of future attacks. However, the motor signs are so minor that they tend to go completely unnoticed by other family members, and the psychological stress caused by this condition is in many cases overlooked. The initial diagnosis is typically made during counseling for learning problems. (You have probably taken an exam after you have missed one or two lectures out of 20. Imagine taking an exam after you have missed every third minute of every lecture.)

Both petit mal and grand mal epilepsy can be treated with barbiturates or other anticonvulsive drugs, such as *phenytoin sodium (Dilantin®)* or *valproic acid (Depakene®).*

 Amnesia   FAP *p. 506*

Amnesia may occur suddenly or progressively, and recovery may be complete, partial, or nonexistent, depending on the nature of the problem. In **retrograde amnesia** (*retro-*, behind), the individual loses memories of past events. Some degree of retrograde amnesia commonly follows a head injury, and accident victims are frequently unable to remember the moments preceding a car wreck or fall. In **anterograde amnesia** (*antero-*, ahead), an individual may be unable to store additional memories, but earlier memories are intact and accessible. The problem appears to involve an inability to generate long-term memories. At least two drugs—*diazepam (Valium)®* and *Halcion®*—have been known to cause brief periods of anterograde amnesia. A person with permanent anterograde amnesia lives in surroundings that are always new. Magazines can be read, chuckled over, and then reread a few minutes later with equal pleasure, as if they had never been seen before. Physicians and nurses must introduce themselves at every meeting, even if they have been visiting the patient for years.

Posttraumatic amnesia (PTA) commonly develops after a head injury. The duration of the amnesia varies with the severity of the injury. PTA combines the characteristics of retrograde and anterograde amnesia; the individual can neither remember the past nor consolidate memories of the present.

Altered States   FAP p. 506

A fine line may separate normal from abnormal states of awareness, and many variations in these states are clinically significant. The state of **delirium** involves wild oscillations in the level of wakefulness. The individual has little or no grasp of reality and often experiences hallucinations. When capable of communication, the person seems restless, confused, and unable to deal with situations and events. Delirium typically develops (1) in individuals with very high fevers; (2) in patients experiencing withdrawal from addictive drugs, such as alcohol; (3) after hallucinogenic drugs, such as LSD, are taken; (4) as a result of brain tumors affecting the temporal lobes; and (5) in the late stages of some infectious diseases, including syphilis and AIDS. The term **dementia** implies a more stable, chronic state characterized by deficits in memory, spatial orientation, language, or personality. We discuss *senile dementia*, a form of dementia that may develop in the elderly, in a later section.

For reference purposes, Table A-14 indicates the entire range of conscious and unconscious states, ranging from *delirium* through *coma*. Assessing the level of consciousness in an awake person involves noting the patient's general alertness, patterns of speech, the content of that speech (as an indication of ongoing thought processes), and general motor abilities. Orientation to three concepts—Who are you? What day is it? and Where are you?—is a basic part of the *mental status* assessment. In an unconscious person, the *pupillary reflex* provides important information about the status of the brain stem. When light is shined into one eye, the pupils of both eyes should constrict. This reflex is coordinated by the superior colliculus in the mesencephalon. If the pupils are unreactive, or *fixed*, serious brain damage has occurred.

A person's level of consciousness is typically reported in terms of the *Glasgow scale*, a classification system detailed in Table A-15. When the rating is completed, a tally is taken. Someone with a total score of 7 or less is probably comatose. Low scores in more than one section indicate that the condition is relatively severe. A comatose patient with a score of 3–5 has probably suffered irreversible brain damage.

Sleep Disorders   FAP p. 507

Sleep disorders include abnormal patterns of REM or deep sleep, variations in the time of onset or the time devoted to sleeping, and unusual behaviors performed while sleeping. Sleep disorders of one kind or another are very common, affecting an esti-

---

Table A-14   States of Awareness

Level or State	Description
Conscious states	
Normal consciousness	Aware of self and external environment; well-oriented; responsive
Delirium	Disorientation, restlessness, confusion, hallucinations, agitation, alternating with other conscious states
Dementia	Difficulties with spatial orientation, memory, language; changes in personality
Confusion	Reduced awareness; easily distracted; easily startled by sensory stimuli; person alternates between drowsiness and excitability; resembles minor form of delirium state
Somnolence	Extreme drowsiness, but response to stimuli is normal
Chronic vegetative state	Conscious but unresponsive; no evidence of cortical function
Unconscious states	
Asleep	Can be aroused by normal stimuli (light touch, sound, etc.)
Stupor	Can be aroused by extreme and/or repeated stimuli
Coma	Cannot be aroused and does not respond to stimuli (coma states can be further subdivided according to the effect on reflex responses to stimuli)

Table A-15   The Glasgow Scale

Area or Aspect Assessed	Response	Score
Motor abilities	None	1
	Decerebrate rigidity (patient is supine with extension of legs, feet in plantar flexion; arms adducted and extended; hands pronated; fingers flexed)	2
	Decorticate rigidity (patient is supine with arms flexed; hands and fingers in a flexed position on the chest; legs rigidly extended with feet in plantar flexion)	3
	Withdrawal reflex to stimulus	4
	Ability to pinpoint painful stimulus	5
	Ability to move body parts according to verbal request	6
Eyes	No reaction	1
	Opens eyes with painful stimulus	2
	Opens eyes upon verbal request	3
	Opens eyes spontaneously	4
Verbal ability	None	1
	Sounds emitted are not understandable	2
	Words used are not appropriate	3
	Able to speak but is not oriented	4
	Converses and is oriented	5

15

mated 25 percent of the U.S. population at any given time.

Many clinical conditions will affect sleep patterns or will be exaggerated by the autonomic changes that accompany the various stages of sleep. Major clinical categories of sleep disorders include *parasomnias, insomnias,* and *hypersomnias.*

Parasomnias are abnormal behaviors performed during sleep. Sleepwalking, sleep talking, teeth-grinding, and so forth commonly involve slow-wave sleep, when the skeletal muscles are not maximally inhibited. Parasomnias may have psychological rather than physiological origins.

Insomnia is characterized by shortened sleeping periods, difficulty in getting to sleep, or early awakening with an inability to return to sleep. This is the most common sleep disorder; an estimated one-third of adults suffer from insomnia. Temporary insomnia may accompany stress, such as family arguments or other crises. Insomnia of longer duration can result from chronic depression, illness, or drug abuse. Treatment varies with the primary cause of the insomnia. For example, treatment of mild, temporary insomnia may involve an exercise program and reduction of caffeine intake. Severe insomnia can be temporarily treated with drugs, usually *benzodiazepines* such as *temazepam, flurazepam,* or *triazolam.* Treatment of underlying depression may also prove helpful.

Hypersomnia involves extremely long periods of otherwise normal sleep. The individual may sleep until noon and nap before dinner, despite an early retirement in the evening. These conditions may have physiological or psychological origins, and successful treatment may involve drug therapy or counseling or both. Two important examples of hypersomnias are *narcolepsy* and *sleep apnea.*

Roughly 0.2–0.3 percent of the population has **narcolepsy,** characterized by dropping off to sleep at inappropriate times. The sleep lasts only a few minutes and is preceded by a period of muscular weakness. Any exciting stimulus, such as laughter or other strong emotion, may trigger the attack. (Interestingly, this condition occurs in other mammals; some dogs will keel over into a sound sleep when presented with their favorite treats.) In some instances, the sleep period is followed by a brief period of amnesia. Drugs that block REM sleep, such as methylphenidate *(Ritalin®),* may reduce or eliminate narcoleptic attacks.

In **sleep apnea,** a sleeping individual stops breathing for short periods of time, probably due to a disturbance or abnormality in the respiratory control mechanism or blockage of the upper airway. The problem can be exaggerated by various drugs (including alcohol), obesity, hypertension, tonsillitis, and other medical conditions that affect either the CNS or cause intermittent obstruction of the respiratory passageways. Roughly half those who experience sleep apnea show simultaneous cardiac arrhythmias, but the link between the two is not understood. Treatment focuses primarily on alleviating potential causes; for example, overweight individuals are put on a diet, and inflamed tonsils are removed. In some cases, air under pressure is administered through the nose to keep the upper airway open. This treatment is called *nasal positive airway pressure.*

✝ Huntington's Disease   FAP *p. 508*

Huntington's disease is an inherited disease marked by a progressive deterioration of mental abilities. There are about 25,000 Americans with this condition. In Huntington's disease, the cerebral nuclei and frontal lobes of the cerebral cortex show degenerative changes. The basic problem is the destruction of ACh-secreting and GABA-secreting neurons in the cerebral nuclei. The cause of this deterioration is not known. The first signs of the disease generally appear in early adulthood. As you would expect in view of the areas affected, the symptoms involve difficulties in performing voluntary and involuntary patterns of movement and a gradual decline in intellectual abilities that eventually lead to dementia and death.

Screening tests can now detect the presence of the gene for Huntington's disease, which is an autosomal dominant gene located on chromosome 4. In people with Huntington's disease, a gene of uncertain function contains a variable number of repetitions of the nucleotide sequence CAG. This DNA segment appears to be unstable, and the number of repetitions can change from generation to generation. The duplication or deletion is thought to occur during gamete formation. The larger the number of repetitions, the earlier in life the symptoms appear, and the more severe the symptoms. The link between the multiple copies of the CAG nucleotide and the disorder has yet to be understood. There is no effective treatment. The children of a person with Huntington's disease have a 50 percent risk of receiving the gene and developing the disease.

℞ Pharmacology and Drug Abuse

FAP *p. 508*

The drug industry certainly qualifies as "big business." Yearly sales approach $10 billion, and each year well over $1 billion are spent on advertising campaigns. If you watch television, you are presented with a dazzling array of advertisements for alcoholic beverages, cold medicines, diet pills, headache remedies, anxiety relievers, and sleep promoters. Although roughly one-third of the drugs sold affect the CNS, few consumers have any understanding of how these drugs exert their effects. Despite an overwhelming interest in medicinal and "recreational" (typically illicit) drugs, we seldom encounter accurate information in the United States about the mechanism of action or the associated hazards. Unfounded and inaccurate rumors are therefore quite common; the concept that "natural" drugs are safer or more effective than "artificial" (synthetic) drugs is an example of the dangerous but popular misconceptions that have appeared in recent years.

The major categories of drugs that affect CNS function are presented in Table A-16. Considerable overlap exists among these categories. For example, any drug that causes heavy sedation will also depress the perception of pain and alter mood at the same time. Well-known prescription, nonprescription, and illicit drugs are included in this table where appropriate. Some of the nonprescription drugs, most of the prescription drugs, and all of the illegal CNS-active drugs are prone to abuse because tolerance and addiction occur.

Sedatives lower the general level of CNS activity, reduce anxiety, and have a calming effect. **Hypnotic drugs** further depress activity, producing drowsiness and promoting sleep. All levels of CNS activity are reduced, so the effects can range from a mild relaxation to a general anesthesia, depending on the drug and the dosage administered. Sedatives and hypnotic drugs are prescribed more often than any others; almost half of the CNS-active drugs sold are sedatives or hypnotics.

Analgesics, the second most popular group, provide relief from pain. Some, like aspirin, act in the periphery by reducing the source of the painful stimulation. (Aspirin slows the release of prostaglandins that promote inflammation and stimulate pain receptors.) Others, especially the drugs structurally related to opium, target the CNS processing of pain sensations. The compounds bind to receptors on the surfaces of CNS neurons, notably in the cerebral nuclei, the limbic system, thalamus, hypothalamus, midbrain, medulla oblongata, and spinal cord. The drugs traditionally administered, such as morphine, mimic the activity and structure of endorphins already present in the CNS.

Psychotropics are used to produce changes in mood and emotional state. Certain forms of **depression** have been shown to result from inadequate production of norepinephrine, dopamine, or serotonin in key nuclei of the brain. Overexcitement, or **mania,** has been correlated with an overproduction of these compounds. Enhancing or blocking the production of these transmitters can often provide the mental stabilization needed to alleviate conditions such as these.

Antipsychotics reduce the hallucinations and behavioral or emotional extremes that characterize the severe mental disorders known as **psychoses,** but they leave other mental functions relatively intact. *Phenothiazines,* notably *chlorpromazine (Thorazine®),* were initially used for control of nausea and vomiting and to promote sedation and relax-

Table A-16   A Simple Classification of Drugs That Affect the CNS

Category (Common name)	Actions	Examples		Comments
		Prescription	Nonprescription	
Sedatives and hypnotics (downers)	Depress CNS activity; may promote sleep, reduce anxiety, create calm	Barbiturates (phenobarbital, Nembutal, amytal); benzodiazepines (Librium, Valium)	Sominex, Ny-tol, Benadryl, alcohol	Those used to depress seizures may be considered as "anticonvulsants"; prescription forms and alcohol can be addictive
Analgesics (pain killers)	Relieve pain at source or along CNS pathways	Opiates (morphine, Demerol, codeine)	Aspirin, Tylenol, ibuprofen	Heroin and cocaine are members of the addictive group of CNS-active drugs
Psychotropics (mood changers)	Alter CNS function and change mental state and/or mood	Antipsychotics (chlorpromazine); antidepressants (imipramine); anti-anxiety (Librium, Valium); mood stabilization (lithium)	Caffeine, alcohol	Many of these are also addictive
Anticonvulsants	Inhibit spread of cortical stimulation	Dilantin; also sedative-hypnotics		
Stimulants	Facilitate CNS activity	Xanthines (caffeine); amphetamines	Diet pills, Sudafed, Actifed, caffeine	Prescription and nonprescription forms are addictive. Typically affect cardiovascular system.

15

ation. However, they ultimately proved more useful in controlling psychotic behavior.

The term *tranquilizer* was first used for the early antipsychotic drugs, such as *Reserpine*®. Later, the category was expanded to include "minor tranquilizers," such as *Valium*®, whose effects more closely resembled those of sedatives. Eventually, it became apparent that "tranquilization" was more useful as a descriptive term than as a classification for specific drugs. For example, a tranquilizing effect can be produced by sedatives (alcohol, Valium), hypnotics (barbiturates), and antipsychotics.

The first **anticonvulsants,** used to control seizures, were sedatives such as phenobarbital. Over time, however, patients required larger and larger doses to achieve the same level of effect. This phenomenon, called **tolerance,** commonly appears when any CNS-active drug is administered. *Dilantin*® and related drugs have powerful anticonvulsant effects without producing tolerance.

Stimulants facilitate activity in the CNS. Few are used clinically, as they may trigger convulsions or hallucinations. Several of the ingredients in coffee and tea are CNS stimulants, caffeine being the most familiar example. **Amphetamines** prompt the release of norepinephrine, stimulating the respiratory and cardiovascular control centers and elevating muscle tone to the point at which tremors may begin. Amphetamines also stimulate dopamine and serotonin release in the CNS and facilitate cranial and spinal reflexes.

Drug abuse has a very hazy definition, for it implies that the individual voluntarily uses a drug in some inappropriate manner. Any drug use that violates medical advice, prevalent social mores, or common law can be included within this definition. **Drug addiction** refers to an overwhelming dependence and compulsion to use a specific drug, despite the medical or legal risks involved. If use is stopped or prevented, the individual will typically suffer physical and psychological symptoms of **withdrawal.** The physiological foundations of drug dependence and addiction vary with the individual compound considered. In general, the mechanisms are poorly understood.

The dangers inherent in recreational drug use have been repeatedly overlooked or ignored. Until recently, the abuse of common addictive drugs such as alcohol and nicotine was considered relatively normal, or at least excusable. The risks of alcohol abuse can hardly be overstated; it has been estimated that 40 percent of young men have problems with alcohol consumption, and 7 percent of the adult population show signs of alcohol abuse or addiction. Long-term abuse increases the risks of diabetes, liver and kidney disorders, cancer, cardiovascular disease, and digestive system malfunctions. In addition, CNS disturbances commonly lead to accidents, due to poor motor control, and violent behaviors, due to interference with normal emotional balance and analytical function. Medications used for other problems, such as those for hypertension or even antibiotics, can also have CNS effects, ranging from drowsiness to hallucinations or seizures. These side effects remind us of the importance of overall homeostasis and our inability to separate the mind from the body.

☤ Alzheimer's Disease   FAP *p. 509*

In its characteristic form, Alzheimer's disease produces a gradual deterioration of mental organization. The afflicted individual loses memories, verbal and reading skills, and emotional control. Initial symptoms are subtle—moodiness, irritability, depression, and a general lack of energy. These symptoms are often ignored, overlooked, or dismissed. Elderly relations are viewed as eccentric or irascible and are humored whenever possible.

As the condition progresses, however, it becomes more difficult to ignore or accommodate. The individual has difficulty making decisions, even minor ones. Mistakes—sometimes dangerous ones—are made, through either bad judgment or forgetfulness. For example, the person might decide to make dinner, light the gas burner, place a pot on the stove top, and go into the living room. Two hours later, the pot, still on the stove, melts into a shapeless blob and starts a fire.

As memory losses continue, the problems become more severe. The individual may forget relatives, his or her home address, or how to use the telephone. The memory loss commonly starts with an inability to store long-term memories, followed by the loss of recently stored memories, and eventually the loss of basic long-term memories, such as the sound of the individual's own name. The loss of memory affects both intellectual and motor abilities, and a patient with severe Alzheimer's disease has difficulty in performing even the simplest motor tasks. Although by this time victims are relatively unconcerned about their mental state or motor abilities, the condition can have devastating emotional effects on the immediate family.

Individuals with Alzheimer's disease show a pronounced decrease in the number of cortical neurons, especially in the frontal and temporal lobes. This loss is correlated with inadequate ACh production in the *nucleus basalis* of the cerebrum. Axons leaving this region project throughout the cerebral cortex, and when ACh production declines, cortical function deteriorates.

Most cases of Alzheimer's disease are associated with unusually large concentrations of *neurofibrillary tangles* and *plaques* in the nucleus basalis, hippocampus, and parahippocampal gyrus. The tangles are intracellular masses of abnormal microtubular proteins. The plaques are extracellular masses that form around a core that consists of an abnormal protein called **beta-amyloid.** Beta-amyloid is also found in other tissues, including the skin, blood vessels, subcutaneous layer, and intestine. Because this protein appears in small quantities in the blood and cerebrospinal fluid of many Alzheimer's patients, a screening test is now being developed to detect the condition before mental deterioration becomes pronounced. Familial cases of Alzheimer's disease are associated with mutations on either chromosome 21 or a small region of chromosome 14. A number of experimental protocols are undergoing clinical trials, but as yet there is no effective treatment for this condition.

☤ Hypersensitivity and Sympathetic Function   FAP *p. 520*

Two interesting clinical conditions result from the disruption of normal sympathetic functions. In **Horner's syndrome,** the sympathetic postganglionic innervation to one side of the face becomes interrupted. This interruption may be the result of an injury, a tumor, or some progressive condition such as multiple sclerosis. The affected side of the face becomes flushed as vascular tone decreases. Sweating stops in the affected region, and the pupil on that side becomes markedly constricted. Other symptoms include a drooping eyelid and an apparent retreat of the eye into the orbit.

Raynaud's disease most commonly affects young women. In this condition, the sympathetic system orders excessive peripheral vasoconstriction. The fingers, toes, ears, and nose may become deprived of their normal circulatory supply and take on pale white or blue coloration. The symptoms may spread to adjacent areas as the disorder progresses. Attempts to block the vasoconstriction with drugs have met with variable success. A regional **sympathectomy** (sim-path-EK-to-mē) may be performed, cutting the fibers that provide sympathetic innervation to the affected area, but results are typically disappointing.

Under normal conditions, sympathetic tone provides the effectors with a background level of stimulation. After the elimination of sympathetic innervation, peripheral effectors may become extremely sensitive to norepinephrine and epinephrine. This hypersensitivity can produce extreme alterations in vascular tone and other functions after stimulation of the adrenal medulla. If the sympathectomy involves cutting the postganglionic fibers, hypersensitivity to circulating norepinephrine and epinephrine may eliminate the beneficial effects. The prognosis improves if the preganglionic fibers are transected, for the ganglionic neurons will continue to release small quantities of neurotransmitter across the neuroeffector junctions. This release keeps the peripheral effectors from becoming hypersensitive.

Ⓡ Pharmacology and the Autonomic Nervous System   FAP *p. 533*

The treatment of many clinical conditions involves the manipulation of autonomic function. As we noted in the text *(p. 533), mimetic drugs* or *blocking agents* can be administered to counteract or reduce symptoms of abnormal ANS activity. Table A-17 relates important mimetic drugs and blocking agents to specific autonomic activities.

Mimetic drugs have advantages over the neurotransmitters or hormones whose actions they simulate. Whereas neurotransmitters must be administered into the bloodstream by injection infusion and their duration of action is limited, mimetic drugs can be administered in a variety of ways, and their actions can be sustained. For example, sympathomimetic drugs can be applied topically, by spray, by inhalation, or in drops. They can also be injected to elevate blood pressure and improve cardiac performance after severe blood loss or heart muscle failure. Sympathomimetic drugs are used to treat a variety of disorders, and we will consider only representative examples here.

Phenylephrine stimulates alpha receptors, causing a constriction of peripheral vessels and elevating blood pressure. It is prescribed in some cases of low blood pressure. Two other sympathomimetic drugs, *ephedrine* and *pseudoephedrine,* stimulate beta receptors. These drugs form the basis of several cold remedies (such as *Actifed*® and *Sudafed*®) that reduce nasal congestion and open respiratory passages through their effects on $ß_2$ receptors. Undesirable side effects, such as jitteriness, anxiety, sleeplessness, and high blood pressure, result from the facilitation of CNS pathways and stimulation of peripheral alpha and $ß_1$ receptors. Recently these drugs have been abused by amphetamine addicts unable to obtain amphetamines. Apparently, these addicts are deliberately seeking the side effects that other people find distressing and objectionable.

Sympathomimetic drugs that selectively target $ß_1$ receptors, such as *dobutamine,* are especially valuable in increasing heart rate and blood pressure. They are commonly prescribed to improve the performance of a failing heart. Sympathomimetic drugs targeting $ß_2$ receptors, such as *albuteral* or *terbutaline,* have been developed to treat the bronchial constriction that accompanies asthma attacks.

Sympathetic blocking agents that prevent a normal response to neurotransmitters or sympathomimetic drugs include *alpha-blockers* and *beta-blockers*. **Alpha-blockers** eliminate the peripheral vasoconstriction that accompanies sympathetic stimulation. The alpha-blockers include *prazosin,* which selectively targets α_1 receptors and is used to reduce high blood pressure. The drug can reduce constriction of the prostatic urethra and is used to treat *prostatitis* and *benign prostatic hypertrophy* (FAP *p. 1050*). **Beta-blockers** are effective and clinically useful for treating chronic high blood pressure and other forms of cardiovascular disease. In general, beta-blockers decrease heart rate and force of contraction, reducing the strain on the heart and simultaneously lowering peripheral blood pressure. *Propranolol* and *metoprolol* are two of the most popular beta-blockers currently on the market. Propranolol affects both $ß_1$ and $ß_2$ receptors, so patients receiving high doses may experience difficulties in breathing as their respiratory passageways constrict. Metoprolol targets $ß_1$ receptors almost exclusively, leaving the respiratory smooth muscles relatively unaffected.

Parasympathomimetic drugs may also be used to increase activity along the digestive tract and encourage defecation and urination. *Physostigmine* and *neostigmine* are important parasympathomimetic drugs that work by blocking the action of acetylcholinesterase. Because this enzyme is rendered inoperative, levels of ACh within the synapses climb, and parasympathetic activity is enhanced. In Chapter 12 we mentioned a blocking agent called *d-tubocurarine* that blocks neuromuscular transmission (FAP *p. 402*). The administration of physostigmine or neostigmine can counteract the paralytic effects of this drug.

Parasympathetic blocking agents such as *atropine* target the muscarinic receptors at the neuroeffector junctions. These drugs have diverse effects, but they are typically used to control the diarrhea and cramps associated with various forms of food poisoning. The drug *Lomotil*®, known as the "traveler's friend," can provide temporary relief from diarrhea for the duration of a plane flight home. Among its other effects, atropine causes an elevation of the heart rate due to a loss of parasympathetic tone. *Scopolamine* has similar effects on peripheral tissues, but it has greater influence on the CNS. Its most useful effects are promoting drowsiness, reducing nausea, and relieving anxiety. As a result, scopolamine is commonly given to a patient who is being prepared for surgery, prior to the administration of the anesthetic agent. (Scopolamine is also administered transdermally to control nausea, as we noted on FAP *pp. 153, 576.*)

🗒 ANALYZING SENSORY DISORDERS   FAP *p. 538*

A recurring theme of the text is that an understanding of how a system works enables you to predict how things might go wrong. You are already

16

Table A-17   Drugs and the ANS

Drug	*Mechanism*	*Action*	*Clinical Uses*
Sympathomimetic			
Phenylephrine (*Neosynephrine*)	Stimulates α_1 receptors	Elevates blood pressure; stimulates smooth muscle	As a nasal decongestant and to elevate low blood pressure
Clonidine	Stimulates α_2 receptors	Lowers blood pressure	Treatment of high blood pressure
Isoproterenol	Stimulates ß receptors	Stimulates heart rate; dilates respiratory passages	Treatment of respiratory disorders and as a cardiac stimulant during cardiac resuscitation
Albuteral, terbutaline	Stimulate $ß_2$ receptors	Dilate respiratory passages	Treatment of asthma, severe allergies, and other respiratory disorders
Ephedrine/ pseudoephedrine	Stimulates NE release at neuroeffector junctions	Similar to effects of epinephrine	As a nasal decongestant and to elevate blood pressure or dilate respiratory passage ways
Parasympathomimetic			
Physostigmine, neostigmine	Block action of acetylcholinesterase	Increase ACh concentrations at parasympathetic neuroeffector junctions	Stimulate digestive tract and smooth muscles of urinary bladder
Pilocarpine	Stimulates muscarinic receptors	Similar to effects of ACh	Applied topically to cornea of eye to cause pupillary contraction
Sympathetic blocking agents			
Prazosin (*Minipress*)	Blocks α_1 receptors	Lowers blood pressure	Treatment of high blood pressure
Propranolol (*Inderal*)	Blocks $ß_1$ and $ß_2$ receptors	Reduces metabolic activity in cardiac muscle but may constrict respiratory passageways; slows heart rate	Treatment of high blood pressure: used to reduce heart rate and force of contraction in heart disease
Metoprolol	Blocks $ß_1$ receptors	Reduces metabolic activity in cardiac muscle	Similar to those of *Inderal* but has less of an effect on respiratory muscles
Parasympathetic blocking agents			
Atropine, related drugs	Block muscarinic receptors	Inhibit parasympathetic activity	Treating diarrhea; dilating pupils; raising heart rate; blocking secretions of digestive and respiratory tracts prior to surgery; used to treat accidental exposure to anticholinesterase drugs, such as pesticides or military nerve gases

familiar with the organization and physiology of sensory systems, and we discussed some of the most important clinical problems in clinical comments on the preceding pages. Placing the entire array into categories provides an excellent example of a strategy that can be used to analyze any system in the body.

Every sensory system contains peripheral receptors, afferent fibers, ascending tracts, nuclei, and areas of the cerebral cortex. Any malfunction affecting the system must involve one of those components. Any clinical diagnosis requires you to seek answers to a series of yes-or-no questions, eliminating one possibility at a time until the nature of the problem becomes apparent. Figure A-28 organizes the disorders considered in this chapter into a "troubleshooting" format similar to that used to diagnose problems with automobiles or other mechanical devices.

A Closer Look: Pain Mechanisms Pathways, and Control *FAP p. 542*

The sensory neurons that bring pain sensations into the CNS release *glutamate* and *Substance P* as neurotransmitters. These two neurotransmitters have very different but complementary effects. Glutamate produces an immediate depolarization of the postsynaptic membrane by binding to *AMPA* receptors,[1] which open channels that permit Na+ entry and K+ departure from the cytoplasm. The excitatory effects are restricted to the postsynaptic cell, because glutamate is rapidly reabsorbed by the synaptic knobs. The binding to AMPA receptors stimulates the interneuron, and pain impulses

[1]AMPA receptors and the NMDA receptors discussed below were originally named after administered compounds that would bind to them. AMPA receptors bind **a**lpha-amino-3-hydroxy-5-**m**ethyl-4-isoxazole **p**ropionic **a**cid; NMDA receptors bind **N**-**m**ethyl-**D**-**a**spartate, a synthetic molecule related to glutamate.

ascend to the thalamus within the spinothalamic tracts.

When the interneuron depolarizes under glutamate stimulation, the depolarization makes another membrane receptor, *NDMA*, available for glutamate binding. The binding process triggers the opening of calcium ion channels in the membrane, and the entry of Ca2+ leads to the activation of second messengers with varied effects on the neuron. The net result is that the stimulated neurons become strongly facilitated, and pain sensitivity increases.

Substance P released by sensory neurons has a more widespread stimulatory effect. This neurotransmitter diffuses through the gray matter of the dorsal gray horn, where it affects large numbers of interneurons. Substance P binds to membrane receptors and is brought into the cytoplasm by means of receptor-mediated endocytosis. The result is the stimulation of some interneurons, producing sensations of pain, and the facilitation of many others.

The ascending pain sensations are widely distributed. Most ascending fibers travel within the lateral spinothalamic tracts for projection to the primary sensory cortex. The thalamus also relays pain sensations to the cingulate gyrus, an emo-

17

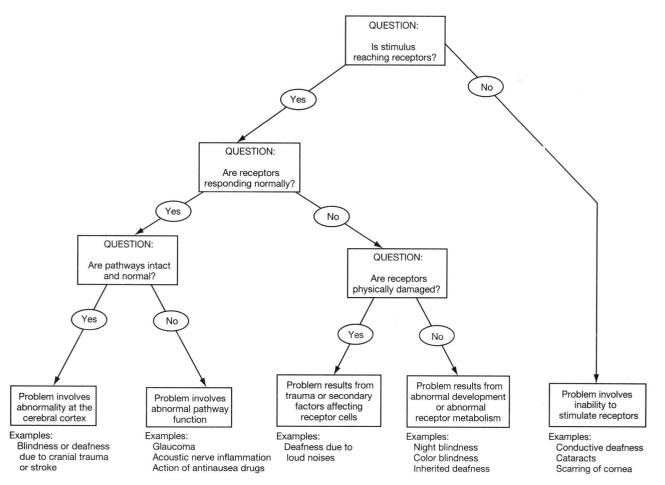

Figure A-28 A Flowchart for the Analysis of Sensory Disorders

tional center of the limbic system. Some of the ascending fibers do not reach the thalamus but synapse in the reticular formation or hypothalamus instead. Pain can thus influence CNS activities at both the conscious and subconscious levels.

Endorphins and enkephalins are neuromodulators whose release inhibits activity along pain pathways in the brain. These compounds, structurally similar to morphine, are found in the limbic system, hypothalamus, and reticular formation. The pain centers in these areas also use Substance P as a neurotransmitter. Endorphins bind to the presynaptic membrane and prevent the release of Substance P, thereby reducing the conscious perception of pain, although the painful stimulus remains.

Due to the facilitation that results from glutamate and Substance P release, the level of pain experienced (especially chronic pain) can be out of proportion to the apparent tissue damage. This effect is probably one reason why people differ so widely in their perception of the pain associated with childbirth or with a particular surgical procedure. This facilitation is also presumed to play a role in phantom limb pain; the sensory neurons may be inactive, but the hyperexcitable interneurons may continue to generate pain sensations. An interesting piece of evidence comes from clinical research: When general anesthesia shuts down the cerebral cortex and prevents conscious perception of pain, the interneurons of the spinal cord are not anesthetized. As a result, they become facilitated, and this effect can exaggerate postoperative pain. In one recent study, roughly 20 percent of patients who had a limb amputated under general anesthesia experienced phantom limb pain. However, when spinal facilitation was prevented, none of the patients experienced phantom limb pain.

ACUTE AND CHRONIC PAIN

Pain management poses a number of problems for clinicians. Painful sensations can result from tissue damage or sensory nerve irritation; it may originate where it is perceived, be referred from another location, or represent a false signal generated along the sensory pathway. The treatment differs in each case, and an accurate diagnosis is an essential first step. Acute pain is the result of tissue injury; the cause is apparent, and treatment is typically effective in relieving the pain. The most effective solution is to stop the damage, end the stimulation, and suppress the painful sensations at the injury site. Pain sensations are suppressed when topical or locally injected anesthetics inactivate nociceptors in the immediate area. Analgesic drugs can also be administered. They work in many different ways; we will consider only a few examples here.

Tissue injury results in damage to cell membranes. A fatty acid called arachidonic acid escapes from injured membranes. Within the interstitial fluid, an enzyme called *cyclo-oxygenase* converts arachidonic acid molecules to prostaglandins, and it is these prostaglandins that stimulate nociceptors in the area. Aspirin and related analgesics reduce inflammation and suppress pain by blocking the action of cyclo-oxygenase.

Chronic pain is more difficult to categorize and treat. It includes (1) pain from an injury that persists after tissue structure has been repaired; (2) pain from a chronic disease, such as cancer; and (3) pain without an apparent cause. Chronic pain in part reflects permanent facilitation of the pain pathways and the creation of a reverberating "pain memory." There are also complex psychological and physiological components. For example, many chronic pain patients develop a tolerance for pain medications, and insomnia and depression are common complaints. Chronic pain can be helped by antidepressants, which affect neurotransmitter levels. Counseling may help the person focus attention outward rather than inward; the outward focus can lessen the perceived level of pain and reduce the amount of pain medication required. Curiously, developing a second, acute source of pain, such as a herpes-zoster infection, can reduce the preception of preexisting chronic pain.

In some cases, chronic pain and severe acute pain can be suppressed by inhibition of the central pain pathway. Analgesics related to morphine reduce pain by mimicking the action of endorphins. The perception of pain may be altered, although the pain remains. For example, patients on morphine report being aware of painful sensations, but they are not affected by them. Surgical steps can be taken to control severe pain; for instance, (1) the sensory innervation of an area can be destroyed by an electrical current, (2) the dorsal roots carrying the painful sensations can be cut (a *rhizotomy*), (3) the ascending tracts in the spinal cord can be severed (a *tractotomy*), or (4) thalamic or limbic centers can be stimulated or destroyed. These options, listed in order of increasing degree of effect, surgical complexity, and associated risk, are used only when other methods of pain control have failed to provide relief. Pain can be remarkably difficult to control; roughly 25 percent of terminal cancer patients are unable to obtain relief from pain.

The Chinese technique of accupuncture to control pain has recently received considerable attention. Fine needles are inserted at specific locations and are either heated or twirled by the therapist. Several theories have been proposed to account for the positive effects, but none is widely accepted. It has been suggested that the pain relief may follow endorphin release. It is not known how acupuncture stimulates endorphin release; the acupuncture points do not correspond to the distribution of any of the major peripheral nerves.

Many other aspects of pain generation and control remain a mystery. Up to 30 percent of patients who receive a nonfunctional medication subsequently experience a significant reduction in pain. It has been suggested that this *placebo effect* results from endorphin release triggered by the expectation of pain relief. Although the medication has no direct effect, the indirect effect is quite significant and complicates the evaluation of analgesic medications.

Assessment of Tactile Sensitivities

FAP *p. 544*

Regional sensitivity to light touch can be checked by gentle contact with a fingertip or a slender wisp of cotton. The **two-point discrimination test** provides a more detailed sensory map of tactile receptors. Two fine points of a drawing compass, a bent paper clip, or another object are applied to the skin surface simultaneously. The subject then describes the contact. When the points fall within a single receptive field, the individual will report only one point of contact. A normal individual loses two-point discrimination at 1 mm (0.04 in.) on the surface of the tongue, at 2–3 mm (0.08–0.12 in.) on the lips, at 3–5 mm (0.12–0.20 in.) on the backs of the hands and feet, and at 4–7 cm (1.6–2.75 in.) over the general body surface.

Vibration receptors are tested by applying the base of a tuning fork to the skin. Damage to an individual spinal nerve produces insensitivity to vibration along the paths of the related sensory nerves. If the sensory loss results from spinal cord damage, the injury site can typically be located by walking the tuning fork down the spinal column, resting its base on the vertebral spines.

Descriptive terms are used to indicate the degree of sensitivity in the area considered. **Anesthesia** implies a total loss of sensation; the individual cannot perceive touch, pressure, pain, or temperature sensations from that area. **Hypesthesia** is a reduction in sensitivity, and **paresthesia** is the presence of abnormal sensations, such as the pins-and-needles sensation when an arm or leg "falls asleep" due to pressure on a peripheral nerve. (We discussed several types of *pressure palsies* that produce temporary paresthesia on p. 78.)

Otitis Media and Mastoiditis

FAP *p. 571*

Otitis media is an infection of the middle ear, most commonly of bacterial origin. *Acute otitis media* typically affects infants and children and is occasionally seen in adults. The pathogens tend to gain access via the pharyngotympanic tube, generally during an upper respiratory infection. As the pathogen population rises in the tympanic cavity, white blood cells rush to the site, and the middle ear becomes filled with pus. Eventually, the tympanic membrane may rupture, producing a char-

acteristic oozing from the external auditory canal. The bacteria can generally be controlled by antibiotics, the pain reduced by analgesics, and the swelling reduced by decongestants. In the United States, it is rare for otitis media to progress to the stage at which rupture of the tympanic membrane occurs.

Otitis media is extremely common in developing countries where medical care and antibiotics are not readily available. Many children and adults in these countries suffer from *chronic otitis media,* a condition characterized by chronic or recurring bouts of infection. This condition produces scarring or perforation of the tympanic membrane, which leads to some degree of hearing loss. Resulting damage to the inner ear or the auditory ossicles may further reduce hearing.

If the pathogens leave the middle ear and invade the air cells within the mastoid process, **mastoiditis** develops. The connecting passageways are very narrow, and as the infection progresses, the subject experiences severe earaches, fever, and swelling behind the ear in addition to symptoms of otitis media. The same antibiotic treatment is used to deal with both conditions; the particular antibiotic selected depends on the identity of the bacterium involved. The major risk of mastoiditis is the spread of the infection to the brain via the connective tissue sheath of the facial nerve (N VII). Should this occur, prompt antibiotic therapy is needed. If the problem remains, the person may have to undergo *mastoidectomy* (opening and drainage of the mastoid sinuses) or *myringotomy* (drainage of the middle ear through a surgical opening in the tympanic membrane).

Vertigo and Ménière's Disease

FAP *p. 576*

The term **vertigo** describes an inappropriate sense of motion. This meaning distinguishes it from "dizziness," a sensation of light-headedness and disorientation that typically precedes a fainting spell. Vertigo can result from abnormal conditions in the inner ear or from problems elsewhere along the sensory pathway that carries equilibrium sensations. It commonly accompanies CNS infection, and many people experience vertigo when they have high fever.

Any event that sets endolymph into motion can stimulate the equilibrium receptors and produce vertigo. Placing an ice pack in contact with the area over the temporal bone or flushing the external auditory canal with cold water may chill the endolymph in the outermost portions of the labyrinth and establish a temperature-related circulation of fluid. A mild and temporary vertigo is the result. The consumption of excessive quantities of alcohol or exposure to certain drugs can also produce vertigo by changing the composition of the endolymph or disturbing the hair cells.

17

Other causes of vertigo include viral infection of the vestibular nerve and damage to the vestibular nucleus or its tracts. Acute vertigo can also result from damage caused by abnormal endolymph production, as in Ménière's disease. Probably the most common cause of vertigo is *motion sickness,* which we discuss in the text (FAP *p. 576*).

In **Ménière's disease,** distortion of the membranous labyrinth of the inner ear by high fluid pressures may rupture the membranous wall and mix endolymph and perilymph together. The receptors in the vestibule and semicircular canals then become highly stimulated, and the individual may be unable to start a voluntary movement because he or she is experiencing intense spinning or rolling sensations. In addition to the vertigo, the victim may "hear" unusual sounds as the cochlear receptors are activated.

Testing and Treating Hearing Deficits   FAP *p. 585*

In the most common hearing test, a subject listens to sounds of varying frequency and intensity generated at irregular intervals. A record is kept of the responses, and the graphed record, or **audiogram,** is compared with that of an individual with normal hearing (Figure A-29). **Bone conduction tests** are used to discriminate between conductive and nerve deafness. If you put your fingers in your ears and talk quietly, you can still hear yourself, because the bones of the skull conduct the sound waves to the cochlea, bypassing the middle ear. In a bone conduction test, the physician places a vibrating tuning fork against the skull. If the subject hears the sound of the tuning fork when it is in contact with the skull but not when it is held next to the entrance to the external auditory canal, the problem must lie within the external or middle ear. If the subject remains unresponsive to both stimuli, the problem must be at the receptors or along the auditory pathway.

Several effective treatments exist for conductive deafness. A hearing aid overcomes the loss in sensitivity by increasing the intensity of stimulation. Surgery may repair the tympanic membrane or free damaged or immobilized ossicles. Artificial ossicles may also be implanted if the original ones are damaged beyond repair.

There are few possible treatments for nerve deafness. Mild conditions may be overcome by the use of a hearing aid if some functional hair cells remain. In a **cochlear implant,** a small, battery-powered device is inserted beneath the skin behind the mastoid process. Small wires run through the round window to reach the cochlear nerve; the implant "hears" a sound and stimulates the nerve directly. Increasing the number of wires and varying their implantation sites make it possible to create a number of different frequency sensations.

Those sensations do not approximate normal hearing, because there is as yet no way to target the specific afferent fibers responsible for the perception of a particular sound. Instead, a random assortment of afferent fibers are stimulated, and the individual must learn to recognize the meaning and probable origin of the perceived sound. Although this technique is obviously not perfect, improvements in electrode placement and computer processing are being made. At present, approximately 2000 people in the United States are using cochlear implants.

A new approach involves inducing the regeneration of hair cells of the organ of Corti. Researchers working with mammals other than humans have been able to induce hair cell regeneration both in cultured hair cells and in live animals. This is a very exciting area of research, and there is hope that it may ultimately lead to an effective treatment for human nerve deafness.

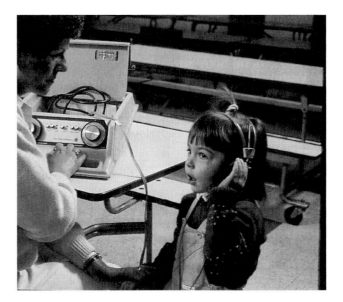

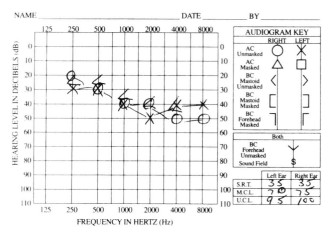

Figure A-29   An Audiogram

CRITICAL-THINKING QUESTIONS

4-1. Ten-year-old Christina falls while climbing a tree and lands on her back. Her frightened parents take her to the local emergency room, where she is examined. Her knee-jerk reflex is normal, and she exhibits a plantar reflex (negative Babinski reflex). These results suggest that:

    a. Christina has injured one of her descending nerve tracts

    b. Christina has injured one of her ascending nerve tracts

    c. Christina has a spinal injury in the lumbar region

    d. Christina has a spinal injury in the cervical region

    e. Christina has suffered no damage to her spinal cord

4-2. Susan brings her husband, Jim, age 32, to the emergency room. Jim has been complaining of a severe headache for the last 12 hours. He has a temperature of 39°C (102°F) and a stiff neck, and he complains of pain when he moves his chin to chest. His reflexes are normal. A lumbar puncture is performed, and the results of CSF analysis are reported as follows:

- Culture: presence of *Streptococcus pneumoniae* (bacteria)
- Pressure of CSF: 201 cm H_2O
- Color of CSF: cloudy
- Glucose: 50 mg/dl
- Protein: 47 mg/dl
- Cell count (lymphocytes): 550 mm^3

What is the likely diagnosis?

    a. brain tumor

    b. meningitis

    c. cerebrovascular accident

    d. multiple sclerosis

4-3. Tapping the calcaneal tendon of a normal individual with a rubber hammer will produce a reflex response. What response would you expect? Which type of reflex is this? Describe the steps involved in the reflex. Discuss other reflexes of this type and what can be learned by testing reflexes.

4-4. Mrs. Glenn, 73 years old, has recently had trouble controlling her movements. Even when resting, she has continual, slight tremors and increased muscle tone to the point of rigidity. She visits her physician, who conducts a series of tests, including an assessment of her reflexes. What would you expect to see in Mrs. Glenn's spinal reflexes?

4-5. Chelsea is mountain climbing with a group of friends when she slips, falls, and bumps the left side of her head on a rock. She gets up slowly and is dazed but otherwise appears unhurt. She feels able to proceed, and the climb continues. An hour later, Chelsea gets a severe headache and experiences a ringing in her ears. She starts having trouble speaking and soon loses consciousness. Before medical personnel can reach the scene, Chelsea dies. What was the likely cause of death?

4-6. Molly comes to lab appearing dazed, and she cannot focus her attention on anything. You suspect that she has taken some type of sympathomimetic drug. What other signs would you expect to observe that would confirm your suspicion?

4-7. Hido has a hypothalamic tumor that compresses the right medial surface of the optic chiasm posterior to the decussation. How would this condition affect his vision?

4-8. M.Romero, 62 years old, has trouble hearing people during conversations, and his family persuades him to have his hearing tested. How can the physician determine whether Mr. Romero's problem results from nerve deafness or conductive deafness?

NOTES:

The Endocrine System

The endocrine system provides long-term regulation and adjustment of homeostatic mechanisms and a variety of body functions. For example, the endocrine system is responsible for the regulation of fluid and electrolyte balance, cell and tissue metabolism, growth and development, and reproductive functions. The endocrine system also assists the nervous system in responding to stressful stimuli.

The endocrine system is composed of nine major endocrine glands and several other organs, such as the heart and kidneys, that have other important functions. The hormones secreted by these endocrine organs are distributed by the circulatory system to target tissues throughout the body. Each hormone affects a specific set of target tissues that may differ from that of other hormones. The selectivity is based on the presence or absence of hormone-specific receptors in the target cell's cell membrane, cytoplasm, or nucleus.

Homeostatic regulation of circulating hormone levels primarily involves negative feedback control mechanisms. The feedback loop involves an interplay between the endocrine organ and its target tissues. An endocrine gland may release a particular hormone in response to one of three different types of stimuli:

1. Some hormones are released in response to variations in the concentrations of specific substances in the body fluids. Parathyroid hormone, for example, is released when calcium levels decline.

2. Some hormones are released only when the gland cells receive hormonal instructions from other endocrine organs. For example, the rate of production and release of T_3 and T_4 by the thyroid gland is controlled by thyroid-stimulating hormone (TSH) from the anterior pituitary gland. The secretion of TSH is in turn regulated by the release of thyrotropin-releasing hormone (TRH) from the hypothalamus.

3. Some hormones are released in response to neural stimulation. The release of epinephrine and norepinephrine from the adrenal medullae during sympathetic activation is an example.

Endocrine disorders can therefore develop due to abnormalities in the endocrine gland, the endocrine or neural regulatory mechanisms, or the target tissues. Figure A-30 provides an overview of the major classes of endocrine disorders. In the discussion that follows, we will use the thyroid gland as an example, because the text introduces major types of thyroid gland disorders. These *primary disorders* may result in overproduction *(hypersecretion)* or underproduction *(hyposecretion)* of hormones. For example, clinicians may categorize a thyroid disorder as *primary hyperthyroidism* or *primary hypothyroidism* if the problem originates within the thyroid gland.

DISORDERS DUE TO ENDOCRINE GLAND ABNORMALITIES

Most endocrine disorders are the result of problems within the endocrine gland itself. Causes of hyposecretion include the following:

*   *Metabolic factors.* Hyposecretion may result from a deficiency in some key substrate needed to synthesize that hormone. For example, hypothyroidism can be caused by inadequate dietary iodine levels or by exposure to drugs that inhibit iodine transport or utilization at the thyroid gland.

*   *Physical damage.* Any condition that interrupts the normal circulatory supply or that physically damages the endocrine cells in some other way will suppress hormone production temporarily. If the damage is severe, the condition may be permanent. Examples of problems that may cause temporary or permanent hypothyroidism include infection or inflammation of the gland *(thyroiditis)*, interruption of normal circulation, and exposure to radiation as part of treatment for cancer of the thyroid or adjacent tissues. The thyroid may also be damaged in an *autoimmune disorder* that results in the production of antibodies that attack and destroy normal follicle cells.

*   *Congenital disorders.* An individual may be unable to produce normal amounts of a particular hormone because (1) the gland itself is too small, (2) the required enzymes are abnormal in some way, (3) the receptors that trigger secretion are relatively insensitive because a mutation affects G-protein structure, or (4) the gland cells lack the receptors normally involved in stimulating secretory activity.

DISORDERS DUE TO ENDOCRINE OR NEURAL REGULATORY MECHANISM ABNORMALITIES

Endocrine disorders can instead result from problems with other endocrine organs involved in the negative feedback control mechanism. For example:

*   *Secondary hypothyroidism* can be caused by inadequate TSH production at the pituitary gland or by inadequate TRH secretion at the hypothalamus.

*   *Secondary hyperthyroidism* can be caused by excessive TRH or TSH production; secondary hyperthyroidism may develop in individuals with tumors of the pituitary gland.

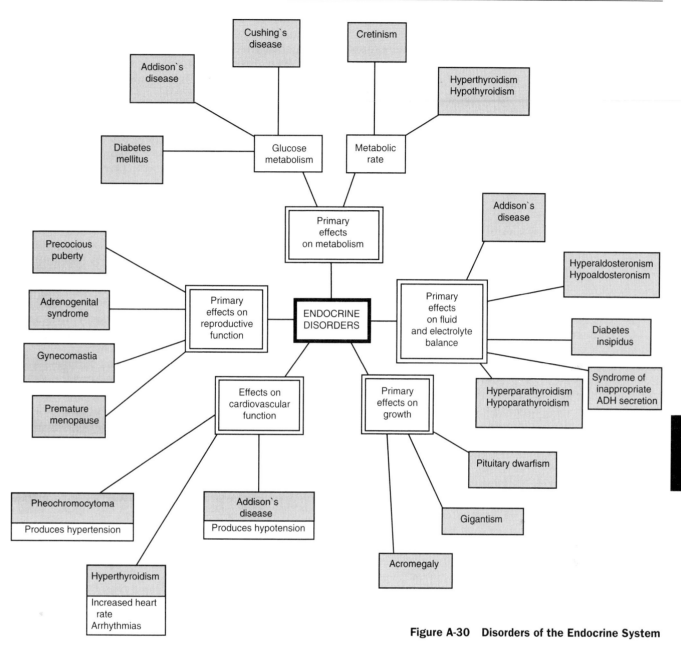

Figure A-30 Disorders of the Endocrine System

DISORDERS DUE TO TARGET TISSUE ABNORMALITIES

Endocrine abnormalities can also be caused by the presence of abnormal hormonal receptors in target tissues. In this case, the gland involved and the regulatory mechanisms may be normal, but the peripheral cells are unable to respond to the circulating hormone. The best example of this type of abnormality is *maturity-onset diabetes* (also known as *Type II diabetes*, NIDDM), in which peripheral cells do not respond normally to insulin. (Maturity-onset diabetes is discussed further on p. 104.)

THE SYMPTOMS OF ENDOCRINE DISORDERS

Knowledge of the individual endocrine organs and their functions makes it possible to predict the symptoms of specific endocrine disorders. For example, thyroid hormones increase basal metabolic rate, body heat production, perspiration, restlessness, and heart rate. An elevated metabolic rate, increased body temperature, weight loss, nervousness, excessive perspiration, and an increased or irregular heartbeat are common symptoms of hyperthyroidism. Conversely, a low metabolic rate, decreased body temperature, weight gain, lethargy, dry skin, and a reduced heart rate typically accompany hypothyroidism. The symptoms associated with over- and underproduction of major hormones are summarized in Table 18-8 in FAP *(p. 635)*.

The Diagnosis of Endocrine Disorders

The first step in the diagnosis of an endocrine disorder is the physical examination. Several disorders produce characteristic physical signs that reflect abnormal hormone activities. A few examples were introduced in the text:

- *Cushing's disease* results from an oversecretion of glucocorticoids by the adrenal cortex. As the condition progresses, there is a shift away from the normal pattern of fat distribution in the body. Adipose tissue accumulates in the abdominal area, the lower cervical area (causing a "buffalo hump"), and the face (a "moonface"), but the extremities become relatively thin.

- *Acromegaly* results from the oversecretion of growth hormone in the adult. In this condition, the facial features become distorted due to excessive cartilage growth, and the lower jaw protrudes, a sign known as *prognathism.* The hands and feet also become enlarged.

- *Adrenogenital syndrome* results from the oversecretion of androgens by the adrenal glands of a female. Hair growth patterns change, and the condition of *hirsutism* (p. 41) develops.

- Hypothyroidism produces a distinctively enlarged thyroid gland, or *goiter.*

- Hyperthyroidism can produce protrusion of the eyes, or *exophthalmos.*

These signs are very useful, but many other signs and symptoms related to endocrine disorders are less definitive. For example, the condition of *polyuria,* or increased urine production, may be the result of hyposecretion of ADH (*diabetes insipidus*), or a form of *diabetes mellitus;* a symptom such as hypertension (high blood pressure) can be caused by a variety of cardiovascular or endocrine problems. In these instances, many diagnostic decisions are based on blood tests, which can confirm the presence of an endocrine disorder by detecting abnormal levels of circulating hormones, followed by tests that determine whether the primary cause of the problem lies within the endocrine gland, the regulatory mechanism(s), or the target tissues. Table A-18 provides an overview of important blood tests and other tests used in the diagnosis of endocrine disorders.

Thyroid Gland Disorders   FAP *p. 613*

Hypothyroidism typically results from some problem that involves the thyroid gland rather than a problem with the pituitary production of TSH. In primary hypothyroidism, TSH levels are elevated, but levels of T_3 (triiodothyronine) and T_4 (tetra-iodothyronine, or *thyroxine*) are depressed. One form of hypothyroidism results from a mutation that affects the structure of the G-proteins at TSH receptors. The structural change reduces the receptors' sensitivity to TSH and depresses thyroid activity. Treatment of chronic hypothyroidism, such as the hypothyroidism that follows radiation exposure, generally involves the administration of synthetic thyroid hormones (thyroxine) to maintain normal blood concentrations.

A **goiter** is an enlargement of the thyroid gland. The enlargement generally indicates increased follicular size, despite a decrease in the rate of thyroid hormone production. Most goiters develop when the thyroid gland is unable to synthesize and release adequate amounts of thyroid hormones. Under continuing TSH stimulation, thyroglobulin production accelerates and the thyroid follicles enlarge. One type of goiter occurs if the thyroid fails to obtain enough iodine to meet its synthetic requirements. (This condition is now rare in the United States, because iodized table salt is widely available.) Administering iodine may not solve the problem entirely, because the sudden availability of iodine may produce symptoms of *hyperthyroidism* as the stored thyroglobulin becomes activated. The usual therapy involves the administration of thyroxine, which has a negative feedback effect on the hypothalamus and pituitary, inhibiting the production of TSH. Over time, the resting thyroid may return to its normal size.

Thyrotoxicosis, or *hyperthyroidism,* occurs when thyroid hormones are produced in excessive quantities. In **Graves' disease,** excessive thyroid activity leads to goiter and the symptoms of hyperthyroidism. Protrusion of the eyes, or **exophthalmos** (eks-ahf-THAL-mōs) may also appear, for unknown reasons. Graves' disease has a genetic autoimmune basis and affects many more women than men. Treatment may involve the use of antithyroid drugs, surgical removal of portions of the glandular mass, or destruction of part of the gland by exposure to radioactive iodine.

Hyperthyroidism may also result from inflammation or, rarely, thyroid tumors. In extreme cases, the individual's metabolic processes accelerate out of control. During a *thyrotoxic crisis,* or "thyroid storm," the subject experiences a dangerously high fever, rapid heart rate, and the malfunctioning of a variety of physiological systems.

Disorders of Parathyroid Function
FAP *p. 613*

When the parathyroid gland secretes inadequate or excessive amounts of parathyroid hormone, calcium concentrations move outside normal homeostatic limits. Hypoparathyroidism with hypocal-

Table A-18   Representative Diagnostic Procedures for Disorders of the Endocrine System

Diagnostic Procedure	Method and Result	Representative Uses	Notes
Pituitary			
Wrist and hand X-ray film	Standard X-rays of epiphyseal growth plate for estimation of bone age	To compare the individual's "bone age" and chronological age; a bone age greater than the 2 years behind the chronological age suggests possible growth hormone deficiency with hypopituitarism or pituitary dwarfism	Determination of bone age is an estimate based on the time of closure of the epiphyseal plates
X-ray study of sella turcica	Standard X-ray of the sella turcica, which houses the pituitary gland	To determine the size and shape of the pituitary gland; to detect a tumor; to detect a tumor or infarction of pituitary tissues	Erosion of the sella turcica suggests pituitary tumor
CT of pituitary gland	Standard cross-sectional CT; contrast media may be used		
MRI of pituitary gland	Standard MRI		
Thyroid			
Thyroid scanning	Radionuclide given by mouth accumulates in the thyroid, giving off radiation captured to create an image of the thyroid	To determine size, shape, and abnormalities of the thyroid gland; to detect presence of nodules that may be tumors or hyperactive or hypoactive areas; to determine cause of a mass in the neck	Nodules are determined as nonfunctional, or "cold" (do not take up radionuclide and appear less dense), or as hyperfunctional, or "hot" (appears darker on scan)
Ultrasound examination of thyroid	Sound waves reflected off internal structures are used to generate a computer image	To detect thyroid cysts or abnormalities in the shape or size of the thyroid gland	
Radioactive iodine uptake test (RAIU)	Radioactive iodine is ingested and trapped by the thyroid; detector determines the amount of radioiodine taken up over a period of time	To determine hyperactivity or hypoactivity of the thyroid gland	Typically is done at the same time as a thyroid scan
Parathyroid			
Ultrasound examination of parathyroid glands	Standard ultrasound	To determine structural abnormalities of the parathyroid gland, such as enlargement	
Intravenous pyelography	Contrast dye filtered at kidneys is followed by an X-ray series of the urinary bladder, ureter, and kidneys	To detect renal calculi in hyperparathyroidism	
Adrenal			
Ultrasound of adrenal gland	Standard ultrasound	To determine abnormalities in adrenal size or shape	
CT scan of adrenal gland	Standard cross-sectional CT	To determine abnormalities in adrenal size or shape	
Adrenal angiography	Injection of radiopaque dye for examination of the vascular supply to the adrenal gland	To detect tumors and hyperplasia	

18

Table A-18 *(continued)*

Laboratory Test	Normal Values in Blood Plasma or Serum	Significance of Abnormal Values	Notes
Pituitary			
Growth hormone (provocative test)	>10 ng/ml	<10 ng/ml upon stimulation for GH suggests deficiency (common in hypopituitarism and pituitary dwarfism)	Provocative tests stimulate or suppress secretions to assess gland function; results are variable and may be difficult to interpret
Plasma ACTH	Morning: 20–80 pg/ml Late afternoon: 10–40 pg/ml	Elevated levels of ACTH could indicate pituitary tumor or hyperpituitarism or Addison's disease as a compensatory mechanism. Decreased levels suggest hypopituitarism.	ACTH suppression and stimulation tests may also be performed
Serum TSH	Adults: <3 ng/ml	Elevated levels indicate hyperpituitarism or hypothyroidism. Decrease could be hypopituitarism or hyperthyroidism.	Ultrasensitive test; provides best measurement of thyroid gland function
Serum LH	Premenopausal females: 3–30 mIU/ml Females, midcycle: 30–100mIU/ml	Elevated levels occur with pituitary tumors and hyperpituitarism. Decreased levels occur in hypopituitarism and adrenal tumors.	Affects ovulation and fertility
Thyroid			
Free serum T_4	Adults: 1.0–2.3 ng/dl	Elevated with hyperthyroidism. Decreased levels with hypothyroidism.	
Total serum T_4 radioimmunoassay	Adults: 5–12 µg/dl	Elevated with hyperthyroidism	Test being replaced by free T_4 test
Calcitonin (plasma)	Adult males: <40 pg/ml Adult females: <20 pg/ml	Elevated levels occur with carcinoma of thyroid	
Thyroid autoantibody test	Negative to 1:20	High titers occur with thyroid tumors and in autoimmune disorders such as Grave's disease	
TSH stimulation test	Patients are given TSH via injection; thyroid gland is monitored for the normal increased function by increased T_4 levels	Patients with a hypoactive thyroid gland who have increased T_4 levels after TSH administration have a pituitary problem with a secondary hypothyroidism. Patients whose thyroid does not respond to TSH have primary hypothyroidism.	Performed to deduce the cause of decreased T_4 levels. If the thyroid is the source of the problem, there will be no response to TSH; if the pituitary is the source, the thyroid will respond normally to TSH.
Parathyroid			
Serum parathyroid hormone	Adults: PTH-N 400–900 pg/ml PTH-C 200–600 pg/ml	Increased in hyperparathyroidism and hypercalcemia. Decreased in hypoparathyroidism (PTH values are used to determine chronic nature).	Two different values are used to monitor acute or chronic conditions.
Serum phosphorus	Adults: 3–4.5 µg/dl	Increased in hypoparathyroidism. Decreased in hyperparathyroidism.	
Serum calcium	Adults: 8.5–10.5 mg/dl	Decreased in hypoparathyroidism. Increased in hyperparathyroidism.	

Table A-18 *(continued)*

Laboratory Test	Normal Values in Blood Plasma or Serum	Significance of Abnormal Values	Notes
Adrenal			
Plasma ACTH	Adults, morning: 20–80 pg/ml Adults, late afternoon: 10–40 pg/ml	Increased in stress, hypofunction of adrenal and pituitary hyperactivity. Decreased in Cushing's syndrome and carcinoma of adrenal gland and with intake of steroids such as prednisone and cortisone.	ACTH stimulation test can be done to determine cause of decreased plasma cortisol as either pituitary or adrenal in origin
Plasma cortisol	Adults, morning: 5–23 µg/dl Adults, afternoon: 3–13 µg/dl	Increased in adrenal hyperactivity, Cushing's syndrome, stress, and steroid use. Decreased in Addison's disease and pituitary hypofunction.	Urine cortisol can be tested also; when plasma cortisol levels are elevated, cortisol is excreted in the urine
Serum aldosterone	Adults: 4–30 ng/dl supine: decreased <1 ng/dl elevated >9 ng/dl	Increased in dehydration, hyperactivity of adrenal, and hyponatremia. Decreased in hypernatremia and adrenal hypoactivity.	Values are given for patient in a sitting position; levels will be lower in a supine position
Serum Na	Adults: 135–145 mEq/l	Increased in dehydration and hypoaldosteronism; decreased with adrenocortical insufficiency and hypoaldosteronism	
Serum K	Adults: 3.5–5.0 mEq/l	Increased with hypoactivity of adrenal glands and hypoaldosteronism. Decreased with hyperactivity of adrenal glands and aldosteronism.	
Urine hydroxycorticosteroid	Adults: 2–12 mg/24 hr	Increased with Cushing's syndrome. Decreased with Addison's disease.	Detects breakdown products of cortisol excreted in the urine
Urine ketosteroids	Adults: 5–25 mg/24 hr	Increased in adrenal hyperactivity and hyperpituitarism. Decreased in adrenal hypoactivity and hypopituitarism.	Detects metabolites of androgens produced by the adrenal cortex and the testes
Pancreas			
Glucose tolerance test	Adults: Results after ingestion of 75–100 g of glucose: 0 hr 75–115 mg/dl 1/2 hr <160 mg/dl 1 hr <170 mg/dl 2 hr <140 mg/dl	Increased in diabetes mellitus, Cushing's syndrome, alcoholism, and infections. Decreased in hyperinsulinism and hypoactivity of adrenal glands.	Monitors serum glucose following ingestion by a patient who has been fasting for 12 hours; 0-hr and 2-hr values are commonly used.
Serum insulin	Adults: 5–25 µU/ml	Increased in early Type II (NIDDM) diabetes mellitus and obesity. Decreased in Type I (IDDM) diabetes mellitus.	
Glycosylated hemoglobin (Alc)	Adults: 4–8%	Increased in diabetes mellitus; a value above 20% indicates poor diabetic control.	

cemia may develop after neck surgery, especially a thyroidectomy, if the blood supply to the parathyroid glands is restricted. In many other cases, the primary cause of the condition is uncertain. Treatment is difficult because at present PTH can be obtained only by extraction from the blood of individuals with normal PTH levels. Thus PTH is extremely costly, and because supplies are very limited, PTH administration is not used to treat this condition, despite its probable effectiveness. As an alternative, a dietary combination of vitamin D and calcium can be used to elevate body fluid calcium concentrations. (As we noted in Chapter 6 of the text, vitamin D stimulates the absorption of calcium ions across the lining of the digestive tract.)

In *hyperparathyroidism* calcium concentrations become abnormally high. Calcium salts in the skeleton are mobilized, and bones are weakened. On X-rays the bones have a light, airy appearance, because the dense calcium salts no longer dominate the tissue. CNS function is depressed, thinking slows, memory is impaired, and the individual often experiences emotional swings and depression. Nausea and vomiting occur, and in severe cases the patient may become comatose. Muscle function deteriorates, and skeletal muscles become weak. Other tissues are typically affected as calcium salts crystallize in joints, tendons, and the dermis; calcium deposits may produce masses called *kidney stones*, which block filtration and conduction passages in the kidney.

Hyperparathyroidism most commonly results from a tumor of the parathyroid gland. Treatment involves the surgical removal of the overactive tissue. Fortunately, there are four parathyroids, and the secretion of even a portion of one gland can maintain normal calcium concentrations.

☤ Disorders of the Adrenal Cortex
FAP pp. 616, 617

Clinical problems related to the adrenal gland vary with the adrenal zone involved. The conditions may result from changes in the functional capabilities of the adrenal cells (primary conditions) or disorders that affect the regulatory mechanisms (secondary conditions).

In **hypoaldosteronism,** the zona glomerulosa fails to produce enough aldosterone, generally either as an early sign of adrenal insufficiency or because the kidneys are not releasing adequate amounts of renin. Low aldosterone levels lead to excessive losses of water and sodium ions at the kidneys, and the water loss in turn leads to low blood volume and a fall in blood pressure. The resulting changes in electrolyte concentrations, including *hyperkalemia* (high extracellular K^+ levels) affect transmembrane potentials, eventually causing dysfunctions in neural and muscular tissues.

Hypersecretion of aldosterone results in **aldosteronism,** or *hyperaldosteronism.* Under contin-

ued aldosterone stimulation, the kidneys retain sodium ions in exchange for potassium ions that are lost in the urine. Hypertension and hypokalemia occur as extracellular potassium levels decline, increasing the concentration gradient for potassium ions across cell membranes. This increase leads to an acceleration in the rate of potassium diffusion out of the cells and into the interstitial fluids. The reduction in intracellular and extracellular potassium levels eventually interferes with the function of excitable membranes, especially cardiac muscle cells and neurons, and kidney cells.

Addison's disease may result from inadequate stimulation of the zona fasciculata by ACTH or from the inability of the adrenal cells to synthesize the necessary hormones, generally due to autoimmune problems or infection. Affected individuals become weak and lose weight owing to a combination of appetite loss, hypotension, and hypovolemia. They cannot adequately mobilize energy reserves, and their blood glucose concentrations fall sharply within hours after a meal. Stresses cannot be tolerated, and a minor infection or injury may lead to a sharp and fatal decline in blood pressure. A particularly interesting symptom is the increased melanin pigmentation in the skin. The ACTH molecule and the MSH molecule are similar in structure, and at high concentrations ACTH stimulates the MSH receptors on melanocytes. President John F. Kennedy suffered from this disorder.

Cushing's disease results from the overproduction of glucocorticoids. The symptoms resemble those of a protracted and exaggerated response to stress. (The stress response is discussed in the text on FAP *p. 629.*) Glucose metabolism is suppressed, lipid reserves are mobilized, and peripheral proteins are broken down. Lipids and amino acids are mobilized in excess of the existing demand. The energy reserves are shuffled around, and the distribution of body fat changes. Adipose tissues in the cheeks and around the base of the neck become enlarged at the expense of other areas, producing a "moon-faced" appearance. The demand for amino acids falls most heavily on the skeletal muscles, which respond by breaking down their contractile proteins. This response reduces muscular power and endurance. The skin becomes thin and may develop *stria*, or stretch marks.

If the primary cause is ACTH oversecretion at the anterior pituitary gland, the most common source is a pituitary *adenoma* (a benign tumor of glandular origin). Microsurgery can be performed through the sphenoid bone to remove the adenomatous tissue. Some oncology centers use pituitary radiation rather than surgery. Several pharmacological therapies act at the hypothalamus, rather than at the pituitary, to prevent release of corticotropin-releasing hormone (CRH). The drugs used are serotonin antagonists, GABA transaminase inhibitors, or dopamine agonists. Alternatively, a bilateral adrenalectomy (removal of the adrenal glands) can be performed, but further complications may arise as the adenoma enlarges. Cushing's disease can also result from the production of ACTH outside the pituitary; for example, the condition may develop with one form of lung cancer (oat cell carcinoma). The removal of the causative tumor can in some cases relieve the symptoms.

The chronic administration of large doses of steroids can produce symptoms similar to those of Cushing's disease, but such treatment is usually avoided. Roughly 75 percent of cases result from an overproduction of ACTH, and afflicted individuals may also show changes in skin pigmentation.

 ## Disorders of the Adrenal Medullae   FAP *p. 618*

The overproduction of epinephrine by the adrenal medullae may reflect chronic sympathetic activation. A **pheochromocytoma** (fē-ō-krō-mō-sī-TŌ-muh) is a tumor that produces catecholamines in massive quantities. The tumor generally develops within an adrenal medulla but may also involve other sympathetic ganglia. The most dangerous symptoms are rapid and irregular heartbeat and high blood pressure; other symptoms include uneasiness, sweating, blurred vision, and headaches. This condition is rare, and surgical removal of the tumor is the most effective treatment.

Diabetes Mellitus   FAP *p. 623*

INSULIN-DEPENDENT DIABETES MELLITUS

The primary cause of **insulin-dependent diabetes mellitus (IDDM),** or **Type I diabetes,** is inadequate insulin production by the beta cells of the pancreatic islets. Glucose transport in most cells cannot occur in the absence of insulin. When insulin concentrations decline, cells can no longer absorb glucose; tissues remain glucose-starved despite the presence of adequate or even excessive amounts of glucose in the circulation.

After a meal rich in glucose, blood glucose concentrations may become so elevated that the kidney cells cannot reclaim all the glucose molecules that enter the urine. The high urinary concentration of glucose limits the ability of the kidneys to conserve water, so the individual urinates frequently and may become dehydrated. The chronic dehydration leads to disturbances of neural function (blurred vision, tingling sensations, disorientation, fatigue) and muscle weakness.

Despite high blood concentrations, glucose cannot enter endocrine tissues, and the endocrine system responds as if glucose were in short supply. Alpha cells release glucagon, and glucocorticoid production accelerates. Peripheral tissues then break down lipids and proteins to obtain the energy needed to continue functioning. The breakdown of large numbers of fatty acids promotes the generation of molecules called **ketone bodies.** These small molecules are metabolic acids whose accumulation in large numbers can cause a dangerous reduction in blood pH. This condition, called **keto-acidosis,** commonly triggers vomiting. In severe cases, it can precede a fatal coma.

If the patient survives (an impossibility without insulin therapy), long-term treatment involves a combination of dietary control and the administration of insulin, either by injection or by infusion, using an **insulin pump.** The treatment is complicated by the fact that tissue glucose demands vary with meals, physical activity, emotional state, stress, and other factors that are hard to assess or predict. Dietary control, including regulation of the type of food, time of meals, and amount consumed, can help reduce oscillations in blood glucose levels. Modern insulin pumps can be programmed by the user to compensate for changes in activity and eating patterns. However, it remains difficult to maintain stable and normal blood glucose levels over long periods of time, even with an insulin pump.

Since 1990, pancreas transplants have been used to treat diabetes in the United States. The procedure is generally limited to gravely ill patients already undergoing kidney transplantation. The graft success rate over 5 years is roughly 50 percent, and the procedure is controversial. Pancreatic islet transplantation has also been attempted but with variable success. The transplantation of fetal pancreatic tissues (performed outside the United States) has not yielded impressive results.

A more promising approach is the use of a **biohybrid artificial pancreas.** This procedure has been used to treat Type I diabetes in dogs. Islet cells can be cultured in the laboratory and inserted within an artificial membrane. The membrane contains pores that allow fluid movement but prevent interactions between the islet cells and white blood cells that would reject them as foreign. The islet cells monitor the blood glucose concentration and secrete insulin or glucagon as needed. The biohybrid artificial pancreas can be located almost anywhere that has an adequate blood supply; in human trials, it is inserted under the skin of the abdomen. In 1994, a human patient became independent of insulin after the insertion of artificially encapsulated human islet cells.

Type I diabetes most commonly appears in individuals under 40 years of age. Because it typically appears in childhood, it has been called **juvenile-onset diabetes.** Roughly 80 percent of people with this type of diabetes have circulating antibodies that target the surfaces of beta cells. The disease may therefore be an *autoimmune disorder,* a condition that results when the immune system attacks normal body cells. (Possible mechanisms responsible for autoimmune disorders are discussed in Chapter 22 of the text.) Consequently, attempts have been made to prevent the appearance of Type I diabetes with *azathioprine (Imuran®),* a drug that suppresses the immune system. This treatment is

18

dangerous, however, because compromising immune function increases the risk of acquiring serious infections or of developing cancer. The disease is complex and probably reflects a combination of genetic and environmental factors. To date, genes associated with the development of Type I diabetes have been localized to chromosomes 6, 11, and 18.

NON-INSULIN-DEPENDENT DIABETES

Non-insulin dependent diabetes mellitus (NIDDM), or **Type II diabetes,** typically affects obese individuals over 40 year of age. Because of the age factor, this condition is also called **maturity-onset diabetes.** Type II diabetes is far more common than Type I diabetes, occurring in an estimated 6.6 percent of the U.S. population. There are 500,000 new cases diagnosed each year in the United States alone, and 90 percent of them involve obese individuals.

In Type II diabetes, insulin levels are normal or elevated, but peripheral tissues no longer respond normally. Treatment consists of weight loss and dietary restrictions that may elevate insulin production and tissue response. The drug *metformin (Glucophage®)*, which received FDA approval in 1995, lowers plasma glucose concentrations, primarily by reducing glucose synthesis and release at the liver. The use of metformin in combination with other drugs that affect glucose metabolism promises to improve the quality of life for many Type II diabetes patients.

The standard testing procedures for Type II diabetes check for primary diabetic signs: high fasting blood glucose concentrations, the appearance of glucose in the urine, and an inability to reduce elevated glucose levels. The latter sign is examined by means of a *glucose tolerance test.* Blood concentrations are monitored after a fasting subject consumes 75–100 g (2.6–3.2 oz) of glucose. In a normal individual, the glucose will enter the circulation, insulin production will rise, and peripheral tissues will absorb the glucose so rapidly that blood concentrations will remain relatively normal. Without adequate insulin production, glucose concentrations skyrocket to more than twice normal levels.

📖 Light and Behavior FAP *p. 625*

Exposure to sunlight can do more than induce a tan or promote the formation of vitamin D_3. There is evidence that daily light/dark cycles have widespread effects on the central nervous system, with melatonin playing a key role. Several studies have indicated that residents of temperate and higher latitudes in the Northern Hemisphere undergo seasonal changes in mood and activity patterns. These people feel most energetic from June through September, and they experience relatively low spirits from December through March. (The opposite situation occurs in the Southern Hemisphere, where winter and summer are reversed relative to the Northern Hemisphere.) The degree of seasonal variation differs from individual to individual: Some people display no symptoms; other people are affected so severely that they seek medical attention. The observed symptoms have recently been termed **seasonal affective disorder,** or **SAD.** Individuals with SAD experience depression and lethargy and have difficultly concentrating. They tend to sleep for long periods, perhaps 10 hours or more a day. They may also go on eating binges and crave carbohydrates.

Melatonin secretion appears to be regulated by exposure to sunlight, not simply by exposure to light. Normal interior lights are apparently not strong enough or do not release the right mixture of light wavelengths to depress melatonin production. Because many people spend very little time outdoors in the winter, melatonin production increases then, and the depression, lethargy, and concentration problems appear to be linked to elevated melatonin levels in the blood. Comparable symptoms can be produced in a normal experimental subject by an injection of melatonin.

Many SAD patients may be successfully treated by exposure to sun lamps that produce full-spectrum light. Experiments are underway to define exactly how intense the light must be and to determine the minimal effective time of exposure. Some people have been using melatonin or MSH obtained (in varying doses and purity) from health food stores to treat SAD, insomnia, and jet lag. Because the health food market is unregulated and few if any controlled studies have been done, it remains unclear whether melatonin and MSH are truly effective therapies.

CRITICAL-THINKING QUESTIONS

5-1. Pheochromocytomas are tumors of the adrenal medulla that cause hypersecretion of the catecholamines produced by this region of the adrenal gland. What symptoms would you associate with a person who suffers from this condition?

  a. hypertension
  b. sweating
  c. nervousness
  d. elevated metabolic rate
  e. all of the above

5-2. Fifty-year-old Barbara B. reports to the emergency room with heart palpitations (rapid, irregular heartbeat). Her EKG recording shows atrial fibrillation. Barbara's medical history includes recent weight loss of 10 pounds and complaints of increased irritability and nervousness within the last month. Some of Barbara's abnormal diagnostic and lab test results are as follows:

Radioactive iodine uptake test (RAIU) at 2 hours: 20% absorption (normal: 1–13%)

Serum thyroxine (T_4) test: 13 ng/dl

Serum triiodothyronine (T_3) test: 210 ng/dl

Serum TSH: <0.1

What is the probable diagnosis?
a. hypothyroidism
b. Graves' disease
c. myxedema
d. goiter

5-3. Bill develops a benign tumor of the parathyroid glands that causes an elevated level of parathyroid hormone in his blood. Which of the following would you expect to occur as a result of this condition?
a. decreases in the blood levels of calcium
b. convulsions
c. decreased muscle strength
d. increased bone density
e. all of the above

5-4. Sixteen-year-old John is a promising athlete who is below the average height for his age. He wants to play football in college but is convinced that he needs to be taller and stronger in order to accomplish his dream. He persuades his parents to visit a physician and inquire about GH treatments. If you were his physician, what would you tell him about the potential risks and benefits of such treatments?

5-5. Angie is diagnosed with diabetic retinopathy. After performing several tests, her physician decides to remove Angie's anterior pituitary gland. Shortly after the surgery, her eyesight returns to normal. What was the apparent cause of Angie's problem?

5-6. A patient is suffering from secondary hypothyroidism. Assuming you had available all the modern testing materials, how could you determine if the problem was due to hypothalamic or pituitary dysfunction?

Clinical Problems

1. The mother of a 5-year-old boy reports to his pediatrician that her son cannot rise from a sitting to a standing position without support and falls frequently. Physical examination reveals muscle weakness, particularly of the lower extremities. The physician orders the following tests:
a. electromyography
b. CT scan of the brain
c. bone X-rays
d. serum aldolase
e. nerve conduction tests

All but two test results are normal. The pediatrician diagnoses the disorder tentatively as muscular dystrophy on the basis of the two abnormal test results. A muscle biopsy confirms his suspicions. Which two tests from the above list were abnormal?

2. Jane, 33, awakens one morning with blurred vision and pain in her right eye. She sees her family physician, who determines that her sclera, conjunctiva, cornea, and ocular tension are normal but that her visual acuity is markedly reduced in that eye. An ophthalmologist is consulted and, after an exam, diagnoses optic neuritis. Jane is then referred to a neurologist, who learns that she had an earlier, previously forgotten episode of weakness and clumsiness of her left arm. This problem persisted for only 2 days, and Jane attributed it to lifting heavy suitcases on a trip. An MRI of her brain reveals several plaque-type lesions in the white matter of the brain but no other abnormalities. A lumbar puncture is performed, and the results are as follows:

Pressure: 150 cm/H_2O

Color: clear, colorless

Glucose: 50 mg/dl

Protein: 50 mg/dl

Cells: no RBCs, WBC count of 6/mm^3

IgG: high ratio of IgG to other proteins

Culture: no bacteria

What is the preliminary diagnosis?

3. Sandra is 50 years old. During a physical examination, her physician notes the following signs:

rounded "moon-shaped" face

obesity of the trunk

slender extremities

muscular weakness

blood pressure: 140/95 (normal = 120/80)

On the basis of these signs, which of the following disorders would you suspect?
a. Addison's disease
b. Cushing's disease
c. pheochromocytoma
d. hypoaldosteronism

Sandra is then sent to an endocrinologist for determination of the cause and treatment of her disorder. Diagnostic and laboratory tests are ordered, and a few of the pertinent test results are listed below:

Plasma cortisol levels: 40 g/dl (taken at 7:00 A.M.)

Plasma ACTH: 100 pg/ml (taken at 7:00 A.M.)

X-ray of skull: erosion of the sella turcica

MRI of pituitary gland: abnormal mass detected

What is the probable cause of Sandra's disorder?

The Cardiovascular System

The components of the cardiovascular system include the blood, heart, and blood vessels. Blood flows through a network of thousands of miles of vessels within the body, transporting nutrients, gases, wastes, hormones, and electrolytes and redistributing the heat generated by active tissues. The exchange of materials between the blood and peripheral tissues occurs across the walls of tiny capillaries that are situated between the arterial and venous systems. The total capillary surface area for exchange is truly enormous, averaging about 6300 square meters—about half the area of a football field.

Because the cardiovascular system plays a key role in supporting all other systems, disorders of this system will affect virtually every cell in the body. One method of organizing the many potential disorders involving the cardiovascular system is by the nature of the primary problem, whether it affects the blood, the heart, or the vascular network. Figure A-31 provides an introductory overview of major cardiovascular disorders that are discussed in the text and in later sections of the *Applications Manual.*

PHYSICAL EXAMINATION AND THE CARDIOVASCULAR SYSTEM

Individuals with cardiovascular problems commonly seek medical attention. In many cases, they present with one or more of the following as chief complaints:

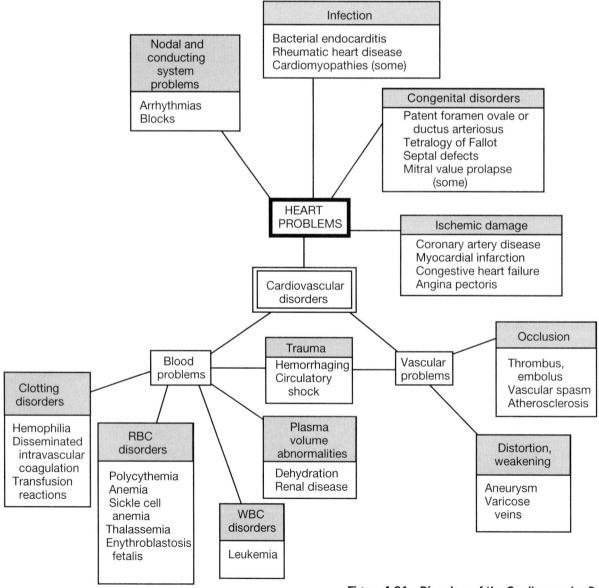

Figure A-31 Disorders of the Cardiovascular System

- *Weakness and fatigue.* These symptoms develop when the cardiovascular system can no longer meet tissue demands for oxygen and nutrients. They may occur because cardiac function is impaired, as in *heart failure* (p. 128) or *cardiomyopathy* (p. 120), or because the blood is unable to carry normal amounts of oxygen, as in the various forms of *anemia* (p. 116). In the early stages of these conditions, the individual feels healthy at rest but becomes weak and fatigued with any significant degree of exertion, because the cardiovascular system cannot keep pace with the rising tissue oxygen demands. In more advanced stages of these disorders, weakness and fatigue are chronic problems that continue, even at rest.

- *Cardiac pain.* This is a deep pressure pain that is felt in the substernal region and typically radiates down the left arm or up into the shoulder and neck. There are two major causes of cardiac pain:

  1. Constant severe pain can result from inflammation of the pericardial sac, a condition known as *pericarditis.* This *pericardial pain* may superficially resemble the pain experienced in a *myocardial infarction (MI)*, or heart attack. Pericardial pain differs from the pain of an MI in that (a) it may be relieved by leaning forward; (b) a fever may be present; and (c) the pain does not respond to the administration of drugs, such as *nitroglycerin,* that dilate coronary blood vessels. Nitroglycerin, which is effective in relieving angina pectoris, does not relieve the pain associated with pericarditis.

  2. Cardiac pain can also result from inadequate blood flow to the myocardium. This type of pain is called *myocardial ischemic pain.* Ischemic pain occurs in *angina pectoris* and in a myocardial infarction. Angina pectoris, discussed in the text *(p. 685),* most commonly results from the narrowing of coronary blood vessels by atherosclerosis. The associated pain appears during physical exertion, when myocardial oxygen demands increase, and the pain is relieved by drugs such as nitroglycerin, which dilate coronary vessels and improve coronary blood flow. The pain associated with a myocardial infarction is felt as a heavyweight or a constriction of the chest. The pain of an MI is also distinctive because (a) it is not linked to exertion; (b) it is not relieved by rest, nitroglycerin, or other coronary vasodilators; and (c) nausea, vomiting, and sweating may occur during the attack.

- *Palpitations.* Palpitations are a person's perception of an altered heart rate. The individual may complain of the heart "skipping a beat" or "racing." The most likely cause of palpitations is an abnormal pattern of cardiac activity known as an *arrhythmia.* The detection and analysis of arrhythmias are considered in a later section (p. 123).

- *Pain on movement.* Individuals with advanced athrosclerosis typically experience pain in the extremities during exercise. The pain may become so severe that the person is unwilling or unable to walk or perform other common activities. The underlying problem is constriction or partial occlusion of major arteries, such as the external iliac arteries to the lower limbs, by plaque formation.

These are only a few of the many symptoms that can be caused by cardiovascular disorders. In addition, the individual may notice the appearance of characteristic signs of underlying cardiovascular problems. A partial listing of important cardiovascular signs includes the following:

- *Edema* is an increase of fluid in the tissues that occurs when (a) the pumping efficiency of the heart is decreased, (b) the plasma protein content of the blood is reduced, or (c) venous pressures are abnormally high. The tissues of the extremities are most commonly affected, and individuals experience swollen feet, ankles, and legs. When edema is so severe that pressing on the affected area leaves an indention, the sign is called *pitting edema.* Edema is discussed in Chapters 21 and 22 of the text *(pp. 726, 773).*

- Breathlessness, or *dyspnea,* occurs when cardiac output is inadequate for tissue oxygen demands. Dyspnea may also occur with *pulmonary edema,* a buildup of fluid within the alveoli of the lungs. Pulmonary edema and dyspnea are typically associated with *congestive heart failure (CHF)* (p. 128).

- *Varicose veins* are dilated superficial veins that are visible at the skin surface. This condition, which develops when venous valves malfunction, can be caused or exaggerated by increased systemic venous pressures. Varicose veins are considered further on p. 126.

- There may be characteristic and distinctive changes in skin coloration. For example, *pallor* is the lack of normal red or pinkish color in the skin of a Caucasian or the conjunctiva and oral mucosa of darker-skinned people. Pallor accompanies many forms of anemia but may also be the result of inadequate cardiac output, shock (p. 129), or circulatory collapse. *Cyanosis* is the bluish color of the skin that occurs with a deficiency of oxygen in the tissues. Cyanosis generally results from either cardiovascular or respiratory disorders.

19

- *Vascular skin lesions* were introduced in the discussion of skin disorders on p. 39. Characteristic vascular lesions may occur in clotting disorders (p. 118) and in several forms of *leukemia* (p. 118). For example, abnormal bruising may be the result of a disorder that affects the clotting system, platelet production, or vessel structure. *Petechiae,* which appear as purple spots on the skin surface, are typically seen in certain types of leukemia or other diseases associated with low platelet counts.

CARDIOVASCULAR DISORDERS AND DIAGNOSTIC PROCEDURES

In many cases, the initial detection of a cardiovascular disorder occurs during the assessment stage of a physical examination:

1. When the vital signs are taken, the pulse is checked for vigor, rate, and rhythm. Weak or irregular heart beats will commonly be noticed at this time.

2. The blood pressure is monitored with a stethoscope, blood pressure cuff, and sphygmomanometer. Unusually high or low readings can alert the examiner to potential problems with cardiac or vascular function. However, a diagnosis of hypotension or hypertension is usually not made on the basis of a single reading but after several readings over a period of time. Hypertension and hypotension are discussed in detail on p. 127.

3. The heart sounds are monitored by auscultation with a stethoscope:

   - Cardiac rate and rhythm can be checked and arrhythmias detected.

   - Abnormal heart sounds, or *murmurs,* may indicate problems with atrioventricular or semilunar valves. Murmurs are noted in relation to their location in the heart (as determined by the position of the stethoscope on the chest wall), the time of occurrence in the cardiac cycle, and whether the sound is low-pitched or high-pitched.

   - Nothing is usually heard during auscultation of normal vessels of the circulatory system. *Bruits* are the sounds that result from turbulent blood flow around an obstruction within a vessel. Bruits are typically heard where large atherosclerotic plaques have formed.

Functional abnormalities of the heart and blood vessels can commonly be detected through physical assessment and the recognition of characteristic signs and symptoms. The structural basis of these problems is generally determined through the use of scans, X-rays, and the monitoring of electrical activity in the heart. For problems with a hema-

tological basis, laboratory tests performed on blood samples normally provide the information necessary to reach a preliminary diagnosis. Table A-19 summarizes information about representative diagnostic procedures used to evaluate the health of the cardiovascular system.

℞ Transfusions FAP *p. 644*

In a **transfusion,** blood components are provided to an individual whose blood volume has been reduced or whose blood is deficient in some way. The blood is obtained under sterile conditions, treated to prevent clotting and to stabilize the red blood cells, and refrigerated. Transfusions of whole blood are most commonly used to restore blood volume after massive blood loss. In an **exchange transfusion,** an individual's entire blood volume is drained off and simultaneously replaced with whole blood from another source.

Chilled whole blood remains usable for about 3.5 weeks. For longer storage, the blood must be fractionated. The red blood cells are separated from the plasma, treated with a special antifreeze solution, and then frozen. The plasma can then be stored chilled, frozen, or freeze-dried. This procedure permits the long-term storage of rare blood types that might not otherwise be available for emergency use.

Fractionated blood has many uses. **Packed red blood cells** (PRBC), with most of the plasma removed, are preferred for cases of anemia, in which the volume of blood may be close to normal but has below-normal oxygen-carrying capabilities. Plasma may be administered when massive fluid losses are occurring, such as after a severe burn. Plasma samples can be further fractionated to yield albumins, immunoglobulins, clotting factors, and plasma enzymes that can be administered separately. White blood cells and platelets can also be collected, sorted, and stored for subsequent transfusion.

Some 12 million pints of blood are used each year in the United States alone, and the demand for blood or blood components commonly exceeds the supply. Moreover, there has been increasing concern over the danger of transfusion recipients' becoming infected with hepatitis viruses or with HIV (the virus that causes AIDS) from contaminated blood. The result has been a number of changes in transfusion practices during recent years. Blood donors and collected blood are screened for these infections. In general, fewer units of blood are now administered. There has also been an increase in **autologous transfusion,** in which blood is removed from a patient (or potential patient), stored, and later transfused back into the original donor when needed—for example, after a surgical procedure. Genetically engineered erythropoietin (EPO) can also be used to help the patient's body quickly restore its full complement of red cells. Moreover,

Table A-19a   Examples of Tests Used in Diagnosing Cardiovascular Disorders

Diagnostic Procedure	Method and Result	Representative Uses	Notes
Electrocardiography (ECG/EKG)	Electrodes placed on the chest register electrical activity of cardiac muscle during a cardiac cycle; information is transmitted to a monitor for graphic recording	Heart rate can be determined through study of the ECG; abnormal wave patterns may occur with cardiac irregularities such as myocardial infarction, chamber hypertrophy, or arrhythmias, such as ectopic pacemaker activity, premature ventricular contractions (PVCs), and defects in the conduction system	Normal ECG pattern is composed of a series of waves called the P, QRS complex, and T waves, which correspond to specific events within the cardiac cycle
Echocardiography	Standard ultrasound examination of the heart	Detection of structural abnormalities of the heart while in motion; useful in the determination of mitral stenosis, cardiomyopathy, and valve dysfunction	
Transesophageal echocardiography	Use of an ultrasound probe attached to an endoscope and inserted into the esophagus		
Phonocardiography	Heart sounds are monitored and graphically recorded	Detection of murmurs and abnormal heart sounds	Used primarily for training medical workers in cardiac auscultation
Exercise stress test	ECG, blood pressure, and heart rate are monitored during exercise on a treadmill	Coronary artery blockage is suspected if ECG patterns change and/or echocardiography results change	
Chest X-ray	Film sheet with radio-dense tissues in white on a negative image	Determination of abnormal shape or size of the heart and abnormalities of the aorta and great vessels	
Cardiac catherization (coronary arteriography)	For study of the left side of the heart, a catheter is inserted into femoral or brachial artery; for study of the right side, the catheter is inserted in the femoral or subclavian vein; the catheter passes along the vessels to the heart. Contrast dye may be injected for better contrast as X-rays are taken.	Most commonly used to detect blockages and spasms in coronary arteries; used also to evaluate congenital defects and ventricular hypertrophy and to determine the severity of a valvular defect; used for monitoring volumes and pressures within the chambers	
Positron emission tomography (PET)	Radionuclides are injected into circulation; accumulation occurs at certain areas of the heart; a computer image for viewing is produced	Allows detection of tissue areas with a reduced metabolic rate due to an infarction and determines the extent of damage; also useful in the determination of blood flow to the heart through the coronary arteries; may detect ischemia	May confirm stress test results without angiography
Pericardiocentesis	Needle aspiration of fluid from the pericardial sac for analysis	*See* pericardial fluid analysis, *Table A-19b*	

19

Table A-19a *(continued)*

Diagnostic Procedure	Method and Result	Representative Uses	Notes
Doppler ultrasound	Transducer is placed over the vessel to be examined, and the echoes are analyzed by computer to provide information on blood velocity and flow direction	Detection of venous occlusion; determination of cardiac efficiency; monitoring of fetal circulation and blood flow in umbilical vessels	
Venography	Radiopaque dye is injected into peripheral vein of a limb, and an X-ray study is done to detect venous occlusions	Detection of venous thrombosis or phlebitis	

new technology permits the reuse of blood "lost" during surgery. The blood is collected and filtered, the platelets are removed, and the remainder of the blood is reinfused into the patient.

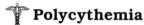 Polycythemia

FAP *p. 646*

An elevated hematocrit with a normal blood volume constitutes **polycythemia** (po-lē-sī-THĒ-mē-uh). There are several different types of polycythemia. We will consider *erythrocytosis* (e-rith-rō-sī-TŌ-sis), a polycythemia that affects only red blood cells, later in the section. **Polycythemia vera** ("true polycythemia") results from an increase in the numbers of all blood cells. The hematocrit may reach 80–90, at which point the tissues become oxygen-starved because red blood cells are blocking the smaller vessels. This condition seldom strikes young people; most cases involve patients age 60–80. There are several treatment options, but none cures the condition. The cause of polycythemia vera is unknown, although there is some evidence that the condition is linked to radiation exposure.

Thalassemia

FAP *p. 649*

The *thalassemias* are a diverse group of inherited blood disorders caused by an inability to produce adequate amounts of alpha or beta hemoglobin chains. A specific condition is categorized as an **alpha-thalassemia** or a **beta-thalassemia** depending on whether the α or β hemoglobin subunits are affected. Normal individuals inherit two copies of alpha-chain genes from each parent, and alpha-thalassemia develops when one or more of these genes are missing or inactive. The severity of the symptoms varies with the number of normal alpha-chain genes that remain functional. For example, an individual with three normal alpha-chain genes will not develop symptoms, but this person can be a carrier, passing the defect to the next generation. A child born of parents who are both carriers is likely to develop a more severe form of the disease:

- Individuals with two copies of the normal alpha-chain gene have somewhat-impaired hemoglobin synthesis. The red blood cells are small and contain less than the normal quantity of hemoglobin.

- Individuals with only one copy of the normal alpha-chain gene have very small (*microcytic*) red blood cells that are relatively fragile.

- Most individuals with no functional copies of the normal alpha-chain gene die shortly after birth, because the hemoglobin synthesized cannot bind and transport oxygen normally.

The genetic mechanisms involved are considered in Chapter 29 of the text.

There are regional and racial differences in the frequency of alpha-thalassemia. For example, roughly 2 percent of African-Americans have only two copies of the normal gene, one from each parent. However, the incidence of severe alpha-thalassemia is highest among Southeast Asians.

Each person inherits only one gene for the beta hemoglobin chain from each parent. If an individual does not receive a copy of the normal gene from either parent, the condition of **beta-thalassemia major**, or *Cooley's disease*, develops. Symptoms of this condition include severe anemia; microcytosis; a low hematocrit (under 20); and enlargement of the spleen, liver, heart, and areas of red bone marrow. Potential treatments for those with severe symptoms include transfusions, splenectomy (to slow the rate of RBC recycling), and bone marrow transplantation. *Beta-thalassemia minor*, or *beta-thalassemia trait*, seldom produces clinical symptoms. The rate of hemoglobin synthesis is depressed by roughly 15 percent, but this decrease does not affect the RBCs' functional abilities, and no treatment is necessary.

Sickle Cell Anemia

FAP *p. 649*

Sickle cell anemia (SCA) results from the production of an abnormal form of hemoglobin. The β

Table A-19b   Laboratory Tests Useful in Diagnosing Cardiovascular Disorders

Laboratory Test	Normal Values in Blood Plasma or Serum	Significance of Abnormal Values	Notes
Creatine phosphokinase (CPK, CK)		Prolonged elevated levels of CK indicate myocardial damage; levels usually do not rise until 6–12 hours following myocardial infarct (MI)	*See* also isoenzymes of CPK
Isoenzymes of CPK (CK-MM) (CK-MB) (CK-BB)	Varies with method; the normal CK-MB is 0–6% of the total	CK-MB percentage rises to 60%. Levels rise within 6–12 hours following MI. Elevation of LDH occurs 24–72 hours after MI.	CK-MM is useful in the diagnosis of muscle disease. CK-BB is used in the diagnosis of brain infarct or stroke. CK-MB isoenzyme provides information about damaged cardiac muscle.
Aspartate aminotransferase (AST or SGOT)	Adults: 7–45 U/l	Elevated in congestive heart failure due to liver damage	
Lactate dehydrogenase (LDH)	Adults: 45–90 U/l	Elevated in roughly 40% of patients with congestive heart failure	
Serum electrolytes			
Sodium	Adults: 135–145 mEq/l	Decreased levels occur with heart failure, hypotension. Increased level occurs with essential hypertension.	
Potassium	Adults: 3.5–5.0 mEq/l	Decreased level when taking certain diuretics	Hypokalemia can cause arrhythmias; hyperkalemia causes bradycardia
Serum lipoproteins	HDL: 29–77 mg/dl LDL: 60–160 mg/dl	Higher risk for coronary heart disease when LDL values are >130 mg/dl and HDL is <35 mg/dl	Electrophoresis is used to separate HDL fraction from LDL
Serum cholesterol	Adults: 150–240 mg/dl (desirable level <200 mg/dl)	High risk when >240 mg/dl for atherosclerosis and coronary artery disease; elevated levels occur in familial hypercholesterolemia	
Serum triglycerides	Adults: 10–160 mg/dl (adults >50 years old may have higher values)	Elevated with acute MI and poorly controlled diabetes mellitus	
Pericardial fluid analysis	Fluid is analyzed for appearance, number and types of blood cells, protein and glucose levels, and is cultured to determine infectious organisms; fluid can also be removed to reduce pressure on the heart (during cardiac tamponade)		

19

Table A-19c   Representative Hematology Studies for Diagnosing Blood Disorders

Laboratory Test	Normal Values in Blood Plasma or Serum	Significance of Abnormal Values	Notes
Complete blood count			
RBC count	Adult males: 4.7–6.1 million/mm^3 Adult females: 4.0–5.0 million/mm^3	Increased RBC count, Hb, and Hct occur in polycythemia vera and in events that induce hypoxia, such as moving to a high altitude or congenital heart disease	
Hemoglobin (Hb, Hgb)	Adult males: 14–18 g/dl Adult females: 12–16 g/dl	Decreased RBC count, Hb, and Hct occur with hemorrhage and the various forms of anemia, including hemolytic, iron deficiency, vitamin B_{12} deficiency, sickle cell anemia, and thalassemia or as a result of bone marrow failure or leukemia	
Hematocrit (Hct)	Adult males: 42–52 Adult females: 37–47	See sections on RBCs and Hb	
RBC Indices			
Mean corpuscular volume (MCV)	Adults: 80–98 μm^3 (measures average volume of single RBC)	Increased MCV and MCH occur in types of macrocytic anemia, including vitamin B_{12} deficiency	
Mean corpuscular hemoglobin (MCH)	Adults: 27–31 pg (measure of the average amount of Hb per RBC)	Decreased MCV and MCH in types of microcytic anemia, such as iron deficiency anemia and thalassemia	
Mean corpuscular hemoglobin concentration (MCHC)	Adults: 32–36% (derived by dividing the total Hb concentration number by the Hct value)	Decreased levels (hypochromic erythrocytes) suggest iron deficiency anemia or thalassemia	
WBC count	Adults: 5000–10,000/mm^3	Increased levels occur with chronic and acute infections; tissue death (MI, burns), leukemia, parasitic diseases and stress. Decreased levels occur in aplastic and pernicious anemias, overwhelming bacterial infection (sepsis), and viral infections.	High-risk values: >30,000/mm^3 and <2500/mm^3
Differential WBC count	Neutrophils: 50–70%	Increased in acute bacterial infection, myelocytic leukemia, rheumatoid arthritis, and stress. Decreased in aplastic and pernicious anemia, viral infections, radiation treatment, and with some medications.	
	Lymphocytes: 20–30%	Increased in lymphocytic leukemia, infectious mononucleosis, and viral infections. Decreased in radiation treatment, AIDS, and corticosteroid therapy.	

Table A-19c *(continued)*

Laboratory Test	Normal Values in Blood Plasma or Serum	Significance of Abnormal Values	Notes
	Monocytes: 2–8%	Increased in chronic inflammation, viral infections, and tuberculosis. Decreased in aplastic anemia and corticosteroid therapy.	
	Eosinophils: 2–4%	Increased in allergies, parasitic infections, and some autoimmune disorders. Decreased with steroid therapy.	
	Basophils: 0.5–1%	Increased in inflammatory processes and during healing. Decreased in hypersensitivity reactions.	
Platelet count	Adults: 150,000–400,000/mm^3	Increased count can cause vascular thrombosis; increase occurs in polycythemia vera. Decreased levels can result in spontaneous bleeding; decrease occurs in different types of anemia and in some leukemias.	High-risk values: <15,000/mm^3 or >1 million/mm^3
Bleeding time (amount of time for bleeding to stop after a small incision is made in skin)	3–7 minutes	Prolongation occurs in patients with decreased platelet count, anticoagulant therapy, aspirin ingestion, leukemia, or clotting factor deficiencies	
Factors assay (coagulation factors, I, II, V, VIII, IX, X, XI, XII)	Factors I–XI are measured for their activity and their minimal hemostatic activity.	Decreased activity of the coagulation factors will result in defective clot formation. Deficiencies can be caused by liver disease or vitamin K deficiencies.	
Plasma fibrinogen (Factor I)	200–400 mg/dl	Elevated values may occur in inflammatory conditions and acute infections and also with medications such as birth control pills. Decreased levels occur in liver disease, leukemia, and DIC (disseminated intravascular coagulation).	
Plasma prothrombin time (PT)	±2 seconds of control (control should be 11–15 seconds)	Elevated levels may occur with trauma, MI, or infection. Decreased levels occur in DIC, liver disease, and medications such as streptokinase or coumadin.	
Hemoglobin electrophoresis			
Hemoglobin A	Adults: within 95–98% of the total Hb		
Hemoglobin F	<2% of the total Hb after age 2 (newborn has 50–80% HbF)	Elevated levels after 6 months of age suggest thalassemia	

19

Table A-19c *(continued)*

Laboratory Test	Normal Values in Blood Plasma or Serum	Significance of Abnormal Values	Notes
Hemoglobin S	0% of the total Hb	Elevated levels occur in sickle cell anemia and the sickle cell trait	
Serum bilirubin	Adults: 0.1–1.2 mg/dl	Increased levels occur with severe liver disease, hepatitis, and liver cancer	
Erythrocyte sedimentation rate (ESR) (measure of the rate at which erythrocytes move or descend in normal saline)	Adult males: 1–15 mm/hr Adult females: 1–20 mm/hr	Increased by disease processes that increase the protein concentration in the plasma, including inflammatory conditions, infections, and cancer	ESR is not useful in the diagnosis of specific disorders
Bone marrow aspiration biopsy	For the evaluation of hematopoiesis	Increased RBC precursors with polycythemia vera; increased WBC precursors with leukemia; radiation or chemotherapy therapy can cause a decrease in all cell populations	Involves laboratory examination of shape and cell size of erythrocytes, leukocytes, and megakaryocytes

chains are involved, and the abnormal subunit is called *hemoglobin S.* Today sickle cell anemia affects 60,000–80,000 African-Americans, or approximately 0.14 percent of the African-American population.

An individual with sickle cell anemia carries two copies of the abnormal gene—one from each parent. If only one copy is present, the individual has a *sickling trait.* One African-American in 12 carries the sickling trait. Although it is now known that the genes are present in Americans of Mediterranean, Middle Eastern, and East Indian ancestry, statistics on the incidence of sickling trait and SCA in these groups are as yet unavailable.

In an individual with the sickling trait, most of the hemoglobin is of the normal form, and the RBCs function normally. But the presence of the abnormal hemoglobin gives the individual the ability to resist the parasitic infection that causes **malaria,** a mosquito-borne illness. The malaria parasites enter the bloodstream when an individual is bitten by an infected mosquito. The microorganisms then invade, and reproduce within, the RBCs. But when they enter the red blood cell of a person with the sickling trait, the cell responds by sickling. Either the sickling itself kills the parasite or the sickling attracts the attention of a phagocyte that engulfs the RBC and kills the parasite. In either event, the individual remains unaffected by the malaria, whereas normal individuals sicken and commonly die of that disease.

Sickled RBCs cells may get stuck in small capillaries and obstruct further blood flow. Symptoms of sickle cell anemia include pain and damage to a variety of organs and systems, depending on the location of the obstructions. In addition, the trapped RBCs eventually die and break down, producing a characteristic anemia. Transfusions of normal blood can temporarily prevent additional complications, and there are experimental drugs that can control or reduce sickling. *Hydroxyurea* is an anticancer drug that stimulates the production of fetal hemoglobin, a slightly different form of hemoglobin produced during development. This drug is effective but has toxic side effects (not surprising in an anticancer drug). The food additive **butyrate,** found in butter and other foods, appears to be even more effective in promoting the synthesis of fetal hemoglobin. In clinical trials, it has been effective in treating sickle cell anemia and other conditions caused by abnormal hemoglobin structure, such as *beta-thalassemia.*

Bilirubin Tests and Jaundice

FAP p. 651

When hemoglobin is broken down, the heme units (minus the iron) are converted to bilirubin. Normal plasma bilirubin concentrations range from 0.5 to 1.2 mg/dl. Of that amount, roughly 85 percent is unconjugated bilirubin that will be removed by the liver. Several different clinical conditions are characterized by an increase in the total plasma bilirubin concentration. In such conditions, bilirubin diffuses into peripheral tissues, giving them a yellow coloration that is most apparent in the skin and over the sclera of the eyes. This combination of signs (yellow skin and eyes) is called **jaundice** (JAWN-dis).

Jaundice can have many different causes, but blood tests that determine the concentration of unconjugated and conjugated forms of bilirubin can provide useful diagnostic clues. For example, **hemolytic jaundice** results from the destruction of large numbers of red blood cells. When this occurs, phagocytes release massive quantities of unconjugated bilirubin into the blood. Because the liver cells then accelerate the secretion of bilirubin in the bile, the blood concentration of conjugated bilirubin does not increase proportionately. A blood test from a patient with hemolytic jaundice would reveal (1) elevated total bilirubin, (2) high concentrations of unconjugated bilirubin, and (3) a conjugated bilirubin contribution of much less than 15 percent of the total bilirubin concentration.

These results are quite different from those seen in **obstructive jaundice.** In this condition, discussed in Chapter 24 of the text, the ducts that remove bile from the liver are constricted or blocked. Liver cells cannot get rid of conjugated bilirubin, and large quantities diffuse into the blood. In this case, diagnostic tests would show (1) elevated total bilirubin, (2) an unconjugated bilirubin contribution of much less than 85 percent of the total bilirubin concentration, and (3) high concentrations of conjugated bilirubin.

⚕ Iron Deficiencies and Excesses

FAP p. 651

If dietary supplies of iron are inadequate, hemoglobin production slows down, and symptoms of *iron deficiency anemia* appear. This form of anemia can also be caused by any condition that produces a blood loss, because the iron in the lost blood cannot be recycled. As the red blood cells are replaced, iron reserves must be mobilized for use in the synthesis of new hemoglobin molecules. If those reserves are exhausted or if dietary sources are inadequate, symptoms of iron deficiency appear. In iron deficiency anemia, the red blood cells are unable to synthesize functional hemoglobin, and they are unusually small when they enter the circulation. The hematocrit declines, and the hemoglobin content and oxygen-carrying capacity of the blood are substantially reduced. Symptoms include weakness and a tendency to fatigue easily.

Women are especially dependent on a normal dietary supply of iron, because their iron reserves are smaller than those of men. A healthy male's body has about 3.5 g of iron in the ionic form Fe^{2+}. Of that amount, 2.5 g is bound to the hemoglobin of circulating red blood cells, and the rest is stored in the liver and bone marrow. In women, the total body iron content averages 2.4 g, with roughly 1.9 g

incorporated into red blood cells. Thus a woman's iron reserves consist of only 0.5 g, half that of a typical man. When the demand for iron increases out of proportion with dietary supplies, iron deficiency develops. An estimated 20 percent of menstruating women in the United States show signs of iron deficiency. Pregnancy also stresses iron reserves, for the woman must provide the iron needed to produce both maternal and fetal erythrocytes.

Good dietary sources of iron include liver, red meats, kidney beans, egg yolks, spinach, and carrots. Iron supplements can help prevent iron deficiency, but too much iron can be as dangerous as too little. Iron absorption across the digestive tract normally keeps pace with physiological demands. When the diet contains abnormally high concentrations of iron, or when hereditary factors increase the rate of absorption, the excess iron gets stored in peripheral tissues. This storage is called *iron loading.* Eventually, cells begin to malfunction as massive iron deposits accumulate in the cytoplasm. For example, iron deposits in pancreatic cells can lead to diabetes mellitus; deposits in cardiac muscle cells lead to abnormal heart contractions and heart failure. (There is evidence that iron deposits in the heart caused by the overconsumption of red meats may contribute to heart disease.) Liver cells become nonfunctional, and liver cancers may develop.

Comparable symptoms of iron loading may follow repeated transfusions of whole blood, because each unit of whole blood contains roughly 250 mg of iron. For example, as we noted previously, the various forms of *thalassemia* result from a genetic inability to produce adequate amounts of one of the four globin chains in hemoglobin. Erythrocyte production and survival are reduced, and so is the blood's oxygen-carrying capacity. Most individuals with severe untreated thalassemia die in their twenties, but not because of the anemia. These patients are treated for severe anemia with frequent blood transfusions, which prolong life, but the excessive iron loading eventually leads to fatal heart problems.

🏃 Erythrocytosis and Blood Doping

FAP p. 652

In **erythrocytosis** (e-rith-rō-sī-TŌ-sis), the blood contains abnormally large numbers of red blood cells. Erythrocytosis generally results from the massive release of EPO by tissues (especially the kidneys) deprived of oxygen. After moving to high altitudes, people typically experience erythrocytosis, because the air there contains less oxygen than it does at sea level. The increased number of red blood cells compensates for the fact that each RBC is carrying less oxygen than it would at sea level. Mountaineers and persons living at altitudes of 10,000–12,000 feet may have hematocrits as high as 65.

19

An individual whose heart or lungs are functioning inadequately may also develop erythrocytosis. For example, this condition is commonly seen in cases of heart failure and emphysema, two conditions we will discuss in later sections. Whether the blood fails to circulate efficiently or the lungs do not deliver enough oxygen to the blood, peripheral tissues remain oxygen-poor despite the rising hematocrit. Having a higher concentration of red blood cells increases the oxygen-carrying capacity of the blood, but it also makes the blood thicker and harder to push around the circulatory system. The work load on the heart increases, making a bad situation even worse.

The practice of **blood doping** has occurred among competitive athletes involved with endurance sports such as cycling. The procedure entails removing whole blood from the athlete in the weeks before an event. The packed red cells are separated from the plasma and stored. By the time of the race, the competitor's bone marrow will have replaced the lost blood. Immediately before the event, the packed red cells are reinfused, increasing the hematocrit. The objective is to elevate the oxygen-carrying capacity of the blood and thereby increase endurance. The consequence is that the athlete's heart is placed under a tremendous strain. The long-term effects are unknown, but the practice obviously carries a significant risk; it has recently been banned in amateur sports. Attempts to circumvent this rule by the use of administered EPO in 1992–1993 resulted in the tragic deaths of 18 European cyclists.

Blood Tests and RBCs   FAP *p. 652*

This section describes several common blood tests that assess circulating RBCs.

Reticulocyte Count. Reticulocytes are immature red blood cells that are still synthesizing hemoglobin. Most reticulocytes remain in the bone marrow until they complete their maturation, but some enter the circulation. Reticulocytes normally account for about 0.8 percent of the erythrocyte population. Values above 1.5 percent or below 0.5 percent indicate that something is wrong with the rates of RBC survival or maturation.

Hematocrit (Hct). The hematocrit value is the percentage of whole blood occupied by cells. Normal adult hematocrits average 46 for men and 42 for women, with ranges of 40–54 for men and 37–47 for women.

Hemoglobin Concentration (Hb). This test determines the amount of hemoglobin in the blood, expressed in grams per deciliter (g/dl). Normal ranges are 14–18 g/dl for males and 12–16 g/dl for females. The differences in hemoglobin concentration reflect the differences in hematocrit. For both genders, a single RBC contains 27–33 picograms (pg) of hemoglobin.

RBC Count. Calculations of the RBC count, the number of RBCs per microliter of blood, are based on the hematocrit and hemoglobin content and can be used to get more information about the condition of the RBCs. Values typically reported in blood screens include:

- *Mean corpuscular volume (MCV),* the average volume of an individual red blood cell, in cubic micrometers. The MCV is calculated by dividing the hematocrit by the RBC count, using the formula

$$MCV = Hct/RBC \text{ count (in millions)} \times 10$$

This is a "cookbook" method that takes advantage of the fact that the hematocrit closely approximates the relative volume of RBCs in any unit sample of whole blood. Normal values range from 82.2 to 100.6. For a representative hematocrit of 46 and an RBC count of 5.2 million, the mean corpuscular volume is

$$MCV = 46/5.2 \times 10 = 88.5 \text{ } \mu m^3$$

Cells of normal size are said to be **normocytic,** whereas larger-than-normal or smaller-than-normal RBCs are called **macrocytic** or **microcytic,** respectively.

- *Mean corpuscular hemoglobin concentration (MCHC),* the amount of hemoglobin within a single RBC, expressed in picograms. Normal values range from 27 to 34 pg. MCHC is calculated as

$$MCHC = Hb/RBC \text{ count (in millions)} \times 10$$

RBCs containing normal amounts of hemoglobin are termed **normochromic,** whereas the names **hyperchromic** and **hypochromic** indicate higher- or lower-than-normal hemoglobin content, respectively.

Anemia (a-NĒ-mē-uh) exists when the oxygen-carrying capacity of the blood is reduced, diminishing the delivery of oxygen to peripheral tissues. Such a reduction causes a variety of symptoms, including premature muscle fatigue, weakness, lethargy, and a lack of energy. Anemia may exist because the hematocrit is abnormally low or because the amount of hemoglobin in the RBCs is reduced. Standard laboratory tests can be used to differentiate among the various forms of anemia on the basis of the number, size, shape, and hemoglobin content of red blood cells. As an example, Table A-20 shows how this information can be used to distinguish among four major types of anemia:

1. **Hemorrhagic anemia** results from severe blood loss. Erythrocytes are of normal size; each contains a normal amount of hemoglobin, and reticulocytes are present in normal concentra-

Table A-20   RBC Tests and Anemias

Anemia Type	Hct	Hb	Reticulocyte Count	MCV	MCHC
Hemorrhagic	Low	Low	Normal	Normal	Normal
Aplastic	Low	Low	Very low	Normal	Normal
Iron deficiency	Low	Low	Normal	Low	Low
Pernicious	Low	Low	Very low	High	High

tions, at least initially. Blood tests would therefore show a low hematocrit and low hemoglobin, but the MCV, MCHC, and reticulocyte counts would be normal.

2. In **aplastic** (ā-PLAS-tik) **anemia,** the bone marrow fails to produce new red blood cells. The 1986 nuclear accident in Chernobyl (of the former USSR) caused a number of cases of aplastic anemia. The condition is fatal unless surviving stem cells repopulate the marrow or a bone marrow transplant is performed. In aplastic anemia, the circulating red blood cells are normal in all respects, but because new RBCs are not being produced, the RBC count, Hct, Hb, and reticulocyte count are low.

3. In **iron deficiency anemia,** normal hemoglobin synthesis cannot occur, because iron reserves are inadequate. Developing red blood cells cannot synthesize functional hemoglobin; as a result, they are unusually small. A blood test shows a low hematocrit, low hemoglobin content, low MCV, and low MCHC but a normal reticulocyte count. An estimated 60 million women worldwide have symptoms of iron deficiency anemia. (See the discussion of iron deficiencies and excesses on p. 115.)

4. In **pernicious** (per-NISH-us) **anemia,** normal red blood cell maturation ceases because the supply of vitamin B_{12} is inadequate. Erythrocyte production declines, and the red blood cells are abnormally large and may develop a variety of bizarre shapes. Blood tests from a person with pernicious anemia indicate a low hematocrit with a very high MCV and a low reticulocyte count.

Ⓡ **Synthetic Blood**   FAP *p. 656*

Shortages of blood and anxieties over the safety of existing stockpiles persist. In addition, some people are unable or unwilling to accept transfusions for medical or religious reasons. Thus there has been widespread interest in recent attempts to develop synthetic blood components. Genetic engineering techniques are now being used to address this problem. For example,

- It is now possible to synthesize one of the subunits of normal human hemoglobin, which can then be introduced into the circulation to increase oxygen transport and total blood volume. The hemoglobin molecules can be attached to inert carrier molecules that will prevent their filtration and loss at the kidneys.

- Hemoglobin can now be obtained from nonhuman sources. Technical progress has been rapid, but practical success has been limited.

- In 1990, the FDA approved testing of a blood substitute called *Hemopure* that contained purified hemoglobin from cattle. Although the hemoglobin was infused without blood cells or plasma, cross-reactions occurred, and the testing was halted.

- In late 1991, researchers created a strain of genetically manipulated pigs that produce human hemoglobin. Because hemoglobin outside RBCs lasts in circulation only minutes or hours, even if this technique is successful (technical and clinical problems must be solved first), it will provide only a short-term solution to blood shortages. However, it would be very useful in the treatment of severe injuries, when liters of blood will be infused and lost through bleeding before the patient can be stabilized.

- In 1992, researchers developed sheep that have human stem cells in their bone marrow. These sheep produce normal human blood cells. This process may eventually provide a safe and reliable source of blood to blood banks.

Whole blood substitutes are highly experimental solutions still undergoing clinical evaluation. In addition to the osmotic agents in plasma expanders, these solutions contain small clusters of synthetic molecules built of carbon and fluorine atoms. The mixtures, known as **perfluorochemical** (PFC) emulsions, can carry roughly 70 percent of the oxygen of whole blood. Animals have been kept alive after an exchange transfusion that completely replaced their circulating blood with a PFC emulsion. PFC solutions have the same advantages of other plasma expanders, plus the added benefits of transporting oxygen. Because no RBCs are involved, the PFC emulsions can carry oxygen to re-

19

gions whose capillaries have been partially blocked by fatty deposits or blood clots. Unfortunately, PFCs do not absorb oxygen as effectively as normal blood does. To ensure that the PFCs deliver adequate oxygen to peripheral tissues, the individual must breathe air rich in oxygen, usually through an oxygen mask. In addition, phagocytes appear to engulf the PFC clusters. These problems have limited the use of PFC emulsions in humans. However, one PFC solution, *Fluosol*, has been used to enhance oxygen delivery to cardiac muscle during heart surgery.

Another approach involves the manufacture of miniature erythrocytes by enclosing small bundles of hemoglobin in a lipid membrane. These **neohematocytes** (nē-ō-hēm-A-tō-sīts) are spherical, with a diameter of less than 1 μm, and they can easily pass through narrow or partially blocked vessels. The major problem with this technique is that phagocytes treat neohematocytes like fragments of normal erythrocytes, so they remain in circulation for only about 5 hours.

☤ The Leukemias
FAP p. 659

Leukemias characterized by the presence of abnormal granulocytes or other cells of the bone marrow are called **myeloid;** leukemias that involve lymphocytes are termed **lymphoid.** The first symptoms appear as immature and abnormal white blood cells appear in the circulation. As the number of white blood cells increases, they travel through the circulation, invading tissues and organs throughout the body.

These cells are extremely active, and they require abnormally large amounts of energy. As in other cancers, described in Chapter 4 and elsewhere in this manual (p. 137), invading leukemic cells gradually replace the normal cells, especially in the bone marrow. Red blood cell, normal WBC, and platelet formation decline, with resulting anemia, infection, and impaired blood clotting. Untreated leukemias are invariably fatal.

Leukemias may be classified as *acute* (short and severe) or *chronic* (prolonged). Acute leukemias may be linked to radiation exposure, hereditary susceptibility, viral infections, or unknown causes. Chronic leukemias may be related to chromosomal abnormalities or immune system malfunctions. Survival in untreated acute leukemia averages about 3 months; individuals with chronic leukemia may survive for years.

Effective treatments exist for some forms of leukemia and not others. For example, when acute lymphoid leukemia is detected early, 85–90 percent of patients can be held in remission for 5 years or longer, but only 10–15 percent of patients with acute myeloid leukemia survive 5 years or more. The yearly mortality rate for leukemia (all types) in the United States has not declined appreciably in the past 30 years, remaining at about 6.8 per 100,000 population. However, new treatments are being developed that show promise when used to combat specific forms of leukemia. For example, the administration of α-*interferon*, a hormone of the immune system, has been very effective in treating hairy cell leukemia and chronic myeloid leukemia.

One option for treating acute leukemias is to perform a bone marrow transplant. In this procedure, massive chemotherapy or radiation treatment is given, enough to kill all the cancerous cells. Unfortunately, this treatment also destroys the patient's blood cells and stem cells in the bone marrow and other blood-forming tissues. The individual then receives an infusion of healthy bone marrow cells that repopulate the blood and marrow tissues.

If the bone marrow is extracted from another person (a **heterologous marrow transplant**), care must be taken to ensure that the blood types and tissue types are compatible (see Chapters 19 and 22 of the text). If they are not, the new lymphocytes may attack the patient's tissues, with potentially fatal results. Best results are obtained when the donor is a close relative. In an **autologous marrow transplant,** bone marrow is removed from the patient, cleansed of cancer cells, and reintroduced after radiation or chemotherapy treatment. Although this method produces fewer complications, the preparation and cleansing of the marrow are technically difficult and time-consuming.

Bone marrow transplants are also performed to treat patients whose bone marrow has been destroyed by toxic chemicals or radiation. For example, heterologous transplants were used successfully in the former USSR to treat survivors of the Chernobyl nuclear reactor accident in 1986.

▢ A Closer Look: The Clotting System
FAP p. 666

Figure A-32a, the clotting cascade, presents a fresh perspective on the interactions of the intrinsic and extrinsic pathways and on the roles of the coagulation factors. Figure A-32b presents the fibrinolytic system, which is activated as thrombin forms. This linked pathway provides a mechanism for the gradual removal of a blood clot during tissue repair.

📋 Testing the Clotting System
FAP p. 667

Several clinical tests check the efficiency of the clotting system.

Bleeding Time. This test measures the time it takes a small skin wound to seal. There are variations on this procedure, with normal values ranging from 1 to 9 minutes. The nonprescription drug *aspirin* prolongs bleeding time by affecting platelet function and suppressing the extrinsic pathway.

Coagulation Time. In this test, a sample of whole blood stands under controlled conditions until a visible clot has formed. Normal values range from 3 to 15 minutes. The test has several poten-

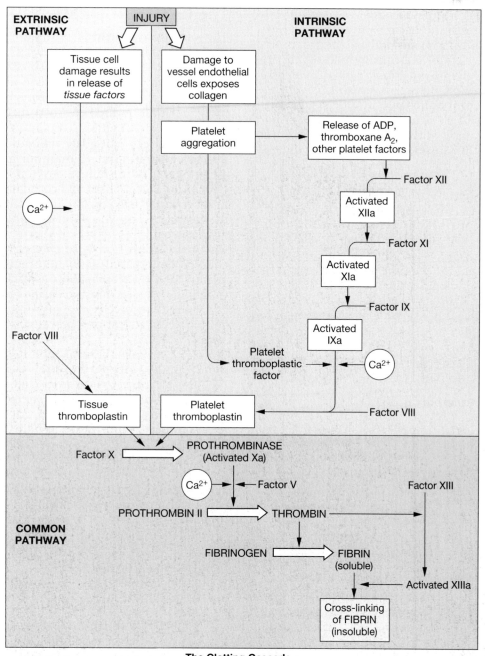

(a)

The Clotting Cascade

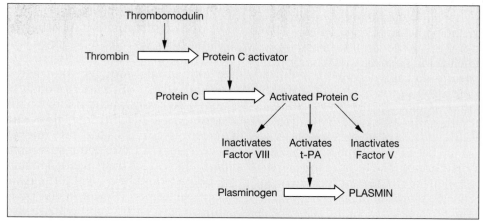

(b)

The Fibrinolytic System

Figure A-32   The Clotting System

20

tial sources of error and so is not very accurate. It is nevertheless of value because it is the simplest test that can be performed on a blood sample. More-sophisticated tests begin by adding citrate ions to the sample. Citrate ions tie up the calcium ions in the plasma and prevent premature clotting.

Partial Thromboplastin Time (PTT). In this test, a plasma sample is mixed with chemicals that mimic the effects of activated platelets. Calcium ions are then introduced, and the clotting time is recorded. Clotting normally occurs in 35–50 seconds if the enzymes and clotting factors of the intrinsic pathway are present in normal concentrations. The PTT is prolonged by heparin therapy.

Plasma Prothrombin Time (Prothrombin Time, PT). This test checks the performance of the extrinsic pathway. The procedure is similar to that in the PTT test, but the clotting process is triggered by exposure to a combination of tissue thromboplastin and calcium ions. Clotting normally occurs in 12–14 seconds. The PT is prolonged by coumadin therapy.

☤ Infection and Inflammation of the Heart FAP *p. 675*

Many different microorganisms may infect heart tissue, leading to serious cardiac abnormalities. **Carditis** (kar-DĪ-tis) is a general term for inflammation of the heart. Clinical conditions resulting from cardiac infection are usually identified by the primary site of infection. For example, those infections that affect the endocardium produce symptoms of **endocarditis.** Endocarditis primarily damages the chordae tendineae and heart valves, and the mortality rate may reach 21–35 percent. The most severe complications result from the formation of blood clots on the damaged surfaces. These clots subsequently break free, entering the circulation as drifting emboli (see *p. 675* of the text) that may cause strokes, heart attacks, or kidney failure. Destruction of heart valves by infection may lead to valve leakage, heart failure, and death.

Bacteria, viruses, protozoans, and fungal pathogens that attack the myocardium produce **myocarditis.** The microorganisms implicated include those responsible for many of the conditions discussed in earlier sections, such as diphtheria, syphilis, polio, and malaria. The membranes of infected heart muscle cells become facilitated, and the heart rate rises dramatically. Over time, abnormal contractions may appear and the heart muscle weakens; these problems may eventually prove fatal.

If the pericardium becomes inflamed or infected, fluid may accumulate around the heart (*cardiac tamponade*) or the elasticity of the pericardium may be reduced (*constrictive pericarditis*). In both, expansion of the heart is restricted, and cardiac output is reduced. Treatment involves draining the excess fluid or cutting a window in the pericardial sac.

The Cardiomyopathies FAP *p. 680*

The **cardiomyopathies** (kar-dē-ō-mī-OP-a-thēz) include an assortment of diseases with a common symptom: the progressive, irreversible degeneration of the myocardium. Cardiac muscle cells are damaged and replaced by fibrous tissue, and the muscular walls of the heart become thin and weak. As muscle tone declines, the ventricular chambers greatly enlarge. When the remaining cells cannot develop enough force to maintain cardiac output, symptoms of heart failure develop.

Chronic alcoholism and coronary artery disease are probably the most common causes of cardiomyopathy in the United States. Infectious agents, including viruses, bacteria, fungi, and protozoans, can also produce cardiomyopathies. Diseases affecting neuromuscular performance, such as muscular dystrophy (discussed elsewhere in this manual), can also damage cardiac muscle cells, as can starvation or chronic variations in the extracellular concentrations of calcium or potassium ions.

There are also several inherited forms of cardiomyopathy. **Hypertrophic cardiomyopathy** *(HCM)* is an inherited disorder that makes the wall of the left ventricle thicken to the point at which it has difficulty pumping blood. Most people with HCM do not become aware of it until relatively late in life. However, HCM can also cause a fatal arrhythmia; it has been implicated in the sudden deaths of several young athletes, and Defense Secretary Les Aspin was hospitalized for this condition in 1993. Implantation of an electronic cardiac pacemaker has proved to be beneficial in controlling these arrhythmias.

Finally, there are a significant number of cases of *idiopathic cardiomyopathy,* a term used when the primary cause cannot be determined.

HEART TRANSPLANTS AND ASSIST DEVICES

Individuals suffering from severe cardiomyopathies may be considered as candidates for heart transplants. This surgery involves the removal of the weakened heart and the replacement with a heart taken from a suitable donor. To survive the surgery, the recipient must be in otherwise satisfactory health. Because the number of suitable donors is limited, the available hearts are generally assigned to individuals younger than age 50. Out of the 8000–10,000 U.S. patients each year who suffer from potentially fatal cardiomyopathies, only about 1000 receive heart transplants. There is an 80–85 percent 1-year survival rate and a 50–70 percent 5-year survival rate after successful transplantation. This rate is quite good, considering that these patients would have died if the transplant had not been performed. However, the procedure remains controversial due to the high cost involved. As health care dollars become managed more closely, society will need to decide whether the probable life extension is worth the expense. Needless to say, this decision will not be an easy one to make.

Many individuals with cardiomyopathy who are initially selected for heart transplant surgery succumb to the disease before a suitable donor becomes available. For this reason, there continues to be considerable interest in the development of an artificial heart. One model, the *Jarvik-7*, had limited clinical use in the 1980s. Attempts to implant it on a permanent basis were unsuccessful, primarily because blood clots formed on the mechanical valves. When these clots broke free, they became drifting emboli that plugged peripheral vessels, producing strokes, kidney failure, and other complications. In 1989, the U.S. government prohibited further experimental use of the Jarvik-7 as a permanent heart implant. Modified versions of this unit and others now under development may still be used to maintain transplant candidates who are awaiting the arrival of a donor organ. These are called *left ventricular assist devices* (LVAD). As the name implies, these devices assist, rather than replace, the damaged heart.

Another interesting approach, called *dynamic cardiomyoplasty*, involves the use of skeletal muscle tissue to apply permanent patches to injured hearts or to build small accessory pumps. For example, one procedure creates a biological LVAD by freeing a portion of the latissimus dorsi muscle from the side and placing this flap, with its circulation intact, into the thoracic cavity. There it is folded to form a sling around the heart, and an electronic pacemaker is used to stimulate its contraction. Each time it contracts, the sling squeezes the heart and helps push blood into the major arteries. These methods are less stressful than heart transplants, because (1) they leave the damaged heart in place, and (2) the transplanted tissue is taken from the same individual, so it will not be attacked by the immune system. Preliminary results have been encouraging, and nationwide clinical trials are now underway. However, until the data are analyzed, the use of skeletal muscle patches will remain an experimental concept rather than a recognized treatment for cardiomyopathy.

In 1996, attention focused on a Brazilian surgeon who has developed a surgical procedure to improve cardiac function in patients with cardiomyopathy. The surgeon removes a portion of the weakened left ventricle. The ventricular muscle that remains is sewn together, forming a smaller chamber. In part because the muscle cells in the dilated heart were excessively stretched, the smaller, remodeled ventricle pumps blood more efficiently. Heart surgeons in the United States have confirmed the effectiveness of this therapy, which may reduce the demand for heart transplants in the years to come.

An experimental approach, which has yet to be tried with human patients, involves the insertion of fetal heart muscle cells in a damaged adult heart.

The fetal cells appear to adapt to their surroundings and differentiate into functional contractile cells.

RHD and Valvular Stenosis

FAP p. 682

Rheumatic (roo-MA-tik) **fever** is an inflammatory condition that may develop after infection by streptococcal bacteria. Rheumatic fever most commonly affects children of ages 5–15; symptoms include high fever, joint pain and stiffness, and a distinctive full-body rash. Obvious symptoms generally persist for less than 6 weeks, although severe cases may linger for 6 months or more. The longer the duration of the inflammation, the more likely it is that carditis will develop. The carditis that does develop in 50–60 percent of patients typically escapes detection, and scar tissue gradually forms in the myocardium and the heart valves. Valve condition deteriorates over time, and valve problems serious enough to affect cardiac function may not appear until 10–20 years after the initial infection.

During the interim, the affected valves become thickened and may also calcify to some degree. This thickening narrows the opening guarded by the valves, producing a condition called **valvular stenosis** (ste-NŌ-sis; *stenos*, narrow). The resulting clinical disorder is known as **rheumatic heart disease (RHD).** The thickened cusps stiffen in a partially closed position, but the valves do not completely block the circulation, because the edges of the cusps are rough and irregular. Regurgitation may occur, and much of the blood pumped out of the heart may flow back in. The abnormal valves are also much more susceptible to bacterial infection, a type of *endocarditis*.

Mitral stenosis and **aortic stenosis** are the most common forms of valvular heart disease. About 40 percent of patients with RHD develop mitral stenosis, and two-thirds of them are women. The reason for the correlation between female gender and mitral stenosis is unknown. In mitral stenosis, blood enters the left ventricle at a slower than normal rate; when the ventricle contracts, blood flows back into the left atrium as well as into the aortic trunk. As a result, the left ventricle has to work much harder to maintain adequate systemic circulation. The right and left ventricles discharge identical amounts of blood with each beat, so as the output of the left ventricle declines, blood "backs up" in the pulmonary circuit. Venous pressures then rise in the pulmonary circuit, and the right ventricle must develop greater pressures to force blood into the pulmonary trunk. In severe cases of mitral stenosis, the ventricular musculature is not up to the task. The heart weakens, and peripheral tissues begin to suffer from oxygen and nutrient deprivation. (We discuss this condition, called heart failure, in more detail in a later section.)

20

Symptoms of aortic stenosis develop in roughly 25 percent of patients with RHD; 80 percent of these individuals are males. Symptoms of aortic stenosis are initially less severe than those of mitral stenosis. Although the left ventricle enlarges and works harder, normal circulatory function can typically be maintained for years. Clinical problems develop only when the opening narrows enough to prevent adequate blood flow. Symptoms then resemble those of mitral stenosis.

One reasonably successful treatment for severe stenosis involves the replacement of the damaged valve with a prosthetic (artificial) valve. Figure A-33a shows a stenotic heart valve; two possible replacements are a valve from a pig (Figure A-33b) and a synthetic valve (Figure A-33c), one of a number of designs that have been employed. Pig valves do not require anticoagulant therapy but may wear out and begin leaking after roughly 10 years in service. The plastic or stainless steel components of the artificial valve are more durable but activate the clotting system of the recipient, leading to inflammation, clot formation, and other potential complications. Synthetic valve recipients must take anticoagulant drugs to prevent strokes and other disorders caused by embolus formation. Valve replacement operations are quite successful, with about 95 percent of the surgical patients surviving for 3 years or more and 70 percent surviving more than 5 years.

Diagnosing Abnormal Heartbeats

FAP pp. 692, 693

Damage to the conduction pathways caused by mechanical distortion, ischemia, infection, or inflammation can affect the normal rhythm of the heart. The resulting condition is called a **conduction deficit,** or **heart block.** Figure A-34a gives the electrocardiograph of a normal heart; heart blocks of varying severity are represented in Figure A-34b–d. In a **first-degree heart block** (Figure A-34b), the AV node and proximal portion of the AV bundle slow the passage of impulses that are heading for the ventricular myocardium. As a result, a pause appears between the atrial and ventricular contractions. Although a delay exists, the regular rhythm of the heart continues, and each atrial beat is followed by a ventricular contraction.

If the delay lasts long enough, the nodal cells will still be repolarizing from the previous beat when the next impulse arrives from the pacemaker. The arriving impulse will then be ignored, the ventricles will not be stimulated, and the normal "atria–ventricles, atria–ventricles" pattern will disappear. This condition is a **second-degree heart block** (Figure A-34c). A mild second-degree block may produce only an occasional skipped beat, but with more substantial delays the ventricles will follow every second atrial beat. The resulting pattern of "atria, atria–ventricles, atria, atria–ventricles" is

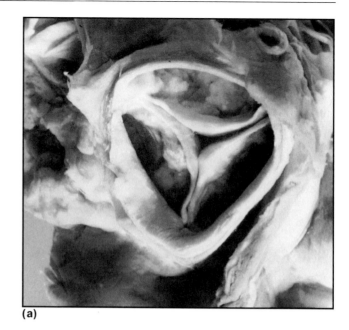

(a)

(b)

(c)

Figure A-33 Artificial Heart Valves.
(a) A stenotic semilunar valve; note the irregular, stiff cusps. (b) Intact Bioprosthetic™ heart valve, which uses the valve from a pig's heart. (c) Medtronic Hall™ prosthetic heart valve.

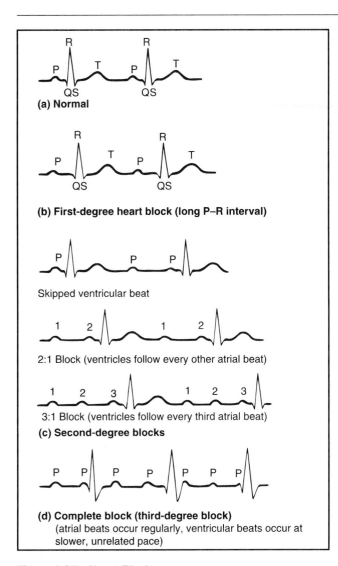

(a) Normal

(b) First-degree heart block (long P–R interval)

Skipped ventricular beat

2:1 Block (ventricles follow every other atrial beat)

3:1 Block (ventricles follow every third atrial beat)

(c) Second-degree blocks

(d) Complete block (third-degree block)
(atrial beats occur regularly, ventricular beats occur at slower, unrelated pace)

Figure A-34   Heart Blocks

known as a **two-to-one (2:1) block.** Three-to-one or even four-to-one blocks are also encountered.

In a **third-degree heart block,** or **complete heart block,** the conducting pathway stops functioning (Figure A-34d). The atria and ventricles continue to beat, but their activities are no longer synchronized. The atria follow the pace set by the SA node, beating 70–80 times per minute, and the ventricles follow the commands of the AV node, beating at a rate of 40–60 beats per minute. A temporary third-degree block can be induced by stimulating the vagus nerve. In addition to slowing the rate of impulse generation by the SA node, such stimulation inhibits the AV nodal cells to the point that they cannot respond to normal stimulation. Comments such as "my heart stopped" or "my heart skipped a beat" generally refer to this phenomenon. The pause typically lasts just a few seconds. Longer delays end when a conducting cell, normally one of the Purkinje fibers, depolarizes to threshold. This phenomenon is called **ventricular escape,** because the ventricles are escaping from the control of the

SA node. Ventricular escape can be a lifesaving event if the conduction system is damaged. Even without instructions from the SA or AV nodes, the ventricles will continue to pump blood at a slow but steady rate.

Tachycardia and Fibrillation

Additional important examples of arrhythmias are shown in Figure A-35. **Premature atrial contractions** *(PACs),* indicated in Figure A-35b, often occur in healthy individuals. In a PAC, the normal atrial rhythm is momentarily interrupted by a "surprise" atrial contraction. Stress, caffeine, and various drugs may increase the frequency of PAC incidence, presumably by increasing the permeabilities of the SA pacemakers. The impulse spreads along the conduction pathway, and a normal ventricular contraction follows the atrial beat.

In **paroxysmal atrial tachycardia** (par-ok-SIZ-mal), or **PAT** (Figure A-35c), a premature atrial contraction triggers a flurry of atrial activity. The ventricles are still able to keep pace, and the heart rate jumps to about 180 beats per minute. In **atrial flutter,** the atria contract in a coordinated manner, but the contractions occur very frequently. During a bout of **atrial fibrillation** (fi-bri-LĀ-shun), the impulses move over the atrial surface at rates of perhaps 500 beats per minute (Figure A-35d). The atrial wall quivers instead of producing an organized contraction. The ventricular rate in atrial flutter or atrial fibrillation cannot follow the atrial rate and may remain within normal limits. Despite the fact that the atria are now essentially nonfunctional, the condition may go unnoticed, especially in older individuals who lead sedentary lives. In chronic atrial fibrillation, blood clots may form by the atrial walls. Clotting promotes the formation of emboli and increases the risks of a stroke. As a result, most people diagnosed with this condition are placed on anticoagulant therapy. PACs, PAT, atrial flutter, and even atrial fibrillation are not considered very dangerous unless they are prolonged or associated with some more serious indications of cardiac damage, such as coronary artery disease or valve problems.

In contrast, ventricular arrhythmias can be serious and even fatal. Because the conduction system functions in one direction only, a ventricular arrhythmia is not linked to atrial activities. **Premature ventricular contractions** (PVCs) occur when a Purkinje cell or ventricular myocardial cell depolarizes to threshold and triggers a premature contraction (Figure A-35e). The cell responsible is called an *ectopic pacemaker.* The frequency of PVCs can be increased by exposure to epinephrine or other stimulatory drugs or to ionic changes that depolarize cardiac muscle cell membranes. Similar factors may be responsible for periods of **ventricular tachycardia,** also known as **VT,** or *V-tach* (Figure A-35f).

20

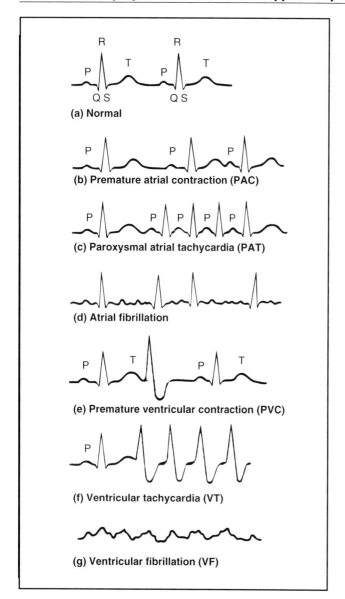

Figure A-35   Cardiac Arrhythmias

administered. The electrical stimulus depolarizes the entire myocardium simultaneously. With luck, after repolarization the SA node will be the first area of the heart to reach threshold. Thus the primary goal of defibrillation is not just to stop the fibrillation but to give the ventricles a chance to respond to normal SA commands.

There are several approaches to treating arrhythmias. Medications can be used to slow down rapid heart rates, or the abnormal portions of the conducting system can be destroyed. Pacemakers are used to accelerate slow heart rates. Implantable pacemakers that are able to sense ventricular fibrillation and deliver an immediate defibrillating shock have been successful in preventing sudden death in patients with previous episodes of ventricular tachycardia and ventricular fibrillation.

Monitoring the Heart   FAP *p. 693*

Many different techniques can be used to examine the structure and performance of the heart. No single diagnostic procedure can provide the complete picture, so the tests used will vary with the suspected nature of the problem. A standard chest X-ray will show the basic size, shape, and orientation of the heart. Additional details require more specialized procedures to enhance the clarity of the images.

Coronary arteriography (ar-tē-rē-OG-ra-fē) is often used to look for abnormalities in the coronary circulation. In this procedure, a catheter is inserted into a major artery in the arm or leg and then maneuvered back along the arterial passageways until its tip reaches the heart. A radiopaque dye can then be released at the openings of the coronary arteries, and its distribution can be followed in a series of high-speed X-rays. The images obtained are called **coronary angiograms** (Figure A-36a). For direct analyses of cardiac performance and the collection of blood samples, a catheter may be introduced into the heart itself. The instrument can enter via the aorta, as we described earlier, or from the venous system by way of the inferior vena cava.

Because the heart is constantly moving, ordinary ultrasound, computerized tomography (CT), and magnetic resonance imaging (MRI) scans create blurred images. Special instruments and computers, however, that can generate images at high speed can be used with these techniques to develop three-dimensional still or moving pictures of the heart as it beats (Figure A-36b–d). These procedures produce dramatic images, but the cost and complexity of the equipment have so far limited their use to major research institutions. Although PET scans can be used to diagnose disorders of coronary circulation, cost factors are also responsible for limiting the clinical use of this technology.

Ultrasound analysis, called **echocardiography** (ek-ō-kar-dē-OG-ra-fē) (Figure A-36b), provides

PVCs and VT often precede the most serious arrhythmia, **ventricular fibrillation** (VF) (Figure A-35g). The resulting condition, also known as **cardiac arrest,** is rapidly fatal, because the heart stops pumping blood. During ventricular fibrillation, the cardiac muscle cells are overly sensitive to stimulation, and the impulses are traveling from cell to cell around and around the ventricular walls. A normal rhythm cannot become established, because the ventricular muscle cells are stimulating one another at such a rapid rate. The problem is exaggerated by a sustained rise in free intracellular calcium ion concentrations, due to massive stimulation of alpha and beta receptors following sympathetic activation.

A **defibrillator** is a device that attempts to eliminate ventricular fibrillation and restore normal cardiac rhythm. Two electrodes are placed in contact with the chest, and a powerful electrical shock is

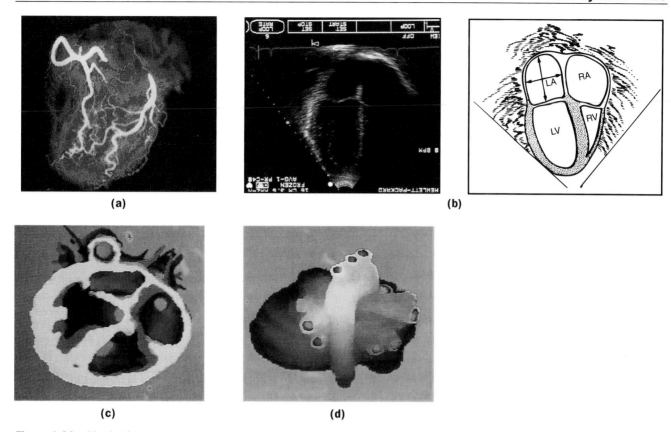

Figure A-36 Monitoring the Heart.
(a) A coronary angiogram. (b) An echocardiogram (left) with interpretive drawing (right). (c) Three-dimensional CT scan of a frontal section and (d) of a posterior/superior view of the heart.

images that lack the clarity of CT or MRI scans, but the equipment is relatively inexpensive and portable. Recent advances in data processing have made the images suitable for following details of cardiac contractions, including valve function and blood flow dynamics. Echocardiography is now an important diagnostic tool.

⚕ Problems with Pacemaker Function
FAP *p. 701*

Symptoms of severe bradycardia (below 50 beats per minute) include weakness, fatigue, fainting, and confusion. Drug therapies are seldom helpful, but artificial pacemakers can be used with considerable success. Wires run to the atria, the ventricles, or both, depending on the nature of the problem, and the unit delivers small electrical pulses to stimulate the myocardium. Internal pacemakers are surgically implanted, batteries and all. These units last 7–8 years or more before another operation is required to change the battery. External pacemakers are used for temporary emergencies, such as immediately after cardiac surgery. Only the wires are implanted, and an external control box is worn on a belt.

More than 50,000 artificial pacemakers are in use at present. The simplest provide constant stim-

ulation to the ventricles at rates of 70–80 per minute. More sophisticated pacemakers vary their rates to adjust to changing circulatory demands, as during exercise. Others are able to monitor cardiac activity and respond whenever the heart begins to function abnormally.

Tachycardia, usually defined as a heart rate of over 100 beats per minute, increases the work load on the heart. Cardiac performance suffers at very high heart rates, because the ventricles do not have enough time to refill with blood before the next contraction occurs. Chronic or acute incidents of tachycardia may be controlled by drugs that affect the permeability of pacemaker membranes or block the effects of sympathetic stimulation.

⚕ Aneurysms
FAP *p. 714*

An **aneurysm** (AN-ū-rizm) is a bulge in the weakened wall of a blood vessel, generally an artery. This bulge resembles a bubble in the wall of a tire. Like a bad tire, the affected artery may suffer a catastrophic blowout. The most dangerous aneurysms are those involving arteries of the brain, where they cause strokes, and of the aorta, where a blowout will cause fatal bleeding in a matter of minutes.

Aneurysms are normally caused by chronic high blood pressure, although any trauma or infec-

20

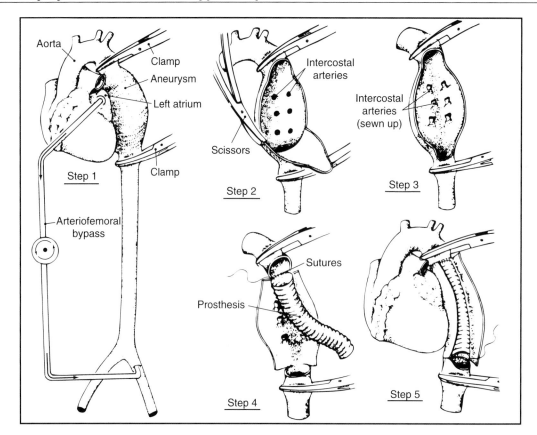

Figure A-37   Repair of an Aneurysm

tion that weakens vessel walls can lead to an aneurysm. In addition, at least some aortic aneurysms have been linked to inherited disorders, such as *Marfan's syndrome,* that have weakened connective tissues in vessel walls. It is not known whether other genetic factors are involved in the development of other aneurysms.

Most aneurysms form gradually as vessel walls become less elastic. When a weak point develops, the arterial pressures distort the wall, creating an aneurysm. Unfortunately, because many aneurysms are painless, they are likely to go undetected.

When aneurysms are detected by ultrasound or other scanning procedures, the risk of rupture can sometimes be estimated on the basis of their size. For example, an aortic aneurysm larger than 6 cm has a 50:50 chance of rupturing within 10 years. Treatment commonly begins with the reduction of blood pressure by means of vasodilators or beta-blockers (see p. 89). An aneurysm in an accessible area, such as the abdomen, may be surgically removed and the vessel repaired. Figure A-37 shows a large aortic aneurysm and the steps involved in its surgical repair with a synthetic patch.

✝ Problems with Venous Valve Function
FAP *p. 718*

Chapter 4 of the text notes that one of the consequences of aging is a loss of elasticity and resilience in connective tissues throughout the body. Blood vessels are no exception; with age, the walls of veins begin to sag. This change generally affects the superficial veins of the legs first, because at these locations gravity opposes blood flow. The situation is aggravated by a lack of exercise and by an occupation requiring long hours of standing or sitting. Because there is no muscular activity to help keep the blood moving, venous blood pools on the proximal (heart) side of each valve. As the venous walls are distorted, the valves become less effective, and gravity can then pull blood back toward the capillaries. This pulling further impedes normal blood flow, and the veins become grossly distended. The sagging, swollen vessels are called **varicose** (VAR-i-kōs) **veins.** Varicose veins are relatively harmless but unsightly. Surgical procedures are sometimes used to remove or constrict the offending vessels.

Varicose veins are not limited to the extremities. Another common site involves a network of veins in the walls of the anus. Pressures within the abdominopelvic cavity rise dramatically when the abdominal muscles are tensed. Straining to force defecation can force blood into these veins, and repeated incidents leave them permanently distended. These distended veins, known as **hemorrhoids** (HEM-o-roydz), can be uncomfortable and in severe cases extremely painful. Hemorrhoids are often associated with pregnancy, because of changes in circulation and abdominal pressures. Minor cases can be treated by the topical application of drugs that promote the contraction of smooth muscles within the venous walls. More severe cases may require the surgical removal or destruction of the distended veins.

Hypertension and Hypotension

FAP *p. 723*

Elevated blood pressure is called **primary hypertension,** or essential hypertension, if no obvious cause can be determined. Known risk factors include a hereditary history of hypertension, gender (males are at higher risk), high plasma cholesterol, obesity, chronic stresses, and cigarette smoking. **Secondary hypertension** appears as the result of abnormal hormonal production outside the cardiovascular system. For example, a condition resulting in excessive production of antidiuretic hormone (ADH), renin, aldosterone, or epinephrine will probably produce hypertension, and many forms of kidney disease will lead to hypertension caused by fluid retention or excessive renin production.

Hypertension significantly increases the work load on the heart, and the left ventricle gradually enlarges. More muscle mass requires a greater oxygen demand. When the coronary circulation cannot keep pace, symptoms of coronary ischemia appear.

Increased arterial pressures also place a physical stress on the walls of blood vessels throughout the body. This stress promotes or accelerates the development of arteriosclerosis and increases the risks of aneurysms, heart attacks, and strokes. Vessels supplying the retinas of the eyes are typically affected, and hemorrhages and associated circulatory changes can produce disturbances in vision. Because a routine physical exam includes the examination of these vessels, retinal changes may provide the first evidence that hypertension is affecting peripheral circulation.

One of the most difficult aspects of hypertension is that in most cases there are no obvious symptoms. As a result, clinical problems do not appear until the condition has reached the crisis stage. There is therefore considerable interest in the early detection and prompt treatment of hypertension.

Treatment consists of a combination of lifestyle changes and physiological therapies. Quitting smoking, getting regular exercise, and restricting dietary intake of salt, fats, and calories will improve peripheral circulation, prevent increases in blood volume and total body weight, and reduce plasma cholesterol levels. These strategies may be sufficient to control hypertension if it has been detected before significant cardiovascular damage has occurred. Most therapies involve antihypertensive drugs, such as calcium channel blockers, beta-blockers, diuretics, and vasodilators, singly or in combination. Beta-blockers reduce the effects of sympathetic stimulation on the heart, and the unopposed parasympathetic system lowers the resting heart rate and blood pressure. Diuretics promote the loss of water and sodium ions at the kidneys, lowering blood volume, and vasodilators further reduce blood pressure. A new class of antihypertensive drugs lowers blood pressure by preventing the conversion of angiotensin I to angiotensin II. These **angiotensin-converting enzyme (ACE) inhibitors,** such as *captopril,* are being used to treat chronic hypertension and congestive heart failure.

In hypotension, blood pressure declines and peripheral systems begin to suffer from oxygen and nutrient deprivation. One clinically important form of hypotension can develop after antihypertensive drugs have been administered. Problems may appear when the individual changes position from lying down to sitting or from sitting to standing. Each time you sit or stand, blood pressure in your carotid sinus drops, for your heart must suddenly counteract gravity to push blood up to your brain. The fall in pressure triggers the carotid reflex, and blood pressure returns to normal. But if the carotid response is prevented by beta-blockers or other drugs, blood pressure at the brain may fall so low that the individual becomes weak, dizzy, disoriented, or unconscious. This condition is known as **orthostatic hypotension** (or-thō-STAT-ik; *orthos,* straight + *statikos,* causing to stand), or simply **orthostasis** (or-thō-STĀ-sis). You may have experienced a brief episode of orthostasis when you stood suddenly after reclining for an extended period. The carotid reflex typically slows with age, so older people must sit and stand more carefully than they used to in order to avoid the effects of orthostatic hypotension.

Ⓡ The Causes and Treatment of Cerebrovascular Disease   FAP *p. 732*

Cerebrovascular disease was introduced in Chapter 14 of the text (see *p. 452*); in this section we consider additional details related to the concepts in that chapter. Most symptoms of cerebrovascular disease appear when atherosclerosis reduces the circulatory supply to the brain. If the circulation to a portion of the brain is completely shut off, a *cerebrovascular accident (CVA),* or *stroke,* occurs. The most common causes of strokes include **cerebral thrombosis** (clot formation at a plaque), **cerebral embolism** (drifting blood clots, fatty masses, or air bubbles), and **cerebral hemorrhages** (rupture of a blood vessel, commonly following the formation of an aneurysm). The observed symptoms and their severity vary with the vessel involved and the

21

location of the blockage. (Examples are included in the discussion on *p. 744* of the text.)

If the circulatory blockage disappears in a matter of minutes, the effects are temporary, and the condition is called a **transient ischemic attack (TIA).** TIAs typically indicate that cerebrovascular disease exists, so preventive measures can be taken to forestall more serious incidents. For example, taking aspirin each day slows blood clot formation in patients who experience TIAs and thereby reduces the risks of cerebral thrombosis and cerebral embolism.

If the blockage persists for a longer period, neurons die and the area degenerates. Stroke symptoms are initially exaggerated by the swelling and distortion of the injured neural tissues; if the individual survives, in many cases, brain function gradually improves. The management and treatment of strokes remain controversial. Surgical removal of the offending clot or blood mass may be attempted, but the results vary. Recent progress in the emergency treatment of cerebral thromboses and cerebral embolisms has involved the administration of clot-dissolving enzymes such as *tissue plasminogen activator* (t-PA; now sold as *Alteplase®*). The best results are obtained if the enzymes are administered within an hour after the stroke, although they may still be of use up to 24 hours after. Subsequent treatment involves anticoagulant therapy, typically with heparin (for one to two weeks) followed by coumadin (for up to a year) to prevent further clot formation. (These fibrinolytic and anticoagulant drugs were introduced in Chapter 19 of the text; see FAP *p. 668*). A more complicated surgical procedure involves the insertion of a transplanted piece of a blood vessel that routes blood around the damaged area. None of these treatments is as successful as preventive surgery, in which plaques are removed before a stroke. It should also be noted that the very best solution is to prevent or restrict plaque formation by controlling the risk factors involved.

℞ Heart Failure   FAP *pp. 698, 732*

A condition of heart failure exists when the cardiac output is insufficient to meet the circulatory demand. The initial symptoms of heart failure depend on whether the problem is restricted to the left ventricle or the right ventricle, or involves both. However, over time these differences are eliminated; for example, the major cause of right ventricular failure is left ventricular failure. Figure A-38 provides a simplified flow chart of heart failure and indicates potential therapies.

Suppose that the left ventricle cannot maintain normal cardiac output due to damage to the ventricular muscle (see the discussion of myocardial infarctions on *p. 698* of the text) or high arterial pressures (hypertension, *p. 732*). In effect, the left ventricle can no longer keep pace with the right

ventricle, and blood backs up into the pulmonary circuit. This venous congestion is responsible for the term **congestive heart failure (CHF).** The right ventricle now works harder, elevating pulmonary arterial pressures and forcing blood through the lungs and into the weakened left ventricle.

At the capillaries of the lungs, arterial and venous pressures are now elevated. This elevated pressure pushes additional fluid out of the capillaries and into the interstitial fluids, most notably at the lungs. The fluid buildup and compression of the airways reduce the effectiveness of gas exchange, leading to shortness of breath, typically the first obvious sign of congestive failure. This fluid buildup begins at a pulmonary postcapillary pressure of about 25 mm Hg. At a capillary pressure of about 30 mm Hg, fluid not only enters the tissues of the lungs but crosses the alveolar walls and begins to fill the air spaces. This condition is called **pulmonary edema.**

Over time, the less muscular right ventricle may become unable to generate enough pressure to force blood through the pulmonary circuit. Venous congestion now occurs in the systemic circuit, and cardiac output declines further. When the reduction in systemic pressures lowers blood flow to the kidneys, renin and erythropoietin are released. This release in turn elevates blood volume as salt and water retention increase at the kidneys and red blood cell production accelerates. This rise in blood volume actually complicates the situation, as it tends to increase venous congestion and cause widespread edema.

The increased volume of blood in the venous system leads to a distension of the veins, making superficial veins more prominent. When the heart contracts, the rise in pressure at the right atrium produces a pressure pulse in the large veins. This venous pulse can be seen and palpated most easily over the right external jugular vein.

Treatment of congestive heart failure commonly includes the following:

- Restriction of salt intake. The expression "water follows salt" applies: When sodium and chloride ions are absorbed across the digestive tract lining, water is also absorbed by osmosis.

- Administration of drugs to promote fluid loss. These drugs, called **diuretics** (dī-ū-RET-iks; *diouretikos*, promoting urine), increase salt and water losses at the kidneys. (The mechanism is described in Chapter 27 of the text.)

- Extended bed rest, to enhance venous return to the heart, coupled with physical therapy to maintain good venous circulation.

- Administration of drugs that enhance cardiac output. These drugs may target the heart, the peripheral circulation, or some combination of the two. When the heart has been weakened, drugs related to digitalis, an extract from the

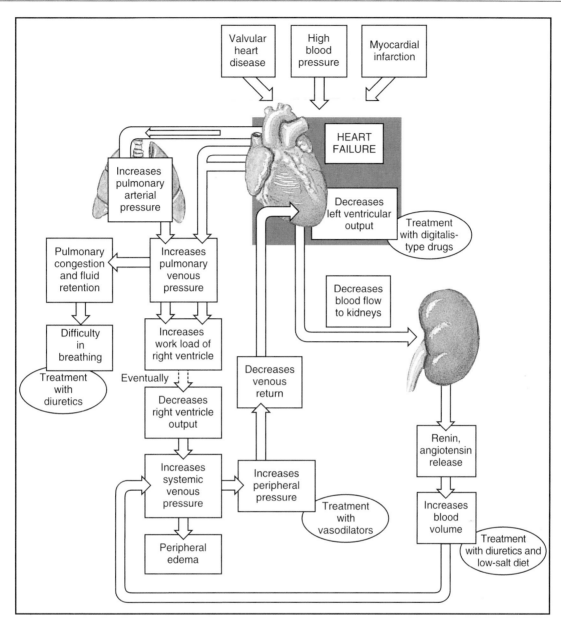

Figure A-38   Heart Failure

leaves of the foxglove plant, are commonly selected. *Digitoxin, digoxin,* and *ouabain* are examples. These compounds increase the force of cardiac muscle cell contractions. When high blood pressure is a factor, some type of vaso-dilator is given to lower peripheral resistance.

- Administration of drugs that reduce peripheral vascular resistance, such as *hydralazine* or *ACE* (angiotensin-converting enzyme) *inhibitors*. The drop in peripheral resistance reduces the work load of the left ventricle.

☤ Other Types of Shock   FAP *p. 737*

Although the text focuses on circulatory shock caused by low blood volume, shock can develop when the blood volume is normal. **Cardiogenic**

(kar-dē-ō-JEN-ik) **shock** occurs when the heart becomes unable to maintain a normal cardiac output. The most common cause is failure of the left ventricle as a result of a myocardial infarction. Between 5 and 10 percent of patients surviving a heart attack must be treated for cardiogenic shock. The use of thrombolytic drugs, such as t-PA, can be very effective in restoring coronary circulation and ventricular function, thereby relieving the periph-eral symptoms. Cardiogenic shock may also be the result of arrhythmias, valvular heart disease, advanced coronary artery disease, cardiomyopathy, or ventricular arrhythmias (see Chapters 20 and 21 of the text and elsewhere in this manual).

In **obstructive shock,** ventricular output is reduced because tissues or fluids are restricting the expansion and contraction of the heart. For example, fluid buildup in the pericardial cavity (cardiac tamponade; see "Infection and Inflammation of the Heart," p. 120) can compress the heart and limit ventricular filling.

Distributive shock is the result of a widespread, uncontrolled vasodilation. The vasodilation produces a dramatic fall in blood pressure that leads to a reduction in blood flow and the onset of shock. Three important examples are *neurogenic shock, septic shock*, and *anaphylactic shock.*

Neurogenic (noo-rō-JEN-ik) **shock** can be caused by general or spinal anesthesia and by trauma or inflammation of the brain stem. The underlying problem is damage to the vasomotor center or to the sympathetic tracts or nerves, leading to a loss of vasomotor tone.

Septic shock results from the massive release of endotoxins, poisons derived from the cell walls of bacteria during a systemic infection. These compounds cause a vasodilation of precapillary sphincters throughout the body, resulting in drops in peripheral resistance and blood pressure. Symp-toms of septic shock generally resemble those of other types of shock, but the skin is flushed, and the individual has a high fever. For this reason septic shock is also known as "warm shock." One form of septic shock, called **toxic shock syndrome (TSS),** results from an infection by the bacterium *Staphylococcus aureus.* This disease was unrecognized before 1978, when it appeared in a group of children. Since that time, there have been roughly 2500 cases in the United States, 95 percent of them affecting women. Although other sources of infection are possible, infection most often appears to occur during menstruation, and the chances of infection are increased with the use of superabsorbent tampons. (The brands involved were taken off the market, and the incidence has declined steadily since 1980.)

Extensive peripheral vasodilation also occurs in **anaphylactic** (an-a-fi-LAK-tik) **shock,** a dangerous allergic reaction. This type of shock, which can be life-threatening, is discussed in Chapter 22 of the text (see FAP *p. 805*).

CRITICAL-THINKING QUESTIONS

6-1. Vickie has a tumor that causes her to release excess amounts of ADH. Possible symptoms arising from this elevated hormone are
   a. decreased blood volume
   b. increased blood pressure
   c. peripheral vasoconstriction
   d. polycythemia
   e. all of the above

6-2. Barbara visits her physician for a routine physical examination. The physician, noting that Barbara is pale and complains of being tired and weak, orders blood tests. The results of the blood work follow:

RBC count: 3.5 million/mm^3

Hematocrit: 32

Hemoglobin: 10 g/dl

MCV: 70

MCH: 20 pg

MCHC: 28

WBC count: 8000/mm^3

Platelet count: 200,000/mm^3

Reticulocyte count: 1% of total erythrocytes

These results would indicate
   a. hemorrhagic anemia
   b. aplastic anemia
   c. iron deficiency anemia
   d. pernicious anemia
   e. macrocytic anemia

6-3. Catherine has just given birth to a baby girl. When the nurse takes the infant back to the nursery and tries to feed her, the baby becomes cyanotic. The episode passes, but when the infant is bathed, she becomes cyanotic again. Blood gas levels show that her arterial blood is only 60 percent saturated. Physical examination indicates that there are no structural deformities involving the respiratory or digestive system. Echocardiography shows a heart defect. What structures may be involved?

The Lymphatic System and Immunity

The lymphatic system consists of the fluid *lymph*, a network of *lymphatic vessels*, specialized cells called *lymphocytes*, and an array of *lymphoid tissues* and *lymphoid organs* scattered throughout the body. This system has two major functions: (1) to protect the body through the *immune response* and (2) to transport fluid from the interstitial fluid to the bloodstream.

The immune response produced by activated lymphocytes is responsible for the detection and destruction of foreign or toxic substances that may disrupt homeostasis. For example, viruses, bacteria, and tumor cells are usually recognized and eliminated by cells of the lymphatic system. Immunity is the specific resistance to disease, and all the cells and tissues involved with the production of immunity are considered to be part of an *immune system*. Whereas the lymphatic system is an anatomically distinct system, the immune system is a physiological system that includes the lymphatic system as well as components of the integumentary, cardiovascular, respiratory, digestive, and other systems.

The role of the lymphatic system in the recirculation of extracellular fluid was detailed in Chapter 21 of the text *(p. 725)*.

PHYSICAL EXAMINATION OF THE LYMPHATIC SYSTEM

Individuals with lymphatic system disorders may experience a variety of symptoms. the pattern observed depends on whether the problem affects the immune or the circulatory functions of the lymphatic system (Figure A-39a). Important symptoms and signs include the following:

- *Enlarged lymph nodes* are typically found during infection. They also develop in cancers of the lymphatic system, such as *lymphoma* (p. 133), or when primary tumors in other tissues have metastasized to regional lymph nodes. The status of regional lymph nodes can therefore be important in the diagnosis and treatment of many different cancers. The onset and duration of swelling; the size, texture, and mobility of the nodes; the number of affected nodes; and the degree of tenderness are all important in diagnosis. For example, nodes containing cancer cells tend to be large, locked in place, and non-

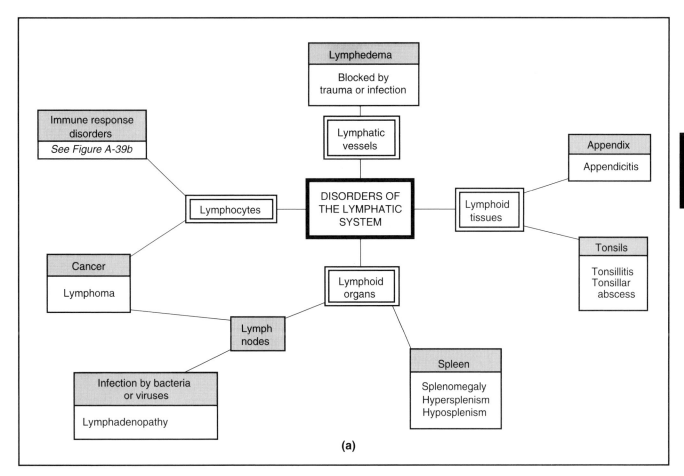

(a)

Figure A-39a   Disorders of the Lymphatic System

tender. On palpation, these nodes feel like dense masses rather than individual lymph nodes. In contrast, infected lymph nodes tend to be large, freely mobile, and very tender.

- *Lymphangitis* consists of *erythematous* (red) streaks on the skin that may develop with an inflammation of superficial lymph vessels. Lymphangitis commonly occurs in the lower limbs; the reddened streaks originate at an infection site. Before the linkage to the lymphatic system was known, this sign was called "blood poisoning."

- *Splenomegaly* (p.138) is an enlargement of the spleen that may result from acute infections such as *endocarditis* (p. 120) and *mononucleosis* (p. 138) or chronic infections such as *malaria* (p. 114) or *leukemia* (p. 118). The spleen can be examined through palpation or percussion (p. 7) to detect splenic enlargement. In percussion, an enlarged spleen produces a distinctive dull sound. The patient history may also reveal important clues. For example, an individual with an enlarged spleen may report a feeling of fullness after eating a small meal, probably because the enlarged spleen limits gastric expansion.

- *Weakness* and *fatigue* typically accompany immunodeficiency disorders (p. 133), *Hodgkins*

disease and other *lymphomas* (p. 136) and *mononucleosis* (p. 138).

- *Skin lesions* such as hives or urticaria (p. 147) can develop during allergic reactions. Immune responses to a variety of allergens, including animal hair, pollen, dust, medications, and some foods may cause such lesions.

- *Respiratory problems,* including rhinitis and wheezing, may accompany the allergic response to allergens such as pollen, hay, dust, and mildew. *Bronchospasms* are smooth muscle contractions that constrict the airways and make breathing difficult. Bronchospasms, which often accompany severe allergic or asthmatic attacks, are a response to the appearance of antigens within the respiratory passageways.

- *Recurrent infections* may occur for a variety of reasons. *Tonsillitis* and *adenoiditis* (p. 133) are common recurrent infections in children. More-serious chronic infections are common among persons with immunodeficiency disorders such as AIDS (p. 142) or *severe combined immunodeficiency disease* (SCID) (p. 30). When the immune response is inadequate, the individual cannot overcome even a minor infection. Infections of the respiratory system are very

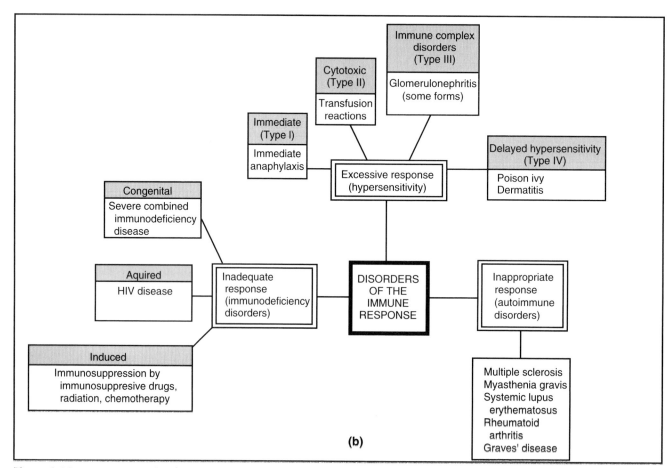

(b)

Figure A-39b   Disorders Affecting the Immune Response

common, and recurring gastrointestinal infections may produce chronic diarrhea. The pathogens involved may not infect persons with a normal immune response. Infections are also a problem for individuals who take medications that suppress the immune response. Examples of immunosuppressive drugs include anti-inflammatories such as the *corticosteroids (Prednisone®)* as well as more specialized drugs such as *FK-506* and *cyclosporin.*

When circulatory functions are impaired, the most common sign is *lymphedema,* a tissue swelling caused by the buildup of interstitial fluid. Lymphedema can result from trauma to a lymphatic vessel or from a lymphatic blockage due to a tumor or infections, including parasitic infection such as *filariasis* (FAP *p. 773*).

Diagnostic procedures and laboratory tests used to detect disorders of the lymphatic system are detailed in Tables A-21 and A-22.

DISORDERS OF THE LYMPHATIC SYSTEM

Disorders of the lymphatic system that affect the immune response can be sorted into three general categories, as diagrammed in Figure A-39b:

1. *Disorders resulting from an insufficient immune response.* This category includes conditions such as AIDS (p. 142) and SCID (p. 30). Individuals with depressed immune defenses may develop life-threatening diseases caused by microorganisms that are harmless to other individuals.

2. *Disorders resulting from an excessive immune response.* Conditions such as *allergies* (FAP *p. 867*) and *immune complex disorders* (p. 147) can result from an immune response that is out of proportion with the size of the stimulus.

3. *Disorders resulting from an inappropriate immune response. Autoimmune disorders* result when normal tissues are mistakenly attacked by T cells or the antibodies produced by activated B cells (FAP *p. 804*).

✝ Infected Lymphoid Nodules
FAP *p. 775*

Lymphoid nodules may be overwhelmed by a pathogenic invasion. The result is a localized infection accompanied by regional swelling and discomfort. An individual with bacterial **tonsillitis** has infected tonsils. Symptoms include a sore throat, high fever, and leukocytosis (an abnormally high white blood cell count). The affected tonsil (normally the pharyngeal) becomes swollen and inflamed, sometimes enlarging enough to partially block the entrance to the trachea. Breathing then becomes

difficult, and in severe cases, impossible. As the infection proceeds, abscesses develop within the tonsilar tissues. The bacteria may enter the bloodstream by passing through the lymphatic capillaries and vessels to the venous system.

In the early stages, antibiotics may control the infection, but once abscesses have formed, the best treatment involves surgical drainage of the abscesses. **Tonsillectomy,** the removal of the tonsil, was once highly recommended and commonly performed to prevent recurring tonsilar infections. The procedure does reduce the incidence and severity of subsequent infections, but questions have arisen concerning the overall cost to the individual. The tonsils are a first line of defense against bacterial invasion of the pharyngeal walls.

Appendicitis generally follows an erosion of the epithelial lining of the appendix. Several factors may be responsible for the initial ulceration—notably bacterial or viral pathogens. Bacteria that normally inhabit the lumen of the large intestine then cross the epithelium and enter the underlying tissues. Inflammation occurs, and the opening between the appendix and the rest of the intestinal tract may become constricted. Mucus secretion and pus formation accelerate, and the organ becomes increasingly distended. Eventually the swollen and inflamed appendix may rupture, or *perforate.* If it does, bacteria will be released into the warm, dark, moist confines of the abdominopelvic cavity, where they can cause a life-threatening peritonitis. The most effective treatment for appendicitis is the surgical removal of the organ, a procedure known as an **appendectomy.**

✝ Lymphomas
FAP *p. 778*

Lymphomas are malignant tumors consisting of cancerous lymphocytes or lymphocytic stem cells. Roughly 80,000 cases of lymphoma are diagnosed in the United States each year, and that number has steadily been increasing. There are many types of lymphoma. One form, called **Hodgkin's disease (HD),** accounts for roughly 40 percent of all lymphoma cases. Hodgkin's disease most commonly strikes individuals at ages 15–35 or those over age 50. The reason for this pattern of incidence is unknown. Although the cause of the disease is uncertain, an infectious agent (probably a virus) is suspected.

Other types of lymphoma are usually grouped together under the heading of **non-Hodgkin's lymphoma (NHL).** They are extremely diverse. More than 85 percent of NHL cases are associated with chromosomal abnormalities. They typically involve *translocations,* in which sections of chromosomes have been swapped from one chromosome to another. The shifting of genes from one chromosome to another interferes with the normal regulatory mechanisms, and the cells become cancerous. The nature of the cancer depends on which of the many

22

Table A-21   Representative Laboratory Tests for Disorders of the Lymphatic and Immune Systems

Laboratory Test	Normal Values	Significance of Abnormal Values	Notes
Complete blood count	See laboratory tests for blood disorders, Table A-19c.		
WBC count	Adults: 5000–10,000/mm^3	Increased in chronic and acute infections, tissue death (MI, burns), leukemia, parasitic diseases, and stress. Decreased in aplastic and pernicious anemias and systemic lupus erythematosus (SLE) and with some medications.	High risk value: >30,000/mm^3 or <2,500/mm^3
Differential WBC count			
	Neutrophils: 50–70%	Increased in acute infection, myelocytic leukemia, and stress. Decreased in aplastic and pernicious anemias, viral infections, and radiation treatment and with some medications.	
	Lymphocytes: 20–40%	Increased in chronic infections, lymphocytic leukemia, infectious mononucleosis, and viral infections. Decreased in radiation treatment, AIDS, and corticosteroid therapy.	
	Monocytes: 2–8%	Increased in chronic inflammation, viral infections, and tuberculosis. Decreased in aplastic anemia and corticosteroid therapy.	
	Eosinophils: 1–4%	Increased in allergies, parasitic infections, and some autoimmune disorders. Decreased in steroid drug therapy.	
	Basophils: 0.5–1%	Increased in inflammatory processes and during healing. Decreased in hypersensitivity reactions and corticosteroid therapy.	
Immunoglobulin electrophoresis (IgA, IgG, IgM)	Adults: IgA: 85–330 mg/dl IgM: 55–145 mg/dl IgG: 565–1765 mg/dl IgD and IgE values should be minimal	Increased levels of IgG occur with infections; IgA levels increase with chronic infections and autoimmune disorders; IgE increases with allergic reactions and skin sensitivities; IgM levels are high in several infectious diseases, including liver disease	Levels of immunoglobulins fall with liver disease, certain types of leukemia, and agammaglobulinemia
Antinuclear antibody	No antinuclear antibodies detected	Positive test occurs in up to 95% of patients diagnosed with systemic lupus erythematosus; false-positive can occur with rheumatoid arthritis (RA) and other autoimmune disorders.	

Table A-21 *(continued)*

Laboratory Test	Normal Values	Significance of Abnormal Values	Notes
Anti-DNA antibody test	Low levels or none	Positive test with high levels of antibodies occur in 40–80% of patients with SLE	
Total complement assay	Total complement: 41–90 hemolytic units C_3: 55–120 mg/dl C_4: 20–50 mg/dl	Total complement and C_3 are decreased in SLE and glomerulo-nephritis and increased in rheumatic fever, rheumatoid arthritis, and certain types of malignancies	
Rheumatoid factor test	Negative	Positive test indicates rheumatoid arthritis, but results may also be positive in SLE and myositis, other inflammatory conditions, and chronic infections	A low positive titer is found in roughly 20% of people over age 70
AIDS serology			
Enzyme-linked immunosorbent assay (ELISA)	Negative	Positive test indicates detection of antibodies against HIV	Test does not detect the the virus itself but the presence of antibodies that appear weeks after infection. Tests given in the early stages of infection yield a negative result; positive results will not develop for several months. HIV-positive status is assigned after 2 different tests are positive for the antibodies and the Western blot is positive.
Western blot	Negative	Positive test	Another test for HIV that monitors IgG and is more specific, because it detects antibodies to specific viral proteins
Coombs direct (blood-RBC)	Negative	Positive test (+1 to +4) results appear in erythroblastosis fetalis, transfusion reactions, and leukemias and during certain types of drug therapy	Coombs indirect test (serum) can be performed to determine the amount of antibodies free in circulation. It is useful to perform this test before transfusions to cross-match blood.
C-reactive protein (serum)	Not usually present	Levels rise after an inflammatory process	Used to measure the amount of damage after an acute attack in patients with rheumatoid arthritis, acute myocardial infarction, SLE, or bacterial infections

22

Table A-22   Representative Diagnostic Tests for Disorders of the Immune and Lymphatic Systems

Diagnostic Procedure	Method and Results	Representative Uses	Notes
Skin tests: Hypersentitivity response	Antigens are extracted and given in a sterile, diluted form; examples include animal dander, pollen, certain foods, medications that cause hypersensitivities (especially penicillin), and insect venom	Aids in diagnosis of cause of hypersensitivity	Useful in detecting specific antigens that may cause hypersensitivity reaction (allergy)
Prick test	A small amount of antigen is applied as a prick to the skin	Erythema, hardening and swelling around puncture area indicates a positive test; usually the area affected is measured and must be of minimal size to qualify for a positive test	Commonly the first attempt to determine sensitivity; less sensitive than the intradermal method
Intradermal test	Antigen is injected into the skin to form a 1–2 mm "bleb"	Positive test: Within 15 minutes, a reddened wheal is produced that is larger than 5 mm in diameter and larger than in the control group	More sensitive than the prick test
Patch test	A patch is impregnated with the antigen and applied to the skin surface	Positive test: The patch provokes an allergic response, usually over several hours to several days	Used to determine the immune system response or allergic reaction (especially useful in contact dermatitis)
Skin Tests: Exposure test	A positive test is usually the production of a wheal of a certain diameter around the injection site within a specified period of time	Diagnostic skin tests can be performed to identify prior exposure to fungal and other parasitic organisms; for example, the test for the organism that causes trichinosis can be performed and read in 15 minutes	
Tuberculin skin test	Injection of tuberculin protein into the skin	Red, hardened area >10 mm wide around injection site 48–72 hours later indicates positive test and antibodies to the organism that causes tuberculosis (infection may be active or dormant)	
Nuclear scan and CT scans of spleen	Scan images provided through energy from radiation emitted from radionuclides or through X-ray waves to reveal position, shape, and size of the spleen	To detect abscesses, tumors of the spleen, and an infarct of splenic tissue	Splenomegaly (enlargement of the spleen) may be the result of the leukemias, polycythemia vera, lymphoma, and Hodgkin's disease

Table A-22   *(continued)*

Diagnostic Procedure	Method and Results	Representative Uses	Notes
Biopsy of lymphoid tissue	Surgical excision of suspicious lymphoid tissue for pathological examination in the laboratory	Determination of potential malignancy and staging of cancers in progress	
Lymphangiography	Dye is injected into extremity and travels through the lymphatic vessels to nodes; X-rays are taken as the dye accumulates in the lymph nodes	Used to identify and determine the staging of Hodgkin's disease and other lymphomas and to detect the cause of lymphedema	See Scan 20, p. S-13.

lymphocyte cell lines are affected. A combination of inherited and environmental factors may be responsible for specific translocations. For example, one form, called **Burkitt's lymphoma,** develops only after genes from chromosome 8 have been translocated to chromosome 14. (There are at least three variations.) Burkitt's lymphoma normally affects male children in Africa and New Guinea. The affected children have been infected with the *Epstein–Barr virus (EBV)*. This highly variable virus is also responsible for infectious mononucleosis (detailed below), and it has been suggested as a possible cause of chronic fatigue syndrome.

The EBV infects B cells, but under normal circumstances the infected cells are destroyed by the immune system. EBV is widespread in the environment, and childhood exposure generally produces lasting immunity. Children who develop Burkitt's lymphoma may have a genetic susceptibility to EBV infection; in addition, the presence of another illness, such as malaria, may weaken their immune system enough that the lymphoma cells are ignored.

The first symptom usually associated with any lymphoma is a painless enlargement of lymph nodes. The involved nodes have a firm, rubbery texture. Because the nodes are pain-free, the condition is typically overlooked until it has progressed far enough for secondary symptoms to appear. For example, patients seeking help for recurrent fevers, night sweats, gastrointestinal or respiratory problems, or weight loss may be unaware of any underlying lymph node changes. In the late stages of the disease, symptoms can include liver or spleen enlargement, central nervous system dysfunction, pneumonia, a variety of skin conditions, and anemia.

In planning treatment, clinicians consider the histological structure of the nodes and the stage of the disease. In a biopsy, the node is described as *nodular* or *diffuse*. A nodular node retains a sem-

blance of normal structure, with follicles and germinal centers. In a diffuse node, the interior of the node has changed, and follicular structure has broken down. In general, the nodular lymphomas progress more slowly than the diffuse forms, which tend to be more aggressive. Conversely, the nodular lymphomas are more resistant to treatment and are more likely to recur even after remission has been achieved.

The most important factor influencing treatment selection is the stage of the disease. Table A-23 presents a simplified staging classification of lymphomas. When the condition is diagnosed early (stage I or II), localized therapies may be effective. For example, the cancerous node(s) may be surgically removed and the region(s) irradiated to kill residual cancer cells. Success rates are very high when a lymphoma is detected in an early stage. For Hodgkin's disease, localized radiation can produce remission that lasts 10 years or more in more than 90 percent of patients. The treatment of localized NHL is somewhat less effective. The 5-year remission rates average 60–80 percent for all types; success rates are higher for nodular forms than for diffuse forms.

Although these results are encouraging, few lymphomas are diagnosed in the early stages. For example, only 10–15 percent of NHL patients are diagnosed at stage I or II. For lymphomas at stages III and IV, most treatments involve chemotherapy. Combination chemotherapy, in which two or more drugs are administered simultaneously, is the most effective treatment. For Hodgkin's disease, a four-drug combination with the acronym **MOPP** (nitrogen **M**ustard, **O**ncovin [*vincristine*], **P**rednisone, and **P**rocarbazine) produces lasting remission in 80 percent of patients.

Bone marrow transplantation is a treatment option for acute, late-stage lymphoma. When suitable donor marrow is available, the patient receives whole-body irradiation, chemotherapy, or some

22

Table A-23   Cancer Staging in Lymphoma

Stage I. Involvement of a single node or region (or of a single extranodal site)

  Typical treatment: surgical removal and/or localized irradiation; in slowly progressing forms of non-Hodgkin's lymphoma, treatment may be postponed indefinitely.

Stage II. Involvement of nodes in two or more regions (or of an extranodal site and nodes in one or more regions) on the same side of the diaphragm.

  Typical treatment: surgical removal and localized irradiation that includes an extended area around the cancer site (the *extended field*).

Stage III. Involvement of lymph node regions on both sides of the diaphragm. This is a large category that is subdivided on the basis of the organs or regions involved. For example, in stage III, the spleen contains cancer cells.

  Typical treatment: combination chemotherapy, with or without radiation; radiation treatment may involve irradiating all thoracic and abdominal nodes plus the spleen *(total axial nodal irradiation,* or *TANI).*

Stage IV. Widespread involvement of extradonal tissues above and below the diaphragm; involvement of bone marrow.

  Treatment: highly variable, depending on the circumstances. Combination chemotheraphy is always used; it may be combined with whole-body irradiation. The "last resort" treatment involves massive chemotherapy followed by a bone marrow transplant.

combination of the two sufficient to kill tumor cells throughout the body. This treatment also destroys normal bone marrow cells. Donor bone marrow is then infused. Within 2 weeks, the donor cells colonize the bone marrow and begin producing red blood cells, granulocytes, monocytes, and lymphocytes.

Potential complications of this treatment include the risk of infection and bleeding while the donor marrow is becoming established. The immune cells of the donor marrow may also attack the tissues of the recipient, a response called **graft-versus-host disease (GVH).** For a patient with stage I or II lymphoma, without bone marrow involvement, bone marrow can be removed and stored (frozen) for over 10 years. If other treatment options fail or the patient comes out of remission at a later date, an autologous marrow transplant can be performed. This procedure eliminates both the need for finding a matched donor and the risk of GVH disease.

⚕ Disorders of the Spleen   FAP p. 780

The spleen responds like a lymph node to infection, inflammation, or invasion by cancer cells. The enlargement that follows is called **splenomegaly** (splen-ō-MEG-a-lē; *megas,* large), and splenic rupture may also occur under these conditions. One relatively common condition that causes splenomegaly is **mononucleosis.** This condition, also known as the "kissing disease," results from infection by the Epstein–Barr virus. In addition to splenic enlargement, symptoms of mononucleosis include fever, sore throat, widespread swelling

of lymph nodes, increased numbers of atypical lymphocytes in the blood, and the presence of circulating antibodies to the virus. The condition typically affects young adults (age 15 to 25) in the spring or fall. Treatment is symptomatic, as no drugs are effective against this virus. The most dangerous aspect of the disease is the risk of rupturing the enlarged spleen, which becomes fragile. Patients are therefore cautioned against heavy exercise and other activities that increase abdominal pressures. If the spleen does rupture, severe hemorrhaging may occur; death will follow unless transfusion and an immediate splenectomy are performed.

An individual whose spleen is missing or nonfunctional has **hyposplenism** (hī-pō-SPLĒN-ism). Hyposplenism usually does not pose a serious problem. Hyposplenic individuals, however, are more prone to some bacterial infections, including infection by *Streptococcus pneumoniae,* than are individuals with normal spleens, so immunization against *S. pneumoniae* is recommended. In **hypersplenism,** the spleen becomes overactive, and the increased phagocytic activities lead to anemia (low RBC count), leukopenia (low WBC count), and thrombocytopenia (low platelet count). Splenectomy is the only known cure for hypersplenism.

📖 A Closer Look: The Complement System   FAP p. 784

There are 11 proteins in the complement system (Figure A-40). The proteins C1q, C1r, and C1s circulate as a single complex designated as C1. The proteins C2–C9 are individual proteins.

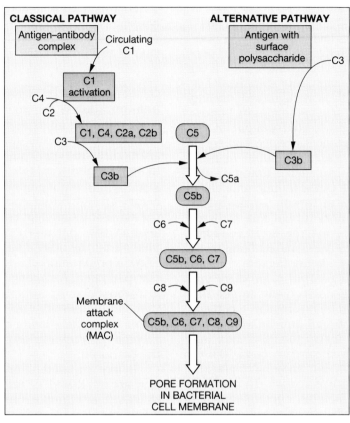

Figure A-40   The Complement Pathways

☤ Complications of Inflammation

FAP *p. 787*

When bacteria invade the dermis of the skin, the production of cellular debris and toxins reinforces and exaggerates the inflammation process. **Pus** is an accumulation of debris, fluid, dead and dying cells, and necrotic tissue components. Pus commonly forms at an infection site in the dermis; an **abscess** is an accumulation of pus in an enclosed tissue space. In the skin, an abscess can form as pus builds up inside the fibrin clot that surrounds the injury site. If the cellular defenses succeed in destroying the invaders, the pus will be either absorbed or surrounded by a fibrous capsule; this capsule is one type of **cyst.** (Cysts can also form in the absence of infection.)

Erysipelas (er-i-SIP-e-las; *erythros*, red + *pella*, skin) is a widespread inflammation of the dermis caused by bacterial infection. If the inflammation spreads into the subcutaneous layer and deeper tissues, the condition is called **cellulitis** (sel-ū-LĪ-tis). Erysipelas and cellulitis develop when bacterial invaders break through the fibrin wall. The bacteria involved produce large quantities of the enzymes *hyaluronidase*, which liquifies the ground

substance, and *fibrinolysin*, which breaks down fibrin and prevents clot formation. These are serious conditions that require prompt antibiotic therapy.

℞ Transplants and Immunosuppressive Drugs FAP *p. 793*

Organ transplantation is an important treatment option for patients with severe disorders of the kidneys, liver, heart, lungs, or pancreas. In 1996, U.S. surgeons transplanted approximately 11,000 kidneys, 4000 livers, 2400 hearts, 900 lungs, 100 heart–lung combinations, 900 kidney–pancreas combinations, and 100 pancreases. These numbers are limited by the availability of suitable organ donors; during the same year, roughly 24,000 patients who could have benefited from a kidney transplant did not receive one.

Graft rejection will not occur if the MHC (major histocompatibility complex) proteins of two individuals are identical, and for this reason most grafts made between identical twins are successful. The greater the difference in MHC structure between the donor and recipient, the greater the likelihood that the graft will be rejected. The process of tissue typing assesses the degree of sim-

22

ilarity between the MHC complexes of two individuals. For this process, lymphocytes are collected and examined, because they can easily be obtained from the blood and bear both Class I and Class II MHC molecules. There are many different MHC proteins, and finding a perfect match among nonidentical siblings, other relatives, or nonrelated persons can be difficult. Full siblings have only a 25 percent chance of being complete matches. For patients with a diverse ethnic background, finding a matched unrelated donor may be almost impossible. With more people being tested and put on the international bone marrow donor registry, there is more hope for these patients. For example, in 1996 an adopted Korean-American soldier found a suitable donor in Korea, and a Chinese-American child from Hawaii found a donor in Taiwan. Unlike organ donors, bone marrow donors do not lose an irreplaceable organ. The bone marrow stem cells left in the donor soon replace the donated bone marrow, and the donor's health is unaffected.

Until recently, the drugs used to produce immunosuppression did not selectively target the immune system. For example, Prednisone, a corticosteroid, was used for its broad anti-inflammatory effects. Two other drugs, **cyclophosphamide** (sī-klō-FOS-fa-mīd) and **azathioprine** (a-za-THĪ-ō-prēn), are more powerful, but they have greater associated risks. These drugs reduce the rates of cellular growth and replication throughout the body. When these drugs are administered, hematopoiesis slows dramatically, and undesirable side effects may develop in the reproductive, nervous, and integumentary systems.

An understanding of the chemical communication between helper T and suppressor T cells, macrophages, and B cells has led to the development of drugs with more selective effects. Cyclosporin A was the most important immunosuppressive drug in the 1980s. In the early 1980s, before the use of cyclosporin, the 5-year survival rate for liver transplants was below 20 percent. In the mid-1990s, the survival rate is approximately 80 percent, and drugs now being tested in clinical trials may further improve survival. One of the most promising drugs, **FK 506,** specifically inhibits lymphocytes.

An obvious problem posed by the use of any immunosuppressive drug is that the individual becomes more susceptible than normal to viral, bacterial, and fungal infections. Immunosuppression may continue indefinitely after a transplant is performed, and the patient may not recover full immune function for months after treatment has been discontinued. Over this period, the individual must be treated with injections of IgG (*gamma globulin*) from pooled sera and, when necessary, with antibiotics.

A more subtle risk of immunosuppression is the reduction in immune surveillance by NK cells. Transplant patients are 100 times more likely to develop cancer than are others in their age group.

Lymphoma-type cancers are the most common; these cancers appear to be linked to posttransplant infection with the Epstein–Barr virus. David, the SCID boy raised in a sterile bubble, died of a B-cell lymphoma after receiving a bone marrow transplant from his sister, who had an asymptomatic case of mononucleosis.

℞ Immunization   FAP *p. 798*

Immunization is the manipulation of the immune system by the administration of antigens under controlled conditions or by the administration of antibodies that can combat an existing infection. In **active immmunization,** a primary response to a particular pathogen is intentionally stimulated before an individual encounters the pathogen in the environment. The result is lasting immunity against that pathogen. Immunization is accomplished by the administration of a **vaccine** (vak-SĒN), a preparation of antigens derived from a specific pathogen. A vaccine may be given orally or via intramuscular or subcutaneous injection. Most vaccines consist of the pathogenic organism in whole or in part—living or dead. In some cases, a vaccine contains one of the metabolic products of the pathogen.

Before live bacteria or viruses are administered, they are weakened, or **attenuated** (a-TEN-ū-Ā-ted), to lessen or eliminate the chance that a serious infection will develop from exposure to the vaccine. The rubella, mumps, measles, smallpox, yellow fever, typhoid, and oral polio vaccines are examples of vaccines that use live attenuated viruses. Despite attenuation, the administration of live microorganisms may produce mild symptoms comparable to those of the disease itself, such as a low-grade fever or rash. However, the likelihood that serious illness will develop as a result of vaccination is very small compared with the risks posed by pathogen exposure *without* prior vaccination.

Inactivated, or "killed," vaccines consist of bacterial cell walls or viral protein coats only. These vaccines have the advantage that they cannot produce even mild symptoms of the disease. Unfortunately, inactivated vaccines may not stimulate as strong an immune response and so may not confer as long-lasting immunity as do live-organism vaccines. In the years after exposure, the antibody titer declines and the system eventually fails to produce an adequate secondary response. As a result, the immune system must be "reminded" of the antigen periodically by the administration of *boosters.* Influenza, typhoid, typhus, plague, and injected polio vaccines use inactivated viruses or bacteria. In some cases, fragments of the bacterial or viral walls, or their toxic products, can be used to produce a vaccine. The tetanus, diphtheria, and hepatitis B vaccines are examples. Data concerning attenuated and inactivated vaccines are presented in Table A-24.

Table A-24 Immunizations Currently Available

Immunization Target	Type of Immunity Provided	Vaccine Type	Remarks
VIRUSES			
Poliovirus	Active	Live, attenuated	Oral
	Active	Killed	Boosters every 2–3 years
Rubella	Active	Live, attenuated	
	Passive	Human antibodies (pooled)	
Mumps	Active	Live, attenuated	
Measles (rubeola)	Active	Live, attenuated	May need second booster
Hepatitis A	Active	Killed	
Hepatitis B	Active	Killed	May need periodic boosters
	Passive	Human antibodies (pooled)	
Smallpox	Active	Live, related virus	Boosters every 3–5 years (no longer required, as disease appears to have been eliminated)
Yellow fever	Active	Live, attenuated	Boosters every 10 years
Herpes zoster	Passive	Human antibodies (pooled)	
Hemophilus influenza B (HIB)	Active	Killed	May need periodic boosters
Rabies	Passive	Human antibodies (pooled)	
	Passive	Horse antibodies	
	Active	Killed	Boosters required
BACTERIA			
Typhoid	Active	Killed or live, attenuated	Boosters every 2, 3, or 5 years, depending on the vaccine type
Tuberculosis	Active	Live, attenuated	
Plague	Active	Killed or live, attenuated	Boosters every 1–2 years
Tetanus	Active	Toxins only	Boosters every 5–10 years
	Passive	Human antibodies (pooled)	
Diphtheria	Active	Toxins only	Boosters every 10 years
	Passive	Horse antibodies	
Streptococcal pneumonia	Active	Bacteria and cell wall components	
Botulism	Passive	Horse antibodies	
Rickettsia: typhus	Active	Killed	Boosters yearly
OTHER TOXINS			
Snake bite	Passive	Horse antibodies	
Spider bite	Passive	Horse antibodies	
Venomous fish spine	Passive	Horse antibodies	

22

Gene-splicing techniques can now be used to incorporate antigenic compounds from pathogens into the cell walls of harmless bacteria. When exposed to these bacteria, the immune system responds by producing antibodies and memory B cells that are equally effective against the engineered bacterium and the pathogen.

Passive immunization is normally used if the individual has already been exposed to a dangerous pathogen or toxin and there is not enough time for active immunization to take effect. In passive immunization, the patient receives a dose of antibodies that will attack the pathogen and overcome the infection, even without the help of the host's own immune system. Passive immunization provides only short-term resistance to infection, for the antibodies are gradually removed from circulation and are not replaced.

The antibodies provided during passive immunization have traditionally been acquired by collecting and combining antibodies from the sera of many other individuals. This *pooled sera* is used to obtain large quantities of antibodies, but the procedure is very expensive. In addition, improper treatment of the sera carries the risk of accidental transmission of an infectious agent, such as the hepatitis virus or the virus responsible for AIDS. Antibodies can also be obtained from the blood of a domesticated animal (typically a horse) exposed to the same antigen. Unfortunately, recipients may suffer allergic reactions to horse serum proteins.

At present, antibody preparations are available to treat hepatitis A, hepatitis B, diphtheria, tetanus, rabies, measles, rubella, botulism, and the venoms of certain fish, snakes, and spiders. Gene-splicing technology can also be used to reproduce pure antibody preparations free from antigenic or viral contaminants; this technology should eventually eliminate the need for pooled or foreign plasma.

Note that passive immunity occurs naturally during fetal development. At that time, maternal IgG antibodies can cross the placental barriers and enter the fetal circulation.

☤ Lyme Disease   FAP *p. 798*

In November 1975, the town of Lyme, Connecticut, experienced an epidemic of adult and juvenile arthritis. Between June and September, 59 cases were reported, 100 times the statistical average for a town of its size. Symptoms were unusually severe; in addition to joint degeneration, victims experienced chronic fever and a prominent rash that began as a red bull's-eye centered around what appeared to be an insect bite (Figure A-41). It took almost 2 years to track down the cause of the problem.

Lyme disease is caused by the bacterium *Borrelia burgdorferi*, which normally lives in white-

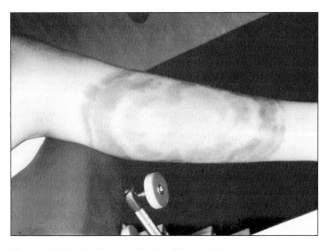

Figure A-41   Bull's Eye Rash of Lyme Disease

footed mice. The disease is transmitted to humans and other mammals through the bite of a tick that harbors that bacterium. Deer, which can carry infected adult ticks without becoming ill, have helped spread infected ticks through populated areas. The high rate of infection among children reflects the fact that they play outdoors during the summer in fields where deer may also be found. Children are thus more likely to encounter—and be bitten by—infected ticks. After 1975, the Lyme disease problem became regional and then national in scope. By 1997, approximately 94,000 people had been diagnosed with Lyme disease in the United States. It has also been found in Europe.

Although some of the joint destruction results from the deposition of immune complexes by a mechanism comparable to that for rheumatoid arthritis, many of the symptoms (fever, pain, skin rash) develop in response to the release of interleukin-1 (Il-1) by activated macrophages. The cell walls of *B. burgdorferi* contain lipid-carbohydrate complexes that stimulate the secretion of Il-1 in large quantities. By stimulating the body's specific and nonspecific defense mechanisms, Il-1 exaggerates the inflammation, rash, fever, pain, and joint degeneration associated with the primary infection. Treatment for Lyme disease consists of the administration of antibiotics and anti-inflammatory drugs; a vaccine is in development.

☤ AIDS   FAP *pp. 799, 806*

Acquired immune deficiency syndrome (AIDS), or *late-stage HIV disease*, develops after infection by the *human immunodeficiency virus (HIV)*. There are at least three different types of HIV, designated *HIV-1*, *HIV-2*, and *HIV-3*. Most people with HIV in the United States are infected with HIV-1. HIV-2 infections are most common in West Africa. Because most of those infected with HIV-1 eventually develop AIDS but not all individuals infected with HIV-2 do so, HIV-2 may be a less-dangerous virus. The distribution and significance of HIV-3 infection remain to be determined.

The discussion that follows, based on information pertaining to HIV-1 infection, expands on the discussion on *p. 806* of the text.

SYMPTOMS OF HIV DISEASE

The initial infection may produce a flulike illness with fever and swollen lymph nodes a few weeks after exposure to the virus. This exposure generally triggers the production of antibodies against the virus. These antibodies appear in the serum within 2–6 months of exposure, and antibody tests can be used to detect infection (see below). Further symptoms may not appear for 5–10 years or more. During this period, the virus content of the blood is very low, but the viruses are at work within lymphoid tissues, especially in the lymph nodes. There is a steady decline in the number of CD4 T cells in the body, and, for reasons as yet unknown, a steady destruction of dendritic cells in lymphoid tissues.

The Centers for Disease Control and Prevention (CDC) in Atlanta, which monitors infectious diseases, now recognizes three categories of HIV disease on the basis of CD4 T cell counts:

1. In *Category 1*, the CD4 T cell count is above 500/µl, and these cells account for at least 29 percent of the total circulating lymphocytes.

2. In *Category 2*, the CD4 T cell count is between 200 and 499/µl, and these cells account for 14–28 percent of the total circulating lymphocytes.

3. In *Category 3*, the CD4 T cell count is below 200/µl, and these cells account for 14 percent of the total circulating lymphocytes. This category, **late-stage HIV disease,** includes all patients with the condition commonly known as AIDS.

Symptoms commonly do not appear until the CD4 T cell concentration falls below 500/µl (Category 2). The symptoms that first appear are typically mild, consisting of lymphadenopathy, chronic nonfatal infections, diarrhea, and weight loss. A person with AIDS (Category 3) develops a variety of life-threatening infections and disorders, and the average life expectancy after diagnosis is 2 years. The life expectancy is short because HIV-1 selectively infects helper T cells. The reduction in helper T cell activity impairs the immune response; the effect is magnified, because suppressor T cells are relatively unaffected by the virus. Over time, circulating antibody levels decline, cellular immunity is reduced, and the body is left without defenses against a wide variety of bacterial, viral, fungal, and protozoan invaders.

This vulnerability is what makes AIDS dangerous. The effects of HIV on the immune system are not by themselves life-threatening, but the infections that result when the immune system is weakened certainly are. With the depression of immune function, ordinarily harmless pathogens can initiate lethal infections known as opportunistic infections. In fact, the most common and dangerous pathogens for an AIDS patient are microorganisms that seldom cause illnesses in humans with normal immune systems. AIDS patients are especially prone to lung infections and pneumonia, commonly caused by infection with the fungus *Pneumocystis carinii.* These patients are also subject to a variety of other fungal infections, such as *cryptococcal meningitis,* and an equally broad array of bacterial and viral infections. Because AIDS patients are virtually defenseless, the symptoms and time course of these infections are very different from those in noninfected individuals.

In addition to problems with pathogenic invasion, immune surveillance is depressed, and the risk of cancer increases. One of the most common cancers in AIDS patients is *Kaposi's sarcoma,* a condition that is extremely rare in uninfected individuals. Kaposi's sarcoma typically begins with rapid cell divisions in endothelial cells of cutaneous blood vessels. Associated lesions, blue or a deep brown-purple, generally appear first in the hands or feet and later occur closer to the trunk. In a small number of patients, the lesions develop in the epithelium of the digestive or respiratory tract. In normal individuals, the tumor usually does not metastasize; in AIDS patients, whose immune systems are relatively ineffective, the tumor typically converts to an aggressive, invasive cancer.

If the AIDS patient survives all these assaults, infection of the CNS by HIV eventually produces neurological disorders and a progressive dementia. So far, AIDS is almost invariably fatal, and most of those who carry HIV will eventually die of AIDS. There are a few long-term survivors who test as HIV-positive but who have no signs of abnormal immune function. Their relative immunity has been linked to the presence of two mutated copies of a gene that directs the synthesis of an integral membrane protein. The protein, called *CC-CKR-5,* is involved in the attachment and penetration of HIV. The most common strains of HIV cannot infect the immune cells of an individual with two mutant copies of this gene. If the individual has one normal gene and one mutant gene, infection occurs, but the disease progresses more slowly than it does in individuals with two normal CC-CKR-5 genes.

INCIDENCE

During 1996, roughly 60,000 cases of AIDS were diagnosed in the United States, and the number of U.S. deaths attributed to this disorder since its discovery in 1982 exceeds 350,000. AIDS is now the leading cause of death in people aged 25–44. The CDC estimates that the number of AIDS cases in the United States will continue to increase rapidly (Figure A-42).

Because HIV can remain in the body for years without producing clinical symptoms, the number of individuals infected and at risk is certain to be far higher than the number of reported cases. An estimated 1–2 million Americans are infected with HIV, and almost all will eventually develop AIDS, probably within the next decade. The numbers worldwide are even more frightening. The World Health Organization (WHO) estimates that as many as 20 million people may be infected with HIV, and the number of AIDS patients worldwide continues to climb rapidly. Several African nations are already on the verge of social and economic collapse due to devastation by AIDS. One-third of the population of Malawi is infected, and 60 percent of pregnant women there carry the virus.

MODES OF INFECTION

Infection with HIV occurs through intimate contact with the body fluids of an infected individual. Although all body fluids, including saliva, tears, and

22

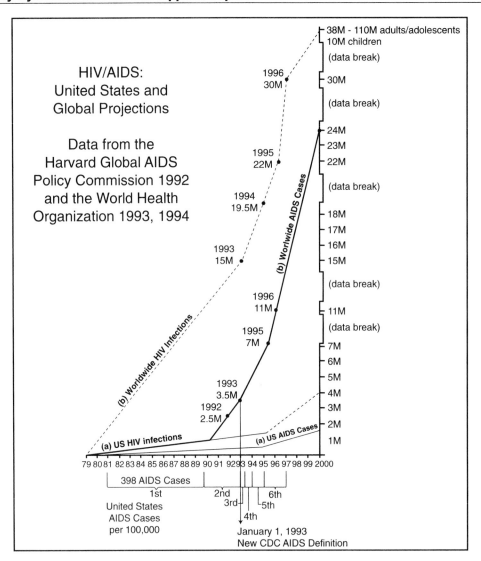

Figure A-42   United States and Global Projections for Total Number of HIV and AIDS Cases.
[From Gerald J. Stine (1997). AIDS Update 1996–1997. © Prentice Hall, Upper Saddle River, NJ.]

breast milk, carry the virus, the major routes of infection involve contact with blood, semen, or vaginal secretions. The transmission pattern has been analyzed for adult and adolescent AIDS patients in the United States. Four major transmission routes have been identified: (1) sexual transmission, (2) intravenous drug use, (3) receipt of blood or tissue products, and (4) prenatal exposure.

Sexual Transmission. In approximately 62 percent of all cases, exposure occurred through sexual contact with an infected individual. Male homosexual contact was involved in 57 percent of all cases (compared with 5 percent for heterosexual contact), and the ratio of male to female AIDS patients is approximately 9 to 1. A comparable transmission pattern is found in Canada, Europe, South America, Australia, and New Zealand. In the Caribbean and Africa, in contrast, AIDS began in the heterosexual community. It affects heterosexual

men and women in roughly equal numbers, and the gender ratio for AIDS patients in the countries is approximately 1 to 1.

Two factors may account for the pattern of transmission observed in the United States. First, the disease appears to have spread through the homosexual community before it entered the heterosexual population via bisexual males. Second, there is statistical evidence that, on a per exposure basis, the risk of male-to-male or male-to-female transmission is many times greater than the risk of female-to-male transmission. As a result, it may take longer to spread the virus through the heterosexual population (male A to female A to male B) than through the homosexual population (male A to male B to male C). Whether homosexual or heterosexual contact is involved, a sex partner whose epithelial defenses are weakened is at increased risk of infection. This factor accounts for the relatively higher rate of transmission by anal intercourse, which tends to damage the delicate lining of the anorectal canal. Genital ulcers from other sexually transmitted diseases, such as syphilis, chancroid, or herpes, also increases the risk of transmission.

Because homosexual transmission is predominant in the United States, many people still consider AIDS to be a disease of homosexuals. It is not. Over time, the number of cases in the heterosexual population has been steadily increasing; since 1987, the percentage of homosexual or bisexual AIDS patients has dropped by roughly 25 percent, and the number of cases transmitted by heterosexual contact has steadily increased. It can be anticipated that the gender ratio will continue to shift toward the 1:1 male-to-female ratio typical of Africa and the Caribbean.

Intravenous Drug Use. Roughly 19 percent of AIDS patients contracted the disease through the shared use of needles. (Another 6 percent were homosexual males who shared needles with other drug users, making it unclear which risk factor was responsible for transmission.) Although only small quantities of blood are inadvertently transferred when needles are shared, this practice injects HIV directly into the bloodstream. It is thus a very effective way to transmit the infection.

Receipt of Blood or Tissue Products. About 3 percent of AIDS patients have become infected with HIV after they received a transfusion of contaminated whole blood or plasma, an infusion of blood products, such as platelets or extracts of pooled sera, or an organ transplant from an infected individual. With careful screening of blood and blood products, the rate of new transmission by this route is now essentially zero in the United States.

Prenatal Exposure. In the United States, approximately 2000 infants are born each year already infected with HIV. Although this number is relatively small compared with the total number of AIDS patients, it is increasing rapidly. An untreated pregnant woman infected with HIV has a 20–33 percent chance of infecting her baby across the placenta. As HIV spreads through the heterosexual population, more pregnant women will become infected, and maternal–fetal transmission will become more common. These unfortunate infants will place social and financial stresses on our society for the rest of the 1990s.

PREVENTION OF AIDS

The best defense against AIDS consists of avoiding exposure to HIV. The most important rule is to avoid sexual contact with infected individuals. All forms of sexual intercourse carry the potential risk of virus transmission. The use of synthetic (latex) condoms has been recommended when you do not know the sexual history of a partner. (Condoms that are not made of synthetic materials are effective in preventing pregnancy but do not block the passage of viruses.) Although condom use does not provide absolute protection, it drastically reduces the risk of infection.

Attempts are underway to ensure that blood and blood products are adequately screened for the presence of HIV-1. A simple blood test exists for the detection of HIV-1 antibodies; a positive reaction indicates previous exposure to the virus. The assay, an example of an **ELISA** (enzyme-linked immunoabsorbent assay) **test,** is now used to screen blood donors, reducing the risk of infection by transfusion or the use of blood products from pooled sera. Pooled sera can also be heat-treated by exposure to temperatures sufficient to kill the virus but too low to denature blood proteins permanently.

Most public health facilities will perform the ELISA test on request for individuals who fear that they may have already been exposed to HIV. Unfortunately, the test is not 100 percent reliable, and false positive reactions occur at a rate of about 0.4 percent. In addition, the ELISA test does not detect HIV-2 or HIV-3. In the event of a positive test result, a retest should be performed by using the more sensitive **Western blot** procedure. Because the incubation period is variable, a positive test for HIV infection does not mean that the individual has AIDS. It does mean that the individual is likely to develop AIDS some time in the future and is now a carrier capable of infecting others. By the time an individual develops AIDS, he or she is obviously sick and usually has little interest in sexual activity. In terms of the spread of this disease, the most dangerous individuals are those who seem perfectly healthy and have no idea that they are carrying the virus.

Despite intensive efforts, a vaccine has yet to be developed that will provide immunity from HIV infection. More than a dozen vaccines have been designed, two have entered advanced clinical trials, and more than 1200 test subjects have been vaccinated. Current research programs are attempting to stimulate antibody production in response to (1) killed but intact viruses; (2) fragments of the viral envelope; (3) HIV proteins on the surfaces of other, less dangerous viruses; or (4) T cell proteins that are targeted by HIV. (The last approach is based on the assumption that the antibodies produced will cover the binding sites, preventing viral attachment and penetration.)

TREATMENT

There is no cure for HIV infection. However, the length of survival for AIDS patients has been steadily increasing, because (1) new drugs are available that slow the progression of the disease and (2) improved antibiotic therapies have helped overcome infections that would otherwise prove fatal. This combination is extending the life of patients as the search for more effective treatment continues.

However, overcoming an infection in an AIDS patient may require antibiotic doses up to 10 times greater than those used to fight infections in healthy individuals. Moreover, once the infection has been overcome, the patient may have to continue taking that drug for the rest of his or her life. As a result, some AIDS patients find themselves taking 50–100 pills a day just to prevent recurrent infections.

The antiviral drug *azidothymidine,* or *AZT* (sold as *Zidovudine* or *Retrovir*), can slow the progression of AIDS. Low doses of AZT have been shown to be effective in delaying the transition of HIV disease to AIDS. Higher doses are used to treat AIDS, but this use can lead to a variety of unpleasant side effects, including anemia and even bone marrow failure. However, AZT is effective in reducing the neurological symptoms of AIDS, because it can cross the blood–brain barrier to reach infected CNS tissues. Side effects are reduced when AZT treatment is alternated with other antiviral drugs, including *ddC (dideoxycytidine), ddI (dideoxyinosine), d4T,* and *3TC.* AZT and these four drugs inhibit the action of *reverse transcriptase,* the enzyme used to insert viral genes into the cell's DNA strands. (See the discussion of viral replication on p. 23 for details.) **Protease inhibitors,** such as *ritonavir, indinavir,* and *saquinavir mesylate,* prevent the assembly and dispersal of new viruses. Alone or in combination with AZT or other drugs, protease inhibitors can dramatically reduce the viral content of the blood, although viruses remain within infected immune cells. There is hope that treatment with protease inhibitors will dramatically improve the life span and quality of life for those with advanced HIV disease. However, because some virus-infected cells are not destroyed, protease inhibitors so far have been unable to cure the disease, although they have dramatically reduced symptoms and extended the lives of individuals with AIDS.

AIDS AND THE COST TO SOCIETY

Treatment of AIDS is complex and expensive. The average cost is $70,000 per patient per year, giving a projected annual cost of roughly $40 billion (1996). As more individuals become infected and more of the 1–2 million people already infected develop AIDS, the costs will continue to escalate rapidly.

There are other, less obvious costs to society. Individuals with depressed immune systems get sick more often and remain sick longer and, as a result, are more likely to spread the infecting pathogen to healthy individuals. Some diseases, such as tuberculosis, that had been present at very low levels in the U.S. population, are now occurring with increased frequency. In part, the reason for this increase is that AIDS patients are succumbing to infection, but it also reflects the spread of these diseases through transmission from AIDS patients into the general population. Long-term use of multiple antibiotics in AIDS patients may also promote the development of more-resistant disease organisms (which is always a risk of chronic antibiotic use; see p. 157).

Ⓡ Technology, Immunity, and Disease FAP *p. 800*

Our understanding of disease mechanisms has been profoundly influenced by recent advances in genetic engineering. The information and technical capabilities developed during the 1980s have already started to affect clinical procedures. This trend is sure to continue, and several lines of research have the potential for broad application. These projects involve a mixture of genetic engineering, computer analysis, and protein biochemistry.

Researchers can identify the individual B cells responsible for producing a given antibody. That B cell can then be isolated in a petri dish and fused with a cultured cancer cell. This technique produces a **hybridoma** (hī-bri-DŌ-ma), a cancer cell that produces large quantities of a single antibody. Because hybridomas, like other cancer cells, undergo rapid mitotic divisions, culturing the original hybridoma cell soon produces an entire population, or **clone,** of genetically identical cells. This particular clone will produce large quantities of a **monoclonal** (mo-nō-KLŌ-nal; *mono,* one) **antibody.** Monoclonal antibodies are useful because they are free from impurities. They are identical molecules, because they are produced from genetically identical cells.

One important use for this technology has been the development of antibody tests for the clinical analysis of body fluids. Labeled antibodies can be used to detect small quantities of specific antigens in a sample of plasma or other body fluids. For example, a popular home pregnancy test relies on monoclonal antibodies that detect small amounts of a placental hormone, *human chorionic gonadotropin (hCG),* in the urine of a pregnant woman. Other monoclonal antibodies are used in standard blood screening tests for venereal diseases and urine tests for ovulation. Monoclonal antibodies can also be used to provide passive immunity to disease. Passive immunizations that include monoclonal antibodies do not cause any of the unpleasant side effects associated with antibodies from pooled sera, for the product will not contain plasma proteins, viruses, or other contaminants. The antibodies can be made to order by exposing a population of B cells to a particular antigen, then isolating any B cells that produce the desired antibodies.

Genetic engineering techniques can be used to promote immunity in other ways as well. One interesting approach involves gene-splicing techniques. The genes coding for an antigenic protein of a viral or bacterial pathogen are identified, isolated, and inserted into a harmless bacterium that can be cultured in the laboratory. The clone that eventually develops will produce large quantities of pure antigen that can then be used to stimulate a primary immune response. Vaccines against malaria and hepatitis were developed in this manner, and a similar strategy may be used to design an AIDS vaccine.

A more controversial experimental technique involves taking a pathogenic organism and adding or removing genes to make it harmless. The modified pathogen can then be used to produce active immunity without the risk of severe illness. Fears that the engineered organism could mutate or regain its pathogenic properties have so far limited the use of this approach, even in animal trials.

Hybridomas that manufacture other products of the immune system can also be produced. Interferons are not effective against all viruses, but interferon nasal sprays appear to provide resistance against the viruses responsible for the common cold. (Unfortunately, these sprays can cause nasal bleeding and other unpleasant side effects, so they have not been approved for sale to the public.) Interferons can also control certain forms of virus-induced cancers and hepatitis. Interleukins may prove useful in increasing the intensity of the immune response.

☤ Fetal Infections   FAP p. 802

Fetal infections are rare, because the developing fetus acquires passive immunity from antibodies produced by the mother. These defenses break down if the maternal antibodies are unable to cope with a bacterial or viral infection. The fetus may then begin producing IgM antibodies. Blood drawn from a newborn infant or taken from the umbilical cord of a developing fetus can be tested for the presence of IgM antibodies. This procedure provides concrete evidence of congenital infection. For example, a newborn infant with congenital syphilis will have IgM antibodies that target the pathogenic bacterium involved (*Treponema pallidum*). Fetal or neonatal (newborn) blood may also be tested for antibodies against the rubella (German measles) virus or other pathogens.

In the case of congenital syphilis, antibiotic treatment of the mother can prevent fetal damage. In the absence of antibiotic treatment, fetal syphilis can cause liver and bone damage, hemorrhaging, and a susceptibility to secondary infections in the newborn infant. There is no satisfactory treatment for congenital rubella infection, which can cause severe developmental abnormalities. For that reason, rubella immunization has been recommended for young children (to slow the spread of the disease in the population) and for women of childbearing age. The vaccination, which contains live attenuated viruses, must be administered before pregnancy to prevent maternal and fetal infection during pregnancy and subsequent resulting fetal damage.

☤ Immune Complex Disorders
FAP pp. 804, 805

Under normal circumstances, immune complexes are promptly eliminated by phagocytosis. But when an antigen appears suddenly in high concentrations, the local phagocytic population may not be able to cope with the situation. The immune complex may then enlarge further, eventually forming insoluble granules that are deposited in the affected area. The presence of these complexes triggers the extensive activation of complement, leading to inflammation and tissue damage at that site (Figure A-43). This condition is known as an **immune complex disorder.** The process of immune complex formation is further enhanced by neutrophils, which release enzymes that attack the inflamed cells and tissues. The most serious immune complex disorders involve deposits within blood vessels and in the filtration membranes of the kidney.

☤ Delayed Hypersensitivity and Skin Tests   FAP pp. 802, 805

Delayed hypersensitivity begins with the sensitization of cytotoxic T cells. At the initial exposure, macrophages called *antigen-presenting cells (APCs)* present antigenic materials to T cells. On subsequent exposures, the T cells respond by releasing cytokines that stimulate macrophage activity and produce a massive inflammatory response in the immediate area (Figure A-44). Examples of delayed hypersensitivity include the many types of contact dermatitis, such as poison ivy, which we considered on p. 39.

Skin tests can be used to check for delayed hypersensitivity. The antigen is administered by shallow injection, and the site is inspected minutes to days later. Most skin tests inject antigens taken from bacteria, fungi, viruses, or parasites. If the individual has previously been exposed to the antigen and has developed antibodies to it, in 2–4 days the injection site will become red, swollen, and invaded by macrophages, T cells, and neutrophils. These signs are considered a positive test, which indicates previous exposure but does not necessarily indicate the presence of the disease. The skin test for tuberculosis, the *ppd,* is the most commonly used test of this sort.

☤ Systemic Lupus Erythematosus
FAP p. 804

Systemic lupus erythematosus (LOO-pus e-rith-ē-ma-TŌ-sis), or **SLE,** appears to result from a generalized breakdown in the antigen recognition mechanism. An individual with SLE manufactures autoantibodies against the body's own nucleic acids, ribosomes, clotting factors, blood cells, platelets, and lymphocytes. The immune complex-

22

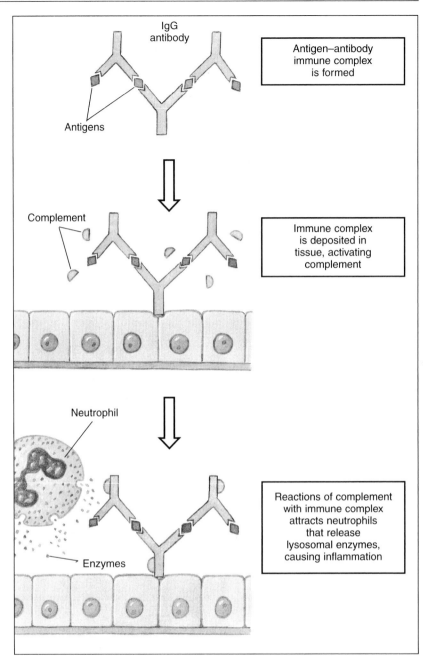

Figure A-43   Immune Complex Hypersensitivity

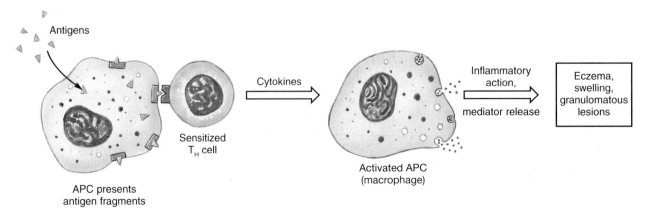

Figure A-44   Delayed Hypersensitivity

es form deposits in peripheral tissues, producing anemia, kidney damage, arthritis, and vascular inflammation. CNS function deteriorates if the blood flow through damaged cranial or spinal vessels slows or stops.

The most obvious sign of this condition is the presence of a butterfly-shaped discoloration of the face, centered over the bridge of the nose (Figure A-45). SLE affects women nine times as often as it affects men, and the incidence in the United States averages 2–3 cases per 100,000 population. There is no known cure, but almost 80 percent of SLE patients survive 5 years or more after diagnosis. Treatment consists of controlling the symptoms and depressing the immune response through the administration of specialized drugs or corticosteroids.

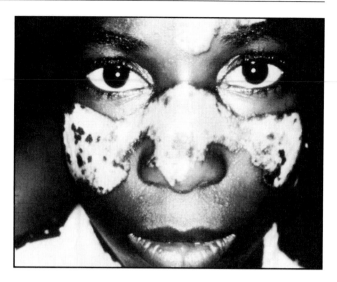

Figure A-45   Butterfly Rash of Systemic Lupus Erythematosus

CRITICAL-THINKING QUESTIONS

7-1. Matthew, a 17-year-old hemophiliac, reports to Dr. M. that he has recently lost weight for no apparent reason and that he has had a flu disorder "on and off" for the last month. Matthew has a fever of 37.8°C (100°F), cervical lymph node swelling, and a persistent cough. Dr. M. is worried about Matthew's immune state and recommends a bronchoscopic analysis to detect the cause of the respiratory problems. The cultured specimen from Matthew's lungs contains the bacterium *Pneumocystis carinii*. Matthew is diagnosed with *Pneumocystis carinii pneumonia (PCP)*. Which further test would you order for Matthew?

7-2. Paula's grandfather is diagnosed as having lung cancer. His physician orders biopsies of several lymph nodes from adjacent regions of the body. Paula wonders why, as her grandfather's cancer is in the lungs. What would you tell her?

7-3. Anne, an 18-year-old college student, goes to her physician with symptoms that have persisted for 2 weeks. Anne has been fatigued and feverish, with a temperature as high as 38.6°C (101°F).

She has inflamed tonsils and enlarged cervical lymph nodes. On palpation, her spleen is slightly enlarged; the physician notes no other signs or symptoms. Her lab results are as follows:

RBC count: 4.2 million/mm³

Hct: 37

Hb: 12 g/dl

MCV: 80 µm³

MCH: 27 pg

Reticulocytes: 1% of total RBCs

WBC count: 7500/mm³

Diff. count: 40% neutrophils

60% lymphocytes

2% monocytes

1% eosinophils

0.5% basophils

Platelets: 200,000/mm³

On the basis of these lab results, what is a likely diagnosis?

Clinical Problems

1. Noticing that her teeth have tartar buildup, Melanie visits her dentist for a cleaning. The dentist reminds Melanie that she has not been in for the last 5 years and recommends that she update her health history form. Melanie, however, is too busy reading the latest magazine and forgets to inform the dentist that she was diagnosed with mitral valve prolapse (MVP) 2 years ago. The dentist cleans her teeth and remarks about how the tartar on her teeth has gone below the gumline, causing bleeding during the procedure. He advises Melanie to keep her teeth clean so that bacteria in her mouth won't produce more plaque. Two weeks later, Melanie becomes very ill and has the following symptoms for several days: night sweats, fever and chills, joint pain, and tachycardia. She is admitted to the hospital for further testing. What might be happening?

2. Bill's father, Emilio, suffers an apparent heart attack and is rushed to the emergency room. A quick medical history from Bill indicates that Emilio was working in the yard when the severe chest pain began several hours ago. The physician on call tells Bill that she has ordered some blood tests from which she will be able to detect if a heart attack has occurred. Why would the levels of the following enzymes help indicate if a person suffered a myocardial infarction?

Total serum CPK: 70 U/ml

Isoenzymes

   CPK-MM: 40%

   CPK-MB: 60%

   CPK-BB: 0%

AST: 20 U/l

Total serum LDH: 90 U/l

Isoenzymes

LDH-1: 27%

3. Teresa is taken to the emergency room after a car accident. She sustained abdominal trauma during the accident and is very sore in the upper left abdominal quadrant. She shows no other signs or symptoms, and her vital signs are within normal limits. Thirty minutes after arrival, she is pale and anxious, and the pain has intensified. She is breathing rapidly, and her hands are clammy and cyanotic. Her pulse is 90 and weak; her blood pressure falls to 80/55, then to 75/50 within minutes. Teresa states that the pain is getting more severe and that it now extends to her left shoulder. While blood tests are being performed, she is put on IV fluids and supplemental oxygen. Inspection of the abdomen reveals abdominal distension and rigidity. Results from the initial blood work and the followup work are as follows:

Test	Upon Arrival	After 30 Minutes
RBC count	3.8 million/mm^3	3.2 million/mm^3
Hct	33	28
Hb	11 g/dl	9 g/dl
MCV	78 mm^3	78 mm^3
MCH	27 pg	27 pg
WBC count	8000/mm^3	8000/mm^3
Platelets	200,000/mm^3	200,000/mm^3

On the basis of this information, Teresa is rushed to surgery. Why?

NOTES:

The Respiratory System

The anatomical components of the respiratory system can be divided into an *upper respiratory system,* which includes the nose, nasal cavity, paranasal sinuses, and pharynx; and a *lower respiratory system,* composed of the larynx, trachea, bronchi, and lungs. The *respiratory tract* consists of the airways that carry air to and from the exchange surfaces of the lungs. The respiratory tract can be divided into a *conducting portion* and a *respiratory portion.* The conducting portion begins at the entrance to the nasal cavity and extends through the pharynx and larynx and along the trachea and bronchi to the terminal bronchioles. The respiratory portion of the tract includes the respiratory bronchioles and the alveoli that are part of the respiratory membrane, where gas exchange occurs.

SYMPTOMS OF RESPIRATORY DISORDERS

When they seek medical attention, individuals with lower respiratory disorders generally do so as a result of one or two major symptoms, specifically *chest pain* and *dyspnea:*

1. The chest pain associated with a respiratory disorder usually worsens when the person takes a deep breath or coughs. This pain with breathing is distinct from the chest pain experienced by individuals with angina (pain appears during exertion) or a myocardial infarction (pain is continuous, even at rest). Several disorders, such as those affecting the pleural membranes, cause chest pain that is localized to specific regions of the thorax. A person with such a condition will usually press against the sensitive area and avoid coughing or deep breathing in an attempt to reduce the pain.

2. *Dyspnea,* or difficulty in breathing, may be a symptom of pulmonary disorders, cardiovascular disorders, metabolic disorders, or environmental factors such as hypoxia at high altitudes. It may be a chronic problem, or it may develop only during exertion or when the person is lying down.

Dyspnea due to respiratory problems generally indicates one of the following classes of disorders:

- *Obstructive disorders* result from increased resistance to airflow along the respiratory passageways. The individual usually struggles to breathe, even at rest, and expiration is more difficult than inspiration. Examples of obstructive disorders include *emphysema* (p. 159) and *asthma* (p. 157).

- *Restrictive disorders* include (a) arthritis (2) paralysis of respiratory muscles caused by trauma and scarring, muscular dystrophy, myasthenia gravis, multiple sclerosis, or polio; (3) physical trauma or congenital structural disorders, such as scoliosis, that limit lung expansion; (4) pulmonary fibrosis, in which abnormal fibrous tissue in the alveolar walls slows oxygen diffusion into the bloodstream. Individuals with restrictive disorders usually experience dyspnea during exertion because pulmonary ventilation cannot increase enough to meet the respiratory demand.

Cardiovascular disorders that produce dyspnea include *heart disease, congestive heart failure,* and *pulmonary embolism.* In *paroxysmal nocturnal dyspnea,* a person awakens at night, gasping for air. In most cases, the underlying cause is a problem with cardiac output due to advanced heart disease or heart failure. *Cheyne–Stokes respiration* consists of cycles of rapid, deep breathing separated by periods of respiratory arrest. This breathing pattern is most commonly seen in persons with CNS disorders or congestive heart failure.

Dyspnea may also be related to metabolic problems, such as the acute acidosis associated with *diabetes mellitus* (p. 103) and with *uremia* (p. 184). The fall in blood pH can trigger *Kussmaul breathing,* which generally consists of rapid, deep respiratory cycles.

THE PHYSICAL EXAMINATION AND THE RESPIRATORY SYSTEM

Several components of the physical examination will detect signs of respiratory disorders:

1. *Inspection* can reveal abnormal dimensions, such as the "barrel chest" that develops in emphysema or other obstructive disorders (p. 159), or *clubbing* of the fingers (p. 37). Clubbing is typically a late sign of disorders such as emphysema or congestive heart failure. *Cyanosis,* a blue color of the skin and mucous membranes, generally indicates hypoxia. Laboratory testing of arterial blood gases will assist in determining the cause and extent of the hypoxia (Table A-25).

2. *Palpation* of the bones and muscles of the thoracic cage can detect structural problems or asymmetry. For example, asymmetrical contraction of respiratory muscles during breathing may indicate a restrictive disorder.

3. *Percussion* on the surface of the thoracic cage over the lungs will yield sharp, resonant sounds. Dull or flat sounds may indicate structural changes in the lungs, such as those that accompany pneumonia, or the collapse of part of a lung (*atelectasis*). Increased resonance may result from obstructive disorders, such as emphysema, due to hyperinflation of the lungs as the individual attempts to improve alveolar ventilation.

23

Table A-25   Representative Diagnostic and Laboratory Tests for Respiratory Disorders

Diagnostic Procedure	Method and Result	Representative Uses	Notes
Pulmonary function studies	A spirometer or other measuring instrument is used to determine lung volumes and capacities, including V_T, IC, ERV, IRV, FRC, VC, TLC on exertion. (See Figure 23-18 in FAP p. 841.) Forced vital capacity (FVC) is the amount of air forcibly expired after maximal inhalation.	Useful in differentiating obstructive from restrictive lung diseases; also used to determine the extent of pulmonary disease. Increased functional residual capacity (FRC) occurs in obstructive diseases such as emphysema and chronic bronchitis; FRC is normal in restrictive diseases such as pulmonary fibrosis.	Can be performed at a resting state or during exercise if dyspnea occurs
Bronchoscopy	Fiber-optic tubing is inserted into oral cavity and further into trachea, larynx, and bronchus for internal viewing	Detects abnormalities such as inflammation and tumors; used to remove aspirated foreign objects, to obtain sample of secretions for analysis, to obtain small specimen of lung tissue for biopsy, and to remove mucus from bronchi	
Mediastinoscopy	A lighted instrument is inserted into an incision made at the jugular notch superior to the manubrium	Detection of abnormalities of mediastinal lymph nodes; biopsy specimen removal for analysis for detection of pulmonary disorders; also useful in detecting metastasis of lung cancer to secondary locations	
Lung biopsy	Lung tissue is removed for pathological analysis via bronchoscopy or during exploratory surgery of the thoracic area	Useful in differentiating pulmonary pathologies and determining the presence of malignancy	
Chest X-ray study	Standard X-ray produces film sheet with radiodense tissues shown in white on a negative image	Detection of abnormalities of lungs, such as tumors, inflammation of the lungs, rib or sternal fractures, or pneumothorax; pulmonary edema (detection of fluid accumulation); pneumonia; atelectasis; determination of heart size	
Thoracentesis	A needle is inserted into the intrapleural space for removal of fluid for laboratory analysis or for relieving pressure due to fluid accumulation	See lab analysis below	
Pulmonary angiography	A catheter is inserted in the femoral vein and threaded through the right ventricle and into the pulmonary arteries. Contrast dye is intermittently injected through the catheter for contrast as X-ray films are taken.	Detection of pulmonary embolism	
Lung scan	Radionuclide is injected intravenously; radiation that is emitted is captured to create an image of the lungs	Determination of areas with decreased blood flow due to pulmonary embolism or pulmonary disease	
Computerized tomography (CT) scan	Standard CT; contrast media are usually used	Detection of tumor, cyst, or other structural abnormality	

Table A-25 *(continued)*

Laboratory Test	Normal Values in Blood Serum or Plasma	Significance of Abnormal Values	Notes
Arterial blood gases and pH			
pH	7.35–7.45	<7.35 indicates acidosis >7.45 indicates alkalosis	See text (Table 27-4, FAP *p. 1031*) for the description and values for compensated and uncompensated acidosis and alkalosis
P_{CO_2}	35–45 mm Hg	>45 mm Hg with pH <7.35 indicates respiratory acidosis present in pulmonary disorders such as emphysema, in chronic obstructive pulmonary disease (COPD), and in CNS depression leading to decrease or irregularity of respiratory rate. <35 mm Hg with a pH >7.45 indicates a respiratory alkalosis that occurs during prolonged hyperventilation; seen in emphysema.	
P_{O_2}	75–100 mm Hg	<75 mm Hg may occur in pneumonia and emphysema	
HCO_3^-	22–28 mEq/l	>28 mEq/l (with elevated P_{CO_2} and decreased pH) indicates renal compensation for respiratory acidosis; <22 mEq/l (with decreased P_{CO_2} and elevated pH) is characteristic of renal compensation for respiratory alkalosis	
Sputum studies			
Cytology	No malignant cells are present	Sloughed malignant cells indicate cancerous process in lungs	Normal result does not rule out lung cancer
Culture and sensitivity (C&S)	Sputum sample is placed on growth medium	Identification of causative pathogenic organism and the organism's susceptibility to antibiotics	Normal test reveals only resident (non-pathogenic) microbes
Acid-fast bacilli	Staining technique reveals no acid-fast bacilli	Presence of stained rod-shaped microbes may indicate tuberculosis	
Alpha$_1$-antitrypsin determination	Adults: >213 mg/dl	Decreased value (<50 mg/dl) indicates a possible genetic predisposition for emphysema	
Tuberculin skin test	Skin wheal produced by intra-dermal application of tuberculin is read at 48–72 hours and should be <10 mm. (Measurement includes the area that is hardened as opposed to the reddened area.)	A skin lesion at injection site that measures >10 mm is a positive test for tuberculosis	Positive test result does not determine whether the TB infection is in an active state
Pleural fluid analysis			
Fluid color and clarity	No pus; fluid is clear.		
Cells	WBC <1000 mm³	WBC >1000 mm³ indicates potential infectious or inflammatory process	Presence of malignant cells indicates lung or pleural malignancy
Culture	Sample of fluid is placed on growth medium	Presence of bacteria indicates infection; fungi can be present in immunocompromised patient	

23

4. *Auscultation* of the lungs with a stethoscope yields the distinctive sounds of inspiration and expiration. These sounds vary in intensity, pitch, and duration. Abnormal breath sounds accompany several pulmonary disorders:

- *Rales* (rahls) are hissing, whistling, scraping, or rattling sounds associated with increased airway resistance. The sounds are created by turbulent airflow past accumulated pus or mucus or through airways narrowed by inflammation or bronchospasms. *Moist rales* are gurgling sounds produced as air flows over fluids within the respiratory tract. They are heard in conditions such as *bronchitis* (p. 159), *tuberculosis*, (p. 157), and *pneumonia* (FAP, *p. 831*). *Dry rales* are produced as air flows over thick masses of mucus, through inflamed airways, or into fluid-filled alveoli. Dry rales are characteristic of *asthma* (p. 157) and *pulmonary edema* (FAP, *p. 733*). *Rhonchi* are loud dry rales produced by mucus buildup in the air passages.

- *Stridor* is a very loud, high-pitched sound that can be heard without a stethoscope. Stridor generally indicates acute airway obstruction, such as the partial blockage of the glottis by a foreign object.

- *Wheezing* is a whistling sound that can occur with inspiration or expiration. It generally indicates airway obstruction due to mucus buildup or bronchospasms.

- *Coughing* is a familiar sign of several respiratory disorders. Although primarily a reflex mechanism that clears the airway, coughing may also indicate irritation of the lining of the respiratory passageways. The duration, pitch, causative factors, and productivity may be important clues in the diagnosis of a respiratory disorder. (A *productive cough* ejects sputum; a *nonproductive cough* does not.) If the cough is productive, the sputum ejected can be analyzed. This analysis will provide information about the presence of epithelial cells, macrophages, blood cells, or pathogens. If pathogens are present, the sputum can be cultured to permit identification of the specific microorganism involved (Table A-25).

- A *friction rub* is a distinctive crackling sound produced by abrasion between abnormal serous membranes. A *pleural rub* accompanies respiratory movements and indicates problems with the pleural membranes, such as *pleurisy* (FAP *p. 834*). A *pericardial rub* accompanies the heartbeat and indicates inflammation of the pericardium, as in *pericarditis* (p. 107).

5. During the assessment of vital signs, the respiratory rate (number of breaths per minute) is recorded, along with notations about the general rhythm and depth of respiration. *Tachypnea* is a respiratory rate faster than 20 breaths per minute in an adult; *bradypnea* is an adult respiratory rate below 12 per min.

Table A-25 introduces important procedures and laboratory tests useful in diagnosing respiratory disorders.

DISORDERS OF THE RESPIRATORY SYSTEM

The respiratory system provides a route for air movement into and out of the lungs and supplies a large, warm, moist surface area for the exchange of oxygen and carbon dioxide between the air and circulating blood. Disorders affecting the respiratory system (Figure A-46) may therefore involve the following three mechanisms:

1. **Interfering with the movement of air along the respiratory passageways.** Internal or external factors may be involved. Within the respiratory tract, the constriction of small airways, as in *asthma* (p. 157), can reduce airflow to the lungs. The blockage of major airways, as in choking, may completely shut off the air supply. External fators that interfere with air movement include (1) the introduction of air *(pneumothorax)* or blood *(hemothorax)* into the plural cavity, with subsequent lung collapse; (2) the buildup of fluid within the plural cavities (a *plural* effusion), which compresses and collapses the lungs; and (3) arthritis, muscular paralysis, or other conditions that prevent the normal skeletal or muscular activities responsible for moving air into and out of the respiratory tract.

2. **Damaging or otherwise impeding the diffusion of gases at the respiratory membrane.** The walls of the alveoli are part of the respiratory membrane, where gas exchange occurs. Any disease process that affects the alveolar walls will reduce the efficiency of gas exchange. In *emphysema* (p. 159) or *lung cancer* (p. 161) alveoli are destroyed. Respiratory exchange can also be disrupted by the buildup of fluid or mucus within the alveoli. This disruption may occur as the result of inflammation or infection of the lungs, as in the various types of pneumonia.

3. **Blocking or reducing the normal circulation of blood through the alveolar capillaries.** Blood flow to portions of the lungs may be pre-

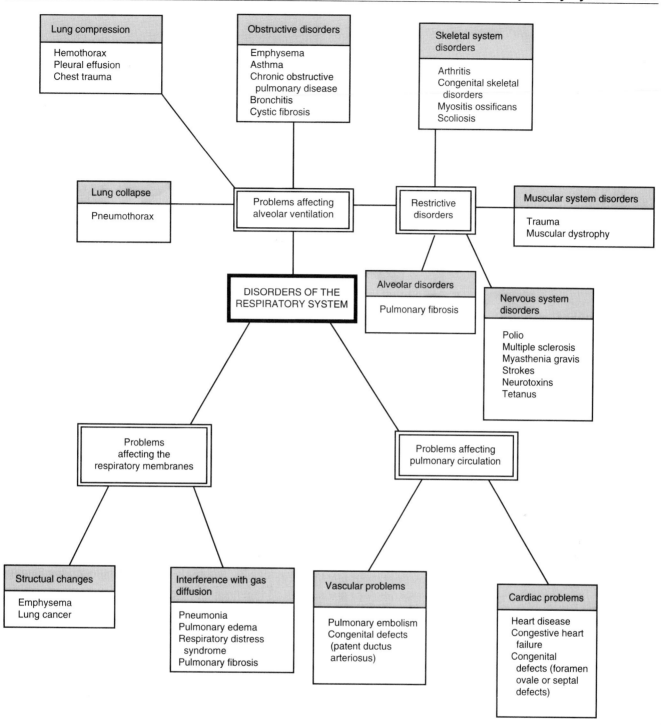

Figure A-46   Disorders of the Respiratory System

vented by a *pulmonary embolism,* a circulatory blockage discussed in the text *(p. 831).* Not only does a pulmonary embolism prevent normal gas exchange in the affected regions of a lung, but also it results in tissue damage and, if the blockage persists for several hours, permanent alveolar collapse. Pulmonary blood pressure may rise (a condition called *pulmonary hypertension)* leading to pulmonary edema and a reduction in alveolar function in other portions of the lungs.

Overloading the Respiratory Defenses
FAP *p. 817*

Large quantities of airborne particles may overload the respiratory defenses and produce a variety of different illnesses. Chemical or physical irritants

23

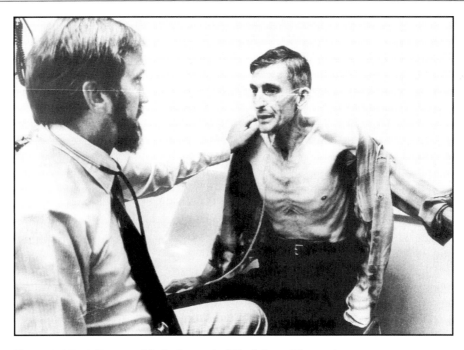

Figure A-47   A Person with Anthracosis (Black Lung Disease)

that reach the lamina propria or underlying tissues promote the formation of scar tissue (*fibrosis*), reducing the elasticity of the lung, and may restrict airflow along the passageway. Irritants or foreign particles may also enter the lymphatics of the lung, producing inflammation of the regional lymph nodes. Chronic irritation and the stimulation of the epithelium and its defenses cause changes in the epithelium that increase the likelihood of lung cancer.

Severe symptoms of such disorders develop slowly; they may take 20 years or more to appear. **Silicosis** (sil-i-KŌ-sis, produced by the inhalation of silica dust), **asbestosis** (as-bes-TŌ-sis, from the inhalation of asbestos fibers), and **anthracosis** (an-thra-KŌ-sis, the "black lung disease" of coal miners) are conditions caused by the overloading of the respiratory defenses (Figure A-47).

Cystic Fibrosis   FAP *p. 817*

Cystic fibrosis (CF) is the most common lethal inherited disease that affects Caucasians of Northern European descent; it occurs at a frequency of 1 in 2500 births. The disease occurs with less frequency in individuals of Southern European ancestry, in the Ashkenazi Jewish population, and in African-Americans. The condition results from a defective gene located on chromosome 7. Within the U.S. Caucasian population, 1 person in 25 carries one copy of the gene for this disorder, and an infant receiving a copy from both parents will develop CF. In the United States, 2000 babies are born with CF each year, and there are roughly 30,000 persons with this condition. Individuals with classic CF seldom survive past age 30; death is generally the result of a massive bacterial infection of the lungs and associated heart failure.

The gene involved carries instructions for a transmembrane protein responsible for the active transport of chloride ions. This membrane protein is abundant in exocrine cells that produce watery secretions. In persons with CF, the protein does not function normally. The secretory cells cannot transport salts and water effectively, and the secretions produced are thick and gooey. Mucous glands of the respiratory tract and secretory cells of the pancreas, salivary glands, and digestive tract are affected.

The most serious symptoms appear because the respiratory defense system cannot transport such dense mucus. The mucus escalator stops working, and mucus plugs block the smaller respiratory passageways. This blockage reduces the diameter of the airways, and the inactivation of the normal respiratory defenses leads to frequent bacterial infections. The most dangerous infections involve a specific bacterium, *Pseudomonas aeruginosa*, which colonizes the stagnant mucus and further stimulates mucus production by epithelial cells.

Treatment has primarily been limited to supportive care and antibiotic therapy to control infections. In a few instances, lung transplants have provided relief, but the technical and logistical problems involved with this approach are formidable. The normal and abnormal gene structures have now been determined, and the current goal is to correct the defect by inserting normal genes within the cells in critical areas of the body. In the meantime, it has been discovered that one of the factors contributing to the thickness of the mucus is the presence of DNA released from degenerating cells within areas of inflammation. Inhaling an aerosol spray containing an enzyme that breaks down DNA has proved to be remarkably effective in improving respiratory performance.

✝ Tuberculosis   FAP *p. 824*

Tuberculosis (TB) is a major health problem throughout the world. With roughly 3 million deaths each year, it is the leading cause of death from infectious diseases. An estimated 2 *billion* people are infected at this time, and 8 million cases are diagnosed each year. Unlike other deadly diseases, such as AIDS, TB is transmitted through casual contact. Anyone who breathes is at risk of contracting this disease; all it takes is exposure to the causative bacterium. Coughing, sneezing, or speaking by an infected individual will spread the pathogen through the air in the form of tiny droplets that can be inhaled by others.

If untreated, the disease progresses in stages. At the site of infection, macrophages and fibroblasts proceed to wall off the area, forming an abscess. If the scar-tissue barricade fails, the bacteria move into the surrounding tissues, and the process repeats itself. The resulting masses of fibrous tissue distort the conducting passageways, increasing resistance and decreasing airflow. In the alveoli, the attacked surfaces are destroyed. The combination severely reduces the area available for gas exchange.

Treatment for TB is complex, because (1) the bacteria can spread to many different tissues, and (2) they can develop a resistance to standard antibiotics relatively quickly. As a result, several drugs are used in combination over a period of 6–9 months. The most effective drugs now available include *isoniazid,* which interferes with bacterial replication, and *rifampin,* which blocks bacterial protein synthesis.

The TB problem is much less severe in developed nations, such as the United States, than in developing nations. However, tuberculosis was extremely common in the United States earlier in the twentieth century. An estimated 80 percent of Americans born around 1900 became infected with tuberculosis during their lives. Although many were able to meet the bacterial challenge, it was the number one cause of death in 1906. These statistics have been drastically changed with the advent of antibiotics and techniques for early detection of infection. Between 1906 and 1984, the death rate fell from 200 deaths per 100,000 population to 1.5 deaths per 100,000 population. From 1984 to 1994, the TB incidence and death rates were on the rise; there were roughly 24,000 TB-related deaths in 1994. Although the death rates in 1995 were slightly lower (22,000), as of 1997 an estimated 10–15 million people in the U.S. are infected with tuberculosis.

Today, tuberculosis is unevenly distributed through the U.S. population, with several groups at relatively high risk of infection. For example, Hispanics, African-Americans, prison inmates, individuals with immune disorders (such as AIDS patients), and hospital employees who work around infected patients are more likely to be infected than are other members of the population. At present 2–5 percent of young American adults have been infected. A growing percentage of these cases are caused by antibiotic-resistant strains. An estimated 14 percent of TB patients diagnosed in the United States each year are resistant to both isoniazid and rifampin. The statistics for some parts of the country are worse; in New York City, 33 percent of new cases are drug-resistant. The most frightening part about this surge in resistant TB is that the fatality rate for infections resistant to two or more antibiotics is 50 percent. Unless efforts are made to combat the spread of TB, over the next decade the United States may face a public health crisis as deadly as HIV infection and even more difficult to control.

✝ Asthma   FAP *p. 827*

Asthma (AZ-muh) affects an estimated 3–6 percent of the U.S. population. There are several forms of asthma, but each is characterized by unusually sensitive and irritable conducting passageways. In many cases, the trigger appears to be an immediate hypersensitivity reaction to an allergen in the inspired air. Drug reactions, air pollution, chronic respiratory infections, exercise, or emotional stress can also induce an asthmatic attack in sensitive individuals.

The most obvious and potentially dangerous symptoms include (1) the constriction of smooth muscles all along the bronchial tree, (2) edema and swelling of the mucosa of the respiratory passageways, and (3) accelerated production of mucus. The combination makes breathing very difficult. Exhalation is affected more than inhalation; the narrowed passageways often collapse before exhalation is completed. Although mucus production increases, mucus transport slows, and fluids accumulate along the passageways. Coughing and wheezing then develop. The bronchoconstriction and mucus production occurs in a few minutes, in response to the release of histamine and prostaglandins by mast cells. The activated mast cells also release interleukins, leukotrienes, and platelet-activating factors. As a result, over a period of hours, neutrophils and eosinophils migrate into the area. The area then becomes inflamed, further reducing airflow and damaging respiratory tissues. Because the inflammation compounds the problem, antihistamines alone are often unable to control a severe asthmatic attack.

When a severe attack occurs, it reduces the functional capabilities of the respiratory system.

23

Peripheral tissues gradually become oxygen starved, a condition that can prove fatal. Asthma fatalities have been increasing in recent years. The annual death rate from asthma in the United States is approximately 4 deaths per million population (for ages 5–34). Mortality among asthmatic African-Americans is twice that among Caucasian Americans.

Treatment of asthma involves the dilation of the respiratory passageways by administering **bronchodilators** (brong-kō-dī-LA-torz) and reducing inflammation and swelling of the respiratory mucosa. Important bronchodilators include *theophylline, epinephrine, albuterol,* and other beta-adrenergic drugs. Although the strongest beta-adrenergic drugs are very useful in a crisis, they are effective only for relatively brief periods, and the individual must be closely monitored due to the potential effects on cardiovascular function. Anti-inflammatory medication with less-acute effects, such as inhaled or ingested steroids, is becoming increasingly important.

Respiratory Distress Syndrome (RDS)   FAP *p. 831*

Surfactant cells begin producing surfactants at the end of the sixth fetal month. By the eighth month, surfactant production has risen to the level required for normal respiratory function. **Neonatal respiratory distress syndrome (NRDS),** also known as *hyaline membrane disease (HMD),* develops when surfactant production fails to reach normal levels. Although there are inherited forms of HMD, the condition most commonly accompanies premature delivery.

In the absence of surfactants, the alveoli tend to collapse during exhalation. Although the conducting passageways remain open, the newborn infant must then inhale with extra force to reopen the alveoli on the next breath. In effect, every breath must approach the power of the first, and the infant rapidly becomes exhausted. Respiratory movements become progressively weaker; eventually the alveoli fail to expand, and gas exchange ceases.

One method of treatment involves assisting the infant by administering air under pressure so that the alveoli are held open. This procedure, known as **positive end-expiratory pressure (PEEP),** can keep the newborn alive until surfactant production increases to normal levels. Surfactant from other sources can also be provided; suitable surfactants can be extracted from cow lungs (*Survanta*), obtained from the liquid (amniotic fluid) that surrounds full-term infants, or synthesized by gene-splicing techniques (*Exosurf*). These preparations are usually administered in the form of a fine mist of surfactant droplets. Clinical trials are now underway using a perfluorocarbon solution (*Liquivent®*) to deliver oxygen to the respiratory membrane.

The lungs are filled with the solution, which readily absorbs oxygen. Because the alveoli are filled with fluid, rather than air, they remain open despite the lack of surfactant. This procedure avoids the necessity of administering air under pressure, which can have undesirable side effects.

Surfactant abnormalities may also develop in adults as the result of severe respiratory infections or other sources of pulmonary injury. Alveolar collapse follows, producing a condition known as **adult respiratory distress syndrome (ARDS).** PEEP is typically used in an attempt to maintain life until the underlying problem can be corrected, but at least 50–60 percent of ARDS cases result in fatalities.

Boyle's Law and Air Over-expansion Syndrome   FAP *p. 835*

Swimmers descending 1 m or more beneath the surface experience significant increases in pressure, due to the weight of the overlying water. Air spaces throughout the body decrease in volume. The pressure increase normally produces mild discomfort in the middle ear, but some people experience acute pain and disorientation. The water pressure first collapses the auditory pharyngotympanic tubes (see Figure 17-23 of the text, *p. 571*). As the volume of air in the middle ear cavities decreases, the tympanic membranes are forced inward. This uncomfortable situation can be remedied by closing the mouth, pinching the nose and exhaling gently, because elevating the pressure in the nasopharynx will force air through the pharyngotympanic tubes and into the middle ear spaces. As the volume of air in each middle ear cavity increases, the tympanic membrane returns to its normal position. When the swimmer returns to the surface, the pressure drops and the air in the middle ear expands. This expansion usually goes unnoticed, for the air simply forces its way along the pharyngotympanic tube and into the nasopharynx.

Scuba divers breathe air under pressure, and that air is delivered at the same pressure as that of their surroundings. A descent to a depth of 10 meters doubles the pressure on a diver due to the weight of the overlying water. Consider what happens if that diver then takes a full breath of air and heads for the surface. As the pressure declines, the volume increases (Boyle's law, FAP *p. 835*). Thus, at the surface, the volume of air in the lungs will have doubled. Such a drastic increase cannot be tolerated, and the usual result is a tear in the wall of the lung. The symptoms and severity of this air overexpansion syndrome depend on where the air ends up. If the air flows into the pleural cavity, the lung may collapse; if it enters the mediastinum, it may compress the pericardium and produce symptoms of cardiac tamponade. Worst of all, the air may rupture blood vessels and enter the circulation. The air bubbles then form emboli that may block blood vessels in the heart or brain, producing a heart attack or stroke. These are all serious conditions, and divers are trained to avoid holding their breath and to exhale when swimming toward the surface.

℞ CPR   FAP *p. 839*

Cardiopulmonary resuscitation, or **CPR,** restores circulation and ventilation to an individual whose heart has stopped beating. Compression applied to the rib cage over the sternum reduces the volume of the thoracic cavity, squeezing the heart and propelling blood into the aorta and pulmonary trunk. When the pressure is removed, the thorax expands and blood moves into the great veins. Cycles of compression are interspersed with cycles of mouth-to-mouth breathing that maintain pulmonary ventilation.

This practice has been credited with saving many thousands of lives. Basic CPR techniques can be mastered in about 8 hours of intensive training, using special equipment. Yearly recertification courses must be taken, for the skills fade with time, and CPR techniques cannot be practiced on a living person without causing severe injuries. Training is available at minimal cost through charitable organizations such as the Red Cross and the American Heart Association.

🏃 Decompression Sickness   FAP *p. 844*

Decompression sickness can develop when an individual experiences a sudden change in pressure. Nitrogen is the gas responsible for this condition. Nitrogen, which accounts for 78.6 percent of the atmospheric gas mixture, has a relatively low solubility in body fluids. Under normal atmospheric pressures, there are few nitrogen molecules in the blood, but at higher than normal pressures, additional nitrogen molecules diffuse across the alveolar surfaces and into the bloodstream.

As more nitrogen enters the blood, the gas is distributed throughout the body. Over time, nitrogen diffuses into peripheral tissues and into body fluids such as the cerebrospinal fluid, aqueous humor, and synovial fluids. If the pressure decreases, the change must occur slowly enough that the excess nitrogen can diffuse out of the tissues, into the blood, and across the alveolar surfaces. If the pressure falls suddenly, this gradual movement of nitrogen from the periphery to the lungs cannot occur. Instead, the nitrogen leaves solution and forms bubbles of nitrogen gas in the blood, tissues, and body fluids.

A few bubbles in peripheral connective tissues may not be particularly dangerous, at least initially. However, these bubbles can fuse together, forming larger bubbles that distort tissues, causing pain. Bubbles typically develop in joint capsules first. These bubbles cause severe pain, and the

afflicted individual tends to bend over or curl up. This symptom accounts for the popular name of this condition, "the bends." Bubbles in the systemic or pulmonary circulation can cause infarcts, and those in the cerebrospinal circulation can cause strokes, leading to sensory losses, paralysis, or respiratory arrest.

Treatment consists of recompression, exposing the individual to pressures that force the nitrogen back into solution and alleviate the symptoms. Pressures are then reduced gradually over a period of 1 or more days. Breathing air with more oxygen and less nitrogen than are in atmospheric air accelerates the removal of excess nitrogen from the blood.

Today most bends cases involve scuba divers who have gone too deep or stayed at depth too long. The condition is not restricted to divers, however, and the first reported cases involved construction crews who worked in pressurized surroundings. Although such accidents are exceedingly rare, the sudden loss of cabin pressure in a commercial airliner can also produce symptoms of decompression sickness.

⚕ Bronchitis, Emphysema, and COPD   FAP *p. 844*

Bronchitis (brong-KĪ-tis) is an inflammation of the bronchial lining. The most characteristic symptom is the overproduction of mucus, which leads to frequent coughing. An estimated 20 percent of adult males have *chronic bronchitis.* This condition is most commonly related to cigarette smoking, but it can also result from other environmental irritants, such as chemical vapors. Over time, the increased mucus production can block smaller airways and reduce respiratory efficiency. This condition is called **chronic airways obstruction.**

Emphysema (em-fi-SĒ-muh) is a chronic, progressive condition characterized by shortness of breath and an inability to tolerate physical exertion. The underlying problem is the destruction of respiratory exchange surfaces. In essence, respiratory bronchioles and alveoli are functionally eliminated. The alveoli gradually expand, capillaries deteriorate, and gas exchange in the affected region comes to a halt.

Emphysema has been linked to the inhalation of air that contains fine particulate matter or toxic vapors, such as those found in cigarette smoke. Early in the disease, local regulation shunts blood away from the damaged areas, and the individual may not notice problems, even with strenuous activity. As the condition progresses, the reduction in exchange surface limits the ability to provide adequate oxygen. However, obvious clinical symptoms typically fail to appear until the damage is extensive.

Alpha-antitrypsin, an enzyme that is normally present in the lungs, helps prevent degenerative

23

changes in lung tissue. Most people requiring treatment for emphysema are adult smokers; this group includes individuals with alpha-antitrypsin deficiency and those with normal tissue enzymes. In the United States, 1 person in 1000 carries two copies of a gene that codes for an abnormal and inactive form of this enzyme. A single change in the amino acid sequence appears responsible for this inactivation. At least 80 percent of nonsmokers with abnormal alpha-antitrypsin will develop emphysema, generally at ages 45–50. *All* smokers will develop at least some emphysema, typically by ages 35–40.

Unfortunately, the loss of alveoli and bronchioles in emphysema is permanent and irreversible. Further progression can be limited by cessation of smoking; the only effective treatment for severe cases is the administration of oxygen. Lung transplants have helped some patients. For persons with alpha-antitrypsin deficiency who are diagnosed early, attempts are underway to provide enzyme supplements by daily infusion or periodic injection.

Two patterns of symptoms may appear in individuals with advanced emphysema. In one group of patients, other aspects of pulmonary structure and function are relatively normal. The respiratory rate in these patients increases dramatically. The lungs are fully inflated at each breath, and expirations are forced. These individuals maintain near-normal arterial P_{O_2}. Their respiratory muscles are working hard, and they use a lot of energy just breathing. As a result, these patients tend to be thin. Because blood oxygenation is near normal, skin color in Caucasian patients will be pink. The combination of heavy breathing and pink coloration has led to the descriptive term *pink puffers* for individuals with this condition.

In the second group, emphysema has been complicated by chronic bronchitis and chronic airways obstruction. This combination, known as **chronic obstructive pulmonary disease** (COPD), is particularly dangerous. Individuals with COPD commonly expand their chests permanently in an effort to enlarge their lung capacities and make the best use of the remaining functional alveoli. This adaptation gives them a distinctive "barrel-chested" appearance. COPD sufferers also have symptoms of heart failure, including widespread edema. Blood oxygenation is low, and the skin has a bluish coloration. The combination of widespread edema and bluish coloration has led to the descriptive term *blue bloaters* for individuals with this condition.

Mountain Sickness   FAP *p. 849*

Mountain sickness commonly develops after an ascent to altitudes of at least 2500 meters. Symptoms, which appear within a day after arrival, include severe headache, insomnia, tachypnea, and nausea. The underlying cause is thought to be a combination of mild cerebral edema and respirato-ry alkalosis from hyperventilation induced by hypoxia. The symptoms, which may persist for a week before subsiding, can be prevented or reduced by the administration of acetazolamide (*Diamox*). This drug opposes the alkalosis by inhibiting carbonic anhydrase activity and promoting bicarbonate loss at the kidneys.

Acute mountain sickness is a life-threatening condition that may affect (1) individuals performing strenuous physical activities shortly after arriving at high altitudes (typically well above 2500 meters) or (2) individuals adapted to high altitudes who descend to sea level for 1–2 weeks and then return. Fatalities result from the combination of cerebral edema and pulmonary edema. Treatment includes rest, breathing pure oxygen, and immediate transport to lower altitudes.

Shallow Water Blackout   FAP *p. 854*

In preparing to dive under water, misguided swimmers may attempt to outwit their chemoreceptor reflexes. The usual method involves taking some extra-deep breaths before submerging. These individuals are intentionally hyperventilating, usually with the stated goal of "taking up extra oxygen."

From the hemoglobin saturation curve, it should be obvious that this explanation is totally incorrect. At a normal alveolar P_{O_2}, the hemoglobin will be 97.5 percent saturated, and no matter how many breaths the swimmer takes, the alveolar P_{O_2} will not rise significantly. But the P_{CO_2} will be affected, because the increased ventilation rate lowers the carbon dioxide concentrations of the alveoli and blood. This lowering produces the desired effect by temporarily shutting off the chemoreceptors that monitor the P_{CO_2}. As long as CO_2 levels remain depressed, the swimmer does not feel the need to breathe, despite the fact that P_{O_2} continues to fall.

Under normal circumstances, breath holding causes a decline in P_{O_2} and a rise in P_{CO_2}. As indicated in Figure A-48, by the time the P_{O_2} has fallen to 60 percent of normal levels, carbon dioxide levels will have risen enough to make breathing an unavoidable necessity. But after hyperventilation, oxygen levels can fall so low that the swimmer becomes unconscious (generally at a P_{O_2} of 15–20 mm Hg) before the P_{CO_2} rises enough to stimulate breathing. Because this "shallow water blackout" has fatal consequences in most cases, swimmers and divers should not hyperventilate.

Chemoreceptor Accommodation and Opposition   FAP *p. 854*

Carbon dioxide receptors show accommodation to sustained P_{CO_2} levels above or below normal. Although these receptors register the initial change quite strongly, over a period of days they adapt to the new level as "normal." As a result, after several days of elevated P_{CO_2} levels, the effects on the res-

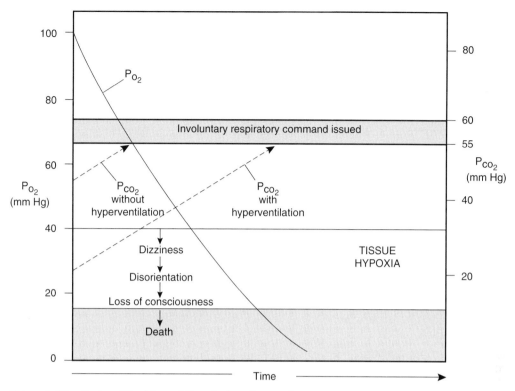

Figure A-48   Carbon Dioxide and Respiratory Demand.
The stimulation of respiration by rising P_{CO_2} normally prevents a dangerous decline in oxygen levels. After hyperventilation, the P_{CO_2} may be so low that respiratory stimulation will not occur before low P_{O_2} impairs neural function.

piratory centers begin to decline. Fortunately, the response to a low arterial P_{O_2} remains intact, and the response to arterial oxygen concentrations becomes increasingly important.

Some patients with severe chronic lung disease are unable to maintain normal P_{O_2} and P_{CO_2} levels in the blood. The arterial P_{CO_2} rises to 50–55 mm Hg, and the P_{O_2} falls to 45–50 mm Hg. At these levels, the carbon dioxide receptors accommodate, and most of the respiratory drive comes from the arterial oxygen receptors. If these patients are given too much oxygen, they may simply stop breathing. Vigorous stimulation to encourage breathing or a mechanical respirator may be required.

☤ Lung Cancer   FAP p. 857

Lung cancers now account for 35 percent of all cancer deaths, making this condition the primary cause of cancer death in the U.S. population. Despite advances in the treatment of other forms of cancer, the survival statistics for lung cancer have not changed significantly. Even with early detection, the 5-year survival rates are only 30 percent for men and 50 percent for women, and most lung cancer patients die within a year of diagnosis. Detailed statistical and experimental evidence has shown that *85–90 percent of all lung cancers are the direct result of cigarette smoking.* Claims to the contrary are simply unjustified and insupportable. The data are far too extensive to detail here, but the

incidence of lung cancer for nonsmokers is 3.4 per 100,000 population, whereas the incidence for smokers ranges from 59.3 per 100,000 for those who smoke between a half-pack and a pack per day to 217.3 per 100,000 for those who smoke one to two packs per day. Before about 1970, this disease primarily affected middle-aged men, but as the number of women smokers has increased, so has the number of women who die from lung cancer.

Smoking changes the quality of the inspired air, making it drier and contaminated with several carcinogenic compounds and particulate matter. The combination overloads the respiratory defenses and damages the epithelial cells throughout the respiratory system. The histological changes that follow were described in Chapter 4 (FAP, p. 142). Whether lung cancer develops appears to be related to the total cumulative exposure to the carcinogenic stimuli. The more cigarettes smoked, the greater the risk, whether those cigarettes are smoked over a period of weeks or years. The histological changes induced by smoking are reversible, and a normal epithelium will return if the stimulus is removed. At the same time, the statistical risks decline to significantly lower levels. Ten years after quitting, a former smoker stands only a 10 percent greater chance of developing lung cancer than does a nonsmoker.

The fact that cigarette smoking typically causes cancer is not surprising in view of the toxic

23

chemicals contained in the smoke. What is surprising is that more smokers do not develop lung cancer. There is evidence that some smokers have a genetic predisposition to developing one form of lung cancer. Dietary factors may also play a role in preventing lung cancer, although the details are controversial. In terms of their influence on the risk of lung cancer, there is a general agreement that (1) vitamin A has no effect, (2) vegetables containing beta-carotene reduce the risk, and (3) a high-cholesterol, high-fat diet increases the risk.

CRITICAL-THINKING QUESTIONS

8-1. During an intramural football game, Joe is tackled so hard that he breaks a rib. On the way to the hospital, he is having a difficult time breathing. Joe may be suffering from
  a.   a collapsed trachea
  b.   an obstruction in the bronchi
  c.   a pneumothorax
  d.   decreased surfactant production
  e.   a bruised diaphragm

8-2. A patient recovering from bacterial pneumonia complains to Dr. Smith of recurrent shortness of breath and pain on breathing. The patient reluctantly admits that he did not finish his antibiotics as prescribed. Dr. Smith auscultates the thorax with a stethoscope and detects a scratching sound during inspiration and decreased breath sounds at the base of one lung. A chest X-ray shows fluid surrounding the base of that lung. Thoracentesis yields cloudy fluid with a large number of WBCs, and bacteria are identified in the cultured pleural fluid specimen. What condition is the patient most likely suffering from?

8-3. Ann, a respiratory care technician, complains of low-grade fever, persistent cough, night sweats, and unexplained weight loss over the last 2 months. Ann does not have a history of cigarette smoking. She has coughed up blood several times in the last 2 days. Her physician orders routine lab tests that are within normal limits. Ann's chest X-ray shows lesions related to prior infection. Her physician is awaiting the results of the sputum culture and has placed a skin test on Ann's forearm. What is a reasonable preliminary diagnosis?
  a.   emphysema
  b.   lung cancer
  c.   tuberculosis
  d.   pleurisy
  e.   pneumonia
What is the significance of the skin test?

NOTES:

The Digestive System

The digestive system consists of the *digestive tract* and *accessory digestive organs.* The digestive tract is divided into the oral cavity, pharynx, esophagus, stomach, small intestine, and large intestine. Most of the absorptive functions of the digestive system occur in the small intestine, with lesser amounts in the stomach and large intestine. The accessory digestive organs provide acids, enzymes, and buffers that assist in the chemical breakdown of food. The accessory organs include the salivary glands, the liver, the gallbladder, and the pancreas. The salivary glands produce saliva, a lubricant that contains enzymes that aid in the digestion of carbohydrates. The liver produces bile, which is concentrated in the gallbladder and released into the small intestine for fat emulsification. The pancreas secretes enzymes and buffers important to the digestion of proteins, carbohydrates, and lipids.

The activities of the digestive system are controlled through a combination of local reflexes, autonomic innervation, and the release of gastrointestinal hormones such as *gastrin, secretin,* and *cholecystokinin.*

SYMPTOMS AND SIGNS OF DIGESTIVE SYSTEM DISORDERS

The functions of the digestive organs are varied, and the symptoms and signs of digestive system disorders are equally diverse. Common symptoms of digestive disorders include the following:

- Pain is a common symptom of digestive disorders. Widespread pain in the oral cavity may result from (1) trauma; (2) infection of the oral mucosa by bacteria, fungi, or viruses; or (3) a deficiency in vitamin C (*scurvy,* p. 4) or one or more of the B vitamins. Focal pain in the oral cavity accompanies (1) infection or blockage of salivary gland ducts; (2) tooth disorders such as tooth fractures, *dental caries, pulpitis,* abscess formation, or *gingivitis* (FAP *p. 872);* and (3) oral lesions, such as those produced by the *herpes simplex* virus.

Abdominal pain is characteristic of a variety of digestive disorders. In most cases the pain is perceived as distressing but tolerable; if the pain is acute and severe, a surgical emergency may exist. Abdominal pain can cause rigidity in the abdominal muscles in the painful area. This rigidity can be easily felt on palpation. The muscle contractions (*guarding*) may be voluntary, in an attempt to protect a painful area, or an involuntary spasm resulting from irritation of the peritoneal lining, as in *peritonitis* (FAP *p. 867).* Persons with peritoneal inflammation also experience *rebound tenderness.* In rebound tenderness, pain appears when fingertip pressure on the abdominal wall is suddenly removed.

Abdominal pain can result from disorders of the digestive, circulatory, urinary, or reproductive system. Digestive tract disorders producing abdominal pain include *appendicitis* (p. 137), *gastric or duodenal ulcers* (p. 165), *pancreatitis* (FAP *p. 891), cholecystitis* (p. 173), *hepatitis* (p. 172), *intestinal obstruction* (p. 164), *diverticulitis* (FAP *p. 901), peritonitis,* and cancers of the digestive tract.

- *Dyspepsia,* or indigestion, is pain or discomfort in the substernal or epigastric region. Digestive tract disorders associated with dyspepsia include *esophagitis* (FAP *p. 877), gastritis* (FAP *p. 881), gastric or duodenal ulcers* (p. 165), *gastroesophageal reflux,* and *cholecystitis* (p. 174).

- *Nausea* is a sensation that usually precedes or accompanies vomiting. Nausea may result from digestive disorders or from disturbances of CNS function.

- *Anorexia* is a decrease in appetite that, if prolonged, is accompanied by weight loss. Digestive disorders that cause anorexia include *gastric cancer* (FAP *p. 884),* pancreatitis (p. 896), hepatitis (p. 172), and several forms of *diarrhea* (p. 174). Anorexia may also accompany disorders that involve other systems. *Anorexia nervosa,* an eating disorder with a psychological basis, is discussed in the text *(p. 950).*

- *Dysphagia* is difficulty in swallowing. This difficulty may result from trauma, infection, inflammation, or blockage of the posterior oral cavity, pharynx, or esophagus. For example, the infections of *tonsillitis, pharyngitis,* and *laryngitis* may cause dysphagia.

THE PHYSICAL EXAMINATION AND THE DIGESTIVE SYSTEM

Physical assessment can provide information useful in the diagnosis of digestive system disorders. The abdominal region is particularly important, because most of the digestive system is located within the abdominopelvic cavity. We will now discuss four methods of physical assessment of the digestive system: *inspection, palpation, percussion,* and *auscultation* (p. 7):

1. *Inspection* can provide a variety of useful diagnostic clues:

    - Bleeding of the gums, as in *gingivitis,* and characteristic oral lesions can be seen on inspection of the oral cavity. Examples of distinctive lesions include those of oral herpes simplex infections and *thrush,* lesions produced by infection of the mouth by *Candida albicans.* This fungus, a normal resident of the digestive tract, causes widespread oral infections in immunodeficient persons, such as AIDS patients or patients undergoing immunosuppressive therapies.

24

(Healthy infants may also get a mild oral fungal infection.)

- Peristalsis in the stomach and intestines may be seen as waves passing across the abdominal wall in persons who do not have a thick layer of abdominal fat. The waves become very prominent during the initial stages of intestinal obstruction.

- A general yellow discoloration of the skin, a sign called *jaundice,* results from excessive levels of bilirubin in body fluids. Jaundice is commonly seen in individuals with *cholecystitis* (p. 174) or liver diseases such as *hepatitis* (p. 172) and *cirrhosis* (p. 173).

- Abdominal distention may be caused by (1) fluid accumulation in the peritoneal cavity, as in *ascites* (FAP *p. 867*); (2) air or gas (flatus) within the digestive tract or peritoneal cavity; (3) obesity; (4) abdominal masses, such as tumors, or enlargement of visceral organs; (5) pregnancy; (6) the presence of an *abdominal hernia* (p. 66); or (7) fecal impaction, as in severe and prolonged constipation.

- *Striae* are multiple scars, 1–6 cm in length, that are visible through the epidermis. Striae develop in damaged dermal tissues after stretching, and they are typically seen in the abdominal region after a pregnancy or other rapid weight gain. Abnormal striae may develop after ascites or in cases of subcutaneous edema, and purple striae are characteristic signs of *Cushing's syndrome* (adrenocortical hypersecretion, p. 97).

2. *Palpation* of the abdomen may reveal specific details about the status of the digestive system, including:

- The presence of abnormal masses, such as tumors, within the peritoneal cavity

- Abdominal distention from (1) excess fluid within the digestive tract or peritoneal cavity or (2) gas within the digestive tract

- Herniation of digestive organs through the inguinal canal or other weak spots in the abdominal wall (p. 66)

- Changes in the size, shape, or texture of visceral organs. For example, in several liver diseases, the liver becomes enlarged and firm, and these changes can be detected on palpation of the right upper quadrant.

- Voluntary or involuntary abdominal muscle contractions

- Rebound tenderness

- Specific areas of tenderness and pain. For example, someone with acute *hepatitis* (p. 172), a liver disease, generally experi-

ences pain on palpation of the upper right quadrant. In contrast, a person with *appendicitis* (p. 137) generally experiences pain when the right lower quadrant is palpated.

3. *Percussion* of the abdomen is less instructive than percussion of the chest, because the visceral organs do not contain extensive air spaces that will reflect the sounds conducted through surrounding tissues. However, the stomach usually contains a small air bubble, and percussion over this area produces a sharp, resonant sound. The sound becomes dull or disappears when the stomach fills, the spleen enlarges, or the peritoneal cavity contains abnormal quantities of peritoneal fluid, as in *ascites* (FAP *p. 867*). Percussion may also detect gas accumulation within the peritoneal cavity or within portions of the colon.

4. *Auscultation* can detect gurgling abdominal sounds, or *bowel sounds,* produced by peristaltic activity along the digestive tract. Increased bowel sounds occur in persons with acute diarrhea, and bowel sounds may disappear in persons with (1) advanced intestinal obstruction; (2) *peritonitis,* an infection of the peritoneal lining; and (3) spinal cord injuries that prevent normal innervation.

Diagnostic procedures, such as endoscopy, are commonly used to provide additional information. Information on representative diagnostic procedures and laboratory tests are given in Tables A-26 and A-27 (p. 168).

DISORDERS OF THE DIGESTIVE SYSTEM

Due to the number and diversity of digestive organs, there are many types of digestive system disorders. One common method of categorizing these disorders (Figure A-49a, p. 171) uses a combination of four anatomical and functional characteristics:

1. *Disorders of the oral cavity.* This category includes disorders that affect the oral mucosa, the teeth, and the salivary glands.

2. *Disorders of the digestive tract.* This category includes the conditions usually called *gastrointestinal disorders.*

3. *Disorders of the digestive viscera.* This category includes the hepatic (liver), biliary (gallbladder and bile ducts), and pancreatic disorders.

4. *Nutritional and metabolic disorders.* These disorders (Figure A-49b, p. 171), which affect the entire body, are often considered with disorders of the digestive system, because they typically accompany digestive system disorders. However, nutritional and metabolic disorders may also reflect (1) disorders involving the endocrine or

Table A-26   Examples of Diagnostic Tests Used in the Diagnosis of Gastrointestinal Disorders

Diagnostic Procedure	Method and Result	Representative Uses	Notes
ORAL CAVITY Sialography	An X-ray film is taken after contrast medium is injected into the salivary ducts while patient's salivary glands are stimulated	Identification of calculi, inflammation and tumors in the salivary duct or gland	
Periapical (PA) X-rays	The periapical film is an X-ray of the crown and root area of several teeth, but the angle of the X-ray cone may cause one tooth to overlap an adjacent tooth	Periapical films allow detection of tooth decay, tooth impactions, fractures, progression of bone loss with periodontal disease, inflammation of the periodontal ligament that surrounds the tooth root, and periapical abscesses (abscess at the end of the tooth root)	
Bitewing X-rays of teeth	The bitewing film is taken interproximally. The X-ray cone is pointed perpendicularly toward the spaces between the crowns of adjacent posterior teeth to provide an X-ray observation of the area where one tooth crown contacts another.	Bitewing films reveal tooth decay between the teeth and early bone loss in peridontal disease	
UPPER GI (esophagus, stomach, and duodenum) Esophagogastroduodenoscopy, esophagoscopy, gastroscopy	A fiber-optic endoscope is inserted into the oral cavity and further into the esophagus, stomach, and duodenum to permit visualization of the lining and lumen of the area of the alimentary tract under question; air is used to prevent lumen collapse	Detection of tumors, ulcerations, polyps, esophageal varices, inflammation and obstructions; to provide biopsy of tissues in the upper GI tract	Endoscopy is also used to obtain biopsy specimens, to remove polyps, and to terminate bleeding along areas of the gastrointestinal tract
Barium swallow	A series of X-rays is taken of the esophagus after barium sulfate is ingested to increase contrast of structures; normally done with the upper GI series	Determination of abnormalities of the pharynx and esophagus, especially to determine the cause of dysphagia; identification of tumors, hiatal hernia, and polyps	
Upper GI series	A series of X-rays is taken of the stomach and duodenum after barium sulfate is ingested to increase contrast of structures	To determine cause of epigastric pain; to detect ulcers, polyps, gastritis, tumors and inflammation within the upper GI tract	

nervous system, (2) congenital metabolic problems, or (3) dietary abnormalities.

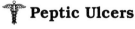

Peptic Ulcers
FAP p. 881

Peptic ulcers are lesions of the gastric or duodenal mucosa caused by bacterial action or erosion by acids and pepsin in gastric juice. Regardless of the primary cause of a gastric ulcer, once gastric juices have destroyed the epithelial layers, the virtually defenseless lamina propria will be exposed to digestive attack. Sharp abdominal pain results. In severe cases, the damage to the mucosa of the stomach may cause significant bleeding, and the acids may even erode through the wall of the stomach and into the peritoneal cavity. This condition, called a **perforated ulcer,** requires immediate surgical correction.

The administration of antacids can typically control gastric or duodenal ulcers by neutralizing the acids and allowing time for the mucosa to

24

Table A-26 *(continued)*

Diagnostic Procedure	Method and Result	Representative Uses	Notes

Esophageal function studies
  Acid reflux
  Acid clearing
	A slender tube is passed from the oral cavity into the stomach. A pH probe is placed in the oral cavity and then passed into the esophagus. A dilute solution of hydrochloric acid is passed through the tube into the stomach. The pH of the esophagus at the level of the gastroesophageal sphincter is measured periodically to detect reflux of the acid. The solution of hydrochloric acid is then given again, and the patient is asked to swallow several times until the acid (as measured by the pH probe) is cleared from the esophagus.	Detection of gastroesophageal reflux
		
		Acid clearing normally occurs in less than 10 swallows. If more are required, the patient may have severe esophagitis from chronic gastric acid reflux.	

Gastric analysis
  Basal gastric
  secretion
	When patient is in fasting state, a nasogastric (NG) tube is inserted and stomach contents are removed. Acid levels are periodically monitored through the NG tube to determine basal gastric secretion.	To identify the cause of ulceration of the gastric lining. Increased acid levels may be the cause of the ulcer. If no increased acidity with ulceration is present, a malignancy may exist.	Also used to detect Zollinger–Ellison syndrome (increased gastrin from pancreatic tumor causes very elevated acid production by stomach)

Gastric acid stimulation	A chemical stimulus is administered and acid levels are monitored to determine the results of the gastric acid stimulation test		
Ateriography (celiac and mesenteric)	A catheter is inserted into the femoral artery and threaded to the celiac or mesenteric artery, where contrast medium is injected; X-ray films are then taken	To determine cause of bleeding in gastrointestinal tract or blockage of the blood supply	
LOWER GI (colon) Barium enema	Barium sulfate enema is administered to provide contrast in intestinal lumen, and an X-ray film series is taken; air may also be administered for better contrast	To determine the cause of abdominal pain and bloody stools; to detect tumors and obstructions of the bowel; to identify polyps and diverticula	
Sigmoidoscopy Colonoscopy	A fiber-optic tube inserted into the rectum for viewing of the anus, rectum, and sigmoid colon; a colonoscopy allows viewing of the entire large intestine to the cecum; biopsy of intestinal tissue can also be performed using these endoscopes	For detection of tumors, polyps, and ulcerations of the intestinal lining. Polyps may be removed with the scope.	

Table A-26 *(continued)*

Diagnostic Procedure	Method and Result	Representative Uses	Notes
LIVER, GALLBLADDER, PANCREAS			
X-ray of abdomen	Radiodense tissues appear in white on a negative film image	For detection of abdominal masses and obstructions	
Computerized tomography of liver, pancreas, gallbladder	Cross-sectional radiation is applied to provide a three-dimensional image of the liver, gallbladder, and pancreas; contrast media is normally used	To identify tumors, pancreatitis (acute, chronic), hepatic cysts and abscesses, and biliary calculi (stones) and to determine the cause of jaundice	
Ultrasonography of liver, pancreas, gallbladder	The liver, pancreas, and gallbladder are examined by means of sound waves emitted by transducer placed on the abdomen. Sound waves deflect off dense structures and produce a series of echoes, which are amplified and graphically recorded.	For detection of polyps, tumors, abscesses, hepatic cysts, and gallstones, and for determination of the cause of jaundice	
Oral cholecystography (gallbladder)	Patient swallows dye tablets the night before the X-ray series is to be taken; calculi appear as dark areas in the radiopaque gallbladder	To detect the presence of calculi and inflammation of the gallbladder; has largely been replaced by ultrasound exams	
Cholangiography Intravenous or percutaneous	Intravenously administered dye is concentrated in the liver and released into the bile duct; a series of X-ray films is taken. Percutaneous method involves the insertion of a needle with catheter into the bile duct, using ultrasonography for guidance; dye is then administered through the catheter.	For identification of biliary calculi obstructing ducts	
Liver biopsy	A needle is inserted through a small skin incision and into the liver; liver tissue is removed by excision or aspiration	Liver tissue is examined for evidence of cirrhosis, hepatitis, tumors, and granuloma	
Endoscopic retrograde cholangiopancreatography	Duodenoscope (fiber-optic tube) is inserted into the oral cavity and threaded through the stomach to the duodenum; through the scope, a catheter is inserted into the duodenal ampulla for dye injection; X-rays are then taken	For detection of calculi, tumors, or cysts of the pancreatic or bile ducts; also useful in the determination of the presence of a pancreatic tumor	

24

regenerate. The drug *cimetidine* (sī-MET-i-dēn), or *Tagamet*®, inhibits the secretion of acid by the parietal cells. Dietary restrictions limit the intake of foods that promote acid production (caffeine and pepper) or that damage unprotected mucosal cells (alcohol and aspirin).

The treatment of peptic ulcers has changed radically since the identification of *Helicobacter pylori* as a likely causative agent. These bacteria are able to resist gastric acids long enough to penetrate the mucous coating of the epithelium. Once within the protective layer of mucus, they bind to the epithelial surfaces, where they are safe from the action of gastric acids and enzymes. Over time, the bacteria release toxins that damage the epithelial lining. These toxins ultimately result in the erosion of the epithelium and the destruction of the lamina propria by gastric juices. Individuals whose stomachs

Table A-27   Examples of Laboratory Tests Used in the Diagnosis of Gastrointestinal Disorders

Laboratory Test	Normal Values in Blood Plasma or Serum	Significance of Abnormal Values	Notes
UPPER AND LOWER GI			
Serum electrolytes			
Potassium Sodium Magnesium	Adults: 3.5–5.0 mEq/l 135–145 mEq/l 1.5–2.5 mEq/l	Vomiting, diarrhea, and naso-gastric intubation can cause electrolyte losses and create imbalances	
Serum gastrin level	Adults: 45–200 pg/ml	Elevated levels occur with pernicious anemia and some gastric ulcers; very high levels in Zollinger–Ellison syndrome (pancreatic tumor that produces excessive gastrin)	
Lactose tolerance test	Fasting patient receives 100 mg of lactose. Blood taken at periodic intervals up to 2 hours should show plasma glucose increasing to >20 mg/dl.	Blood glucose levels do not rise to normal levels after lactose ingestion because the patient lacks the necessary enzyme lactase, which breaks down lactose into galactose and glucose	Patients who are lactose-intolerant frequently have abdominal cramping and diarrhea after ingesting milk products
Carcinoembryonic antigen (CEA) (in plasma)	Adults: Nonsmokers <2.5 ng/ml Smokers <3.5 ng/ml	Colorectal cancer is a possibility with elevated levels; this antigen is not specific for colorectal cancer; elevated levels may occur in other types of carcinomas and in ulcerative colitis	CEA levels are useful in determining the effectiveness of colorectal cancer treatment
Stool culture			
Culture and sensitivity (C&S). A sample of the stool is cultured for bacterial content.	Only normal intestinal flora such as *E. coli* should be isolated	Bacteria such as *Shigella*, *Campylobacter,* and *Salmonella* can cause acute diarrhea	
Ova and parasites (O&P). A stool sample is microscopically examined for parasite eggs or adults.	None found	Typical parasites found in feces that cause intestinal disturbances are tapeworms, entamoeba, and some protozoans, such as *Giardia*	
Fecal analysis			
Occult blood	None found	Hidden (occult) blood in the feces can occur when bleeding from GI tract is minimal due to inflammation, ulceration, or with a tumor that is causing small amounts of bleeding	
Fat content	Adults: <6 g/24 hours	Fat content >6 g/24 hours, a condition called steatorrhea, often indicates malabsorption syndrome; may also occur in cystic fibrosis and pancreatic diseases	

Table A-27 (*continued*)

Laboratory Test	Normal Values in Blood Plasma or Serum	Significance of Abnormal Values	Notes
LIVER and GALLBLADDER			
Serum bilirubin (total)	Total serum bilirubin: 0.1–1.2 mg/dL	Can be elevated with hepatitis (infectious or toxic) and biliary obstruction	
	Indirect bilirubin: 0.2–1.0 mg/dL (unconjugated)	Excessive hemolysis as in transfusion reactions or erythroblastosis fetalis; elevation also occurs with hepatitis	
	Direct bilirubin: 0.1–0.3 mg/dl (conjugated)	Elevated levels occur with obstructions of the bile duct system, hepatitis, cirrhosis of the liver, and liver cancer	Direct bilirubin is water-soluble; when levels are elevated, it can be released in the urine
	Infant (newborn) total serum bilirubin: 1–12 mg/dl	Bilirubin >15 mg/dl can result in serious neurological problems	
Urine bilirubin	None	Presence of detectable bilirubin occurs when the direct bilirubin is elevated as in obstruction of the bile duct system, hepatitis, cirrhosis of the liver, and liver cancer	
Liver Enzyme Tests			These enzymes are also produced in other areas of the body; elevations are not specific to liver disease
Aspartate aminotransferase (AST) (or SGOT)	Adults: 0–35 U/l	Elevated levels occur with hepatitis, acute pancreatitis, and cirrhosis	
Alanine aminotransferase (ALT or SGPT)	Adults: 0–35 IU/l	Elevated levels occur with hepatitis and other liver diseases	Along with 5'-nucleotidase, this test is more specific than the others for diseases of the liver
Alkaline phosphatase (isoenzyme ALP$_1$)	Adults: 30–120 mU/mL	Elevated levels occur in biliary obstructions, hepatitis, and liver cancer	
5'-nucleotidase	Adults: 0–1.6 units	Elevated levels are useful for early detection of liver disease	
Gamma-glutamyl transferase/transpeptidase (GGT or GGTP)	Adults: 8–38 U/l	Elevated levels occur with liver disease; levels also become elevated after the ingestion of alcohol	
Serum protein	Adults: Albumin 3.5–5.5 g/dl (or 52–68% of the total protein)	Decreased levels of albumin and increased levels of globulins occur in chronic liver disease	Liver production of albumin may be reduced with liver disorders
	Adults: Globulin 2.0–3.0 g/dl (or 32–48% of the total protein)		

harbor *H. pylori* are also at higher than normal risk of gastric cancer, although the reason is not known.

Current treatment for *H. pylori*-related ulcers consists of a combination of up to three antibiotics (*tetracycline, amoxycillin,* and *metronidazole*), Pepto-Bismol™, and treatment with drugs that suppress acid production. Because strains of *H. pylori* resistant to at least one of these antibiotics have already appeared, research is under way to find alternative methods of controlling these infections.

24

Table A-27 *(continued)*

Laboratory Test	Normal Values in Blood Plasma or Serum	Significance of Abnormal Values	Notes
Ammonia (plasma)	Adult: 15–45 µg/dl	Elevated level occurs with liver failure and hepatic encephalopathy	
Hepatitis virus studies	None	Antibodies can be detected against the virus that causes hepatitis A (HAV); hepatitis B (HBV) can be detected by testing for both the viral antigen and antibodies; hepatitis C (HCV) can be detected by testing for antibodies against the virus	
Prothrombin time (PT) test	Adults: 11–15 seconds	Prolonged time occurs with liver disease, vitamin K deficiency, and warfarin (coumadin), administration	Reduced liver production of clotting factors I, II, VII, X will cause prolonged PT
PANCREAS			
Amylase			
Urine	Adult: 35–260 U/l hour	Elevated levels occur for 7–10 days after pancreatic disease begins	
Serum	Adult: 60–180 U/l	Elevated levels occur with pancreatic disease and with obstruction of the pancreatic duct	
Lipase (serum)	Adult: 0–110 units/l	Elevated levels most commonly occur in acute pancreatitis	
Sweat electrolytes test			
Sodium	Children: <70 mEq/l	Elevated levels of both sodium and chloride occur in cystic fibrosis; does not reflect severity of the disorder	
Chloride	Children: <50 mEq/l		

 Drastic Weight-Loss Techniques

FAP *p. 887*

At any moment, an estimated 20 percent of the U.S. population is dieting to promote weight loss. In addition to the appearance of "fat farms" and exercise clubs, there has been an increase in the use of surgery to promote weight loss. Many of the techniques involve the surgical remodeling of the gastrointestinal tract. **Gastric stapling** attempts to correct an overeating problem by reducing the size of the stomach. A large portion of the gastric lumen is stapled shut, leaving only a small pouch in contact with the esophagus and duodenum. After this surgery, the individual will be able to eat only a small amount before the stretch receptors in the gastric wall become stimulated and a feeling of fullness results. Gastric stapling is a major surgical procedure with many potential complications. In addition, the smooth muscle in the wall of the functional portion of the stomach gradually becomes increasingly tolerant of distension, and the operation may have to be repeated.

A more drastic approach involves the surgical removal or bypass of a large portion of the jejunum. This procedure reduces the effective absorptive area, producing a marked weight loss. After the operation, the individual must take dietary supplements to ensure that all the essential nutrients and vitamins can be absorbed before the chyme enters the large intestine. Chronic diarrhea and serious liver disease are potential complications.

 Liver Disease FAP *p. 894*

The liver is the largest and most important visceral organ, and liver disorders affect almost every other vital system in the body. A variety of clinical tests are used to check the functional and physical state of the liver (see Tables A-26 and A-27):

- **Liver function tests** assess specific functional capabilities.

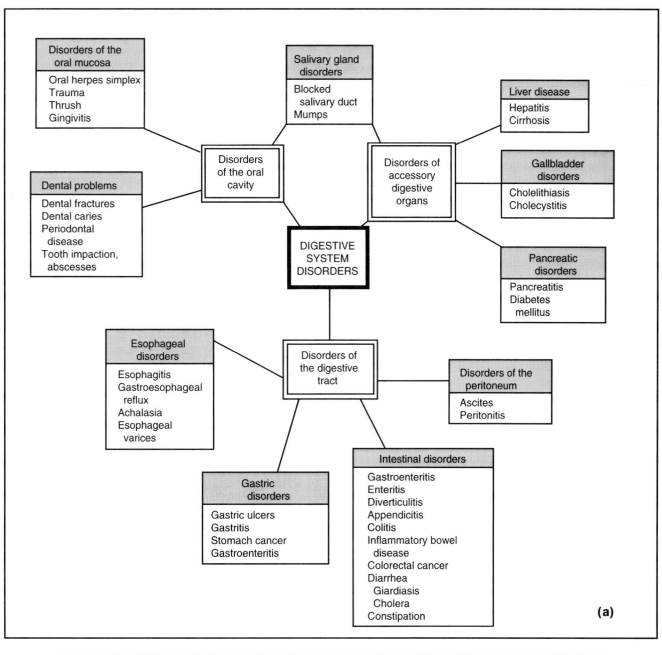

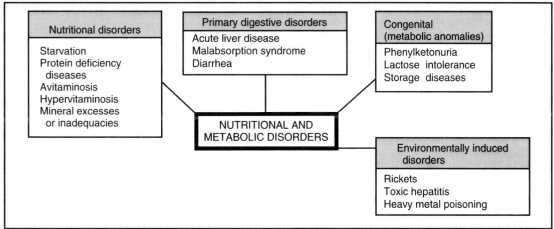

Figure A-49   Disorders of the Digestive System.

(a) Major categories of digestive system disorders. (b) Nutritional and metabolic disorders.

- A **serum bilirubin assay** indicates how efficiently the liver has been able to extract and excrete bilirubin.

- Serum and plasma protein assays can detect changes in the liver's rate of plasma protein synthesis, and serum enzyme tests can reveal liver damage by detecting intracellular enzymes in the circulating blood.

- **Liver scans** involve the injection of radioisotope-labeled compounds into the circulation. Compounds are chosen that will be selectively absorbed by either Kupffer cells, liver cells, or abnormal liver tissues.

- CT scans of the abdominal region are commonly used to provide information about cysts, abscesses, tumors, or hemorrhages in the liver.

- A **liver biopsy** can also be taken by laparoscopy, in which a long needle is inserted through the abdominal wall. Laparoscopic examination can also reveal gross structural changes in the liver or gallbladder.

The term *liver disease* includes a variety of disorders. We will focus here on two major liver diseases, *hepatitis* and *cirrhosis*.

HEPATITIS

Hepatitis (hep-a-TĪ-tis) is defined as inflammation of the liver. Viruses that target the liver are responsible for most cases of hepatitis, although some environmental toxins can cause similar symptoms. There are at least six forms of viral hepatitis, A, B, C, D, E, and G:

1. **Hepatitis A,** or *infectious hepatitis,* typically results from the ingestion of good, water, milk, or shellfish contaminated by infected fecal wastes. It has a relatively short incubation period of 2–6 weeks. The disease generally runs its course in a matter of months, and fatalities are rare among individuals under 40 years of age.

2. **Hepatitis B,** or *serum hepatitis,* is transmitted by the exchange of body fluids during intimate contact. For example, infection may occur through the transfusion of blood products, through a break in the skin or mucosa, or by sexual contact. The incubation period ranges from 1 to 6 months. If a pregnant woman is infected, the newborn baby may become infected at birth.

3. **Hepatitis C,** originally designated *non-A, non-B hepatitis,* is most commonly transferred from individual to individual through the collection and transfusion of contaminated blood. Since 1990, screening procedures have been used to lower the incidence of transfusion-related hepatitis C. Hepatitis C can also be transmitted among intravenous drug users through shared needles, and there is evidence that hepatitis C

may also be sexually transmitted. Chronic hepatitis C infections produce significant liver damage in roughly half the individuals infected with the virus. Interferon treatment early in the disease can slow the progression of the disease in some patients.

4. **Hepatitis D** is caused by a virus that produces symptoms only in persons already infected with hepatitis B. The transmission of hepatitis D resembles that of hepatitis B. In the United States, the disease is most common among intravenous drug users. The combination of hepatitis B and hepatitis D causes progressive and severe liver disease.

5. **Hepatitis E** resembles hepatitis A in that it is transmitted by the ingestion of contaminated food or water. Hepatitis E is the most common form of hepatitis worldwide, but cases seldom occur in the United States. For reasons that are as yet unknown, hepatitis E infections are most acute and severe among pregnant women, and the disease is typically fatal.

6. **Hepatitis G,** the most recently described form, appears in the same populations as hepatitis C. Little else is known about this type of hepatitis.

The hepatitis viruses disrupt liver function by attacking and destroying liver cells. An infected individual may develop a high fever, and the liver may become inflamed and tender. As the disease progresses, several hematological parameters change markedly. For example, enzymes normally confined to the cytoplasm of functional liver cells appear in the circulating blood. Normal metabolic regulatory activities become less effective, and blood glucose declines. Plasma protein synthesis slows, and the clotting time becomes unusually long. The injured hepatocytes stop removing bilirubin from the circulating blood, and symptoms of jaundice appear.

Hepatitis may be *acute* or *chronic.* Acute hepatitis is characteristic of hepatitis forms A and E. Almost everyone who contracts hepatitis A or hepatitis E eventually recovers, although full recovery may take several months. Symptoms of acute hepatitis include severe fatigue and jaundice. In chronic hepatitis, fatigue is less pronounced and jaundice is rare. *Chronic active hepatitis* is a progressive disorder that can lead to severe medical problems as liver function deteriorates and *cirrhosis* develops. Common complications include the following:

- The formation of *varices* (FAP *p. 876*) due to portal hypertension

- *Ascites* (FAP *p. 867*) due to increased peritoneal fluid production at the elevated venous pressures

- *Bacterial peritonitis,* which may recur for unknown reasons

- *Hepatic encephalopathy*, which is characterized by disorientation and confusion. These symptoms are the result of the disruption of CNS function due to some combination of factors in the blood. High ammonia levels and the presence of abnormal concentrations of fatty acids, amino acids, and waste products have been implicated.

Hepatitis B, C, D, and E infections may produce either acute or chronic hepatitis. The chronic forms may eventually lead to liver failure and death. Roughly 10 percent of hepatitis B patients develop potentially dangerous complications; the percentage may be higher for hepatitis C. The prognosis is significantly worse for those infected with both hepatitis B and hepatitis D. Chronic active hepatitis infections (especially B or C) may also lead to liver cancer. The long-term effects of hepatitis G infection remain to be determined.

Passive immunization with pooled immunoglobulins is available for both the hepatitis A and B viruses. (Active immunization for hepatitis A and B is also available.) No vaccines are available to stimulate immunity to hepatitis forms C, D, E, or G.

CIRRHOSIS

The underlying problem in **cirrhosis** (sir-Ō-sis) appears to be the widespread destruction of hepatocytes by exposure to drugs (especially alcohol), viral infection, ischemia, or blockage of the hepatic ducts. Two processes are involved in producing the symptoms. Initially, the damage to hepatocytes leads to the formation of extensive areas of scar tissue that branch throughout the liver. The surviving hepatocytes then undergo repeated cell divisions, but the fibrous tissue prevents the new hepatocytes from achieving a normal lobular arrangement. As a result, the liver gradually converts from an organized assemblage of lobules to a fibrous aggregation of poorly functioning cell clusters. Jaundice, ascites, and other symptoms may appear as the condition progresses.

Cholecystitis   FAP *p. 896*

An estimated 16–20 million people in the United States have gallstones that go unnoticed; small stones are commonly flushed down the bile duct and excreted. If gallstones enter and jam the cystic or bile duct, the painful symptoms of **cholecystitis** (kō-lē-sis-TĪ-tis) appear. Approximately 1 million people develop acute symptoms of cholecystitis each year. The gallbladder becomes swollen and inflamed, infections may develop, and symptoms of *obstructive jaundice* develop.

A blockage that does not work its way down the duct to the duodenum must be removed or destroyed. Small gallstones can be chemically dissolved. One chemical now under clinical review is *methyl tert-butyl ether* (MTBE). When introduced into the gallbladder, MTBE dissolves gallstones in a matter of hours; testing to date has not shown many undesirable side effects. (You may have heard of this compound in another context: It is a gasoline additive used to prevent engine knocking.)

In most cases, surgery is required to remove large gallstones. (The gallbladder is also removed to prevent recurrence.) Many surgeons are now using a laparoscope, inserted through two or three small abdominal incisions, to perform this surgery. A *lithotripter* may also be used to shatter the gallstones with focused sound waves.

Colon Inspection and Cancer

FAP *p. 899*

Each year roughly 152,000 new cases of colorectal cancer are diagnosed in the United States, and 57,000 deaths result from this condition. This deadly form of cancer is normally diagnosed in persons over 50 years of age. Primary risk factors for colorectal cancer include (1) a diet rich in animal fats and low in fiber, (2) *inflammatory bowel disease* (p. 174), and (3) a number of inherited disorders that promote epithelial tumor formation along the intestines.

Successful treatment of colorectal cancer depends on the early identification of the condition. It is believed that colorectal cancers begin as small, localized mucosal tumors, or **polyps** (POL-ips), that grow from the intestinal wall. The prognosis improves dramatically if cancerous polyps are removed before metastasis has occurred. If the tumor is restricted to the mucosa and submucosa, the 5-year survival rate is higher than 90 percent. If it extends into the serosa, the survival rate drops to 70–85 percent; after metastasis to other organs, the rate drops to about 5 percent.

One of the early signs of polyp formation is the appearance of blood in the feces. Unfortunately, many people ignore small amounts of blood in fecal materials, because they attribute the bleeding to "harmless" hemorrhoids. This offhand diagnosis should always be professionally verified.

When blood is detected in the feces, X-ray techniques are commonly used as a first step in diagnosis. In the usual procedure, a large quantity of a liquid barium solution is introduced by enema. Because this solution is radiopaque, the X-rays will reveal any intestinal masses, such as a large tumor, blockages, or structural abnormalities.

Typically, the most precise surveys can be obtained with the aid of a flexible **colonoscope** (ko-LON-o-skōp). This instrument permits direct visual inspection of the lining of the large intestine. Variations on this procedure permit the collection of tissue samples. The colonoscope can also be used to remove polyps, and this removal avoids the potential complications of traditional surgery.

24

Inflammatory Bowel Disease

FAP p. 901

The general term **colitis** (ko-LĪ-tis) may be used to indicate a condition characterized by inflammation of the colon. **Irritable bowel syndrome** is characterized by diarrhea, constipation, or an alternation between the two. When constipation is the primary problem, this condition may be called a *spastic colon,* or *spastic colitis.* Irritable bowel syndrome may have a partly psychological basis; the mucosa is usually normal in appearance. **Inflammatory bowel disease,** such as *ulcerative colitis,* involves chronic inflammation of the digestive tract, most commonly affecting the colon. The mucosa becomes inflamed and ulcerated, extensive areas of scar tissue develop, and colonic function deteriorates. Acute bloody diarrhea and cramps are common symptoms. Fever and weight loss are also typical complaints. Treatment of inflammatory bowel disease normally involves anti-inflammatory drugs and corticosteroids that reduce inflammation. In severe cases, oral or intravenous fluid replacement is required. In cases that do not respond to other therapies, immunosuppresive drugs, such as *cyclosporine* may be used to good effect.

Treatment of severe inflammatory bowel disease may also involve a **colectomy** (ko-LEK-to-mē), the removal of all or a portion of the colon. If a large part or even all of the colon must be removed, normal connection with the anus cannot be maintained. Instead, the end of the intact digestive tube is sutured to the abdominal wall, and wastes then accumulate in a plastic pouch or sac attached to the opening. If the attachment involves the colon, the procedure is a **colostomy** (ko-LOS-to-mē); if the ileum is involved, it is an **ileostomy** (il-ē-OS-to-mē).

 Diarrhea

FAP p. 903

Diarrhea is characterized as the production of copious, watery stools. Many different conditions will result in diarrhea, and we will consider only a representative sampling. Most infectious diarrhea involves organisms that are shed in the stool and then spread from person to person (or from person to food to person). Proper hand washing after defecation, clean drinking water, and good sewage disposal are the best preventive measures.

GASTROENTERITIS

An irritation of the small intestine may lead to a series of powerful peristaltic contractions that eject the contents of the small intestine into the large intestine. An extremely powerful irritating stimulus will produce a "clean sweep" of the absorptive areas of the digestive tract. Vomiting clears the stomach, duodenum, and proximal jejunum, and peristaltic contractions evacuate the distal jejunum and ileum. Bacterial toxins, viral infections, and various poisons may produce these extensive gas-

trointestinal responses. Conditions affecting primarily the small intestine are usually referred to as a form of **enteritis** (en-ter-Ī-tis). If both vomiting and diarrhea are present, the term **gastroenteritis** (gas-trō-en-ter-Ī-tis) may be used instead.

TRAVELER'S DIARRHEA

Traveler's diarrhea, a form of infectious diarrhea generally caused by a bacterial or viral infection, develops because the irritated or damaged mucosal cells are unable to maintain normal absorption levels. The irritation stimulates the production of mucus, and the damaged cells and mucous secretions add to the volume of feces produced. Despite the inconvenience, this type of diarrhea is usually temporary, and mild diarrhea is probably a reasonably effective method of rapidly removing an intestinal irritant. Drugs, such as *Lomotil®,* that prevent peristaltic contractions in the colon relieve the diarrhea but leave the irritant intact, and the symptoms may return with a vengeance when the drug effects fade. A 5-day course of antibiotics may be effective in controlling diarrhea due to bacterial infection.

GIARDIASIS

Giardiasis is an infection caused by the protozoans *Giardia intestinalis* and *G. lamblia.* The pathogens can colonize the duodenum and jejunum and interfere with the normal absorption of lipids and carbohydrates. Many people do not develop acute symptoms but instead act as carriers who can spread the disease. When acute symptoms develop, they usually appear within 3 weeks of initial exposure. Diarrhea, abdominal pains, cramps, nausea, and vomiting are the primary complaints. These symptoms persist for 5–7 days, although some patients are subject to relapses. Treatment typically consists of the oral administration of drugs, such as *quinacrine* or *metronidazole,* that can kill the protozoan.

The transmission of giardiasis requires that food or water be contaminated with feces that contain *cysts,* resting stages of the protozoan that are produced during passage through the large intestine. Rates of infection are highest (1) in developing countries with poor sanitation, (2) among campers drinking surface water, (3) among individuals with impaired immune systems (as in AIDS), and (4) among toddlers and young children. The cysts can survive in the environment for months, and they are not killed during the chlorine treatment used to kill bacteria in drinking water. Travelers are advised to boil or ultrafilter water and to boil food before eating it, as these preventive measures will destroy the cysts.

CHOLERA EPIDEMICS

Cholera epidemics are most common in areas where sanitation is poor and where drinking water is contaminated by fecal wastes. After an incubation period of 1–2 days, the symptoms of nausea, vomiting, and diarrhea persist for 2–7 days. Fluid loss during the worst stage of the disease can approach 1 liter per hour. This dramatic loss causes a rapid drop in blood volume, leading to acute hypovolemic shock and damage to the kidneys and other organs.

Treatment consists of oral or intravenous fluid replacement while the disease runs its course. Antibiotic therapy may also prove beneficial. A vaccine is available, but its low success rate (40–60 percent) and short duration (4–6 month protection) make it relatively ineffective in preventing or controlling cholera outbreaks. A cholera epidemic began in Peru in 1991 and has since spread throughout South America. More than 500,000 cases have been reported, with a death rate of 0.5 percent. This outbreak has had a remarkably low mortality rate; death rates in other outbreaks in the twentieth century have been as high as 60 percent.

⚕ Lactose Intolerance   FAP *p. 904*

Lactose intolerance is a malabsorption syndrome that results from the lack of the enzyme *lactase* at the brush border of the intestinal epithelium. This condition poses more of a problem than would be expected if the only important outcome were the inability to make use of this particular disaccharide. The clinical symptoms result because undigested lactose provides a particularly stimulating energy source for the bacterial inhabitants of the colon. Increased gas generation, cramps, and diarrhea often bring extreme discomfort and distress after the individual drinks a single glass of milk or eats small amounts of any other dairy product.

Lactose intolerance appears to have a genetic basis. Infants produce lactase to digest milk, but older children and adults may stop producing this enzyme. In some populations, lactase production continues throughout adulthood. Only about 15 percent of Caucasians develop lactose intolerance, whereas estimates ranging from 80 to 90 percent have been suggested for the adult African and Asian populations. These differences affect dietary preferences in these groups. Food relief efforts must take such preferences into account; for example, shipping powdered milk to starvation areas in Africa may make matters worse if supplies are distributed to adults rather than to children.

📖 A Closer Look: Aerobic Metabolism   FAP *p. 922*

Mitochondria use organic substrates, such as the pyruvic acid produced by glycolysis (Figure A-50a), from the surrounding cytoplasm. Aerobic metabolism includes the TCA cycle (Figure A-50b) and the oxidation–reduction reactions of the electron transport system (FAP *p. 925*).

⚕ Phenylketonuria   FAP *p. 935*

Phenylketonuria (fen-il-kē-tō-NOO-rē-uh), or **PKU,** is one of about 130 disorders that have been traced to the lack of a specific enzyme. Individuals with PKU are deficient in a key enzyme, *phenylalanine hydroxylase,* responsible for the conversion of the amino acid phenylalanine to tyrosine. This reaction is a necessary step in the synthesis of tyrosine, an important component of many proteins and the structural basis for a pigment (melanin), two hormones (epinephrine and norepinephrine), and two neurotransmitters (dopamine and norepinephrine). In addition, this conversion must occur before the carbon chain of a phenylalanine molecule can be recycled or broken down in the TCA cycle.

If PKU is undetected and untreated, plasma concentrations of phenylalanine gradually escalate from normal (about 3 mg/dl) to levels above 20 mg/dl. High plasma concentrations of phenylalanine affect overall metabolism, and a number of unusual byproducts are excreted in the urine. The synthesis and degradation of proteins and other amino acid derivatives are affected. Developing neural tissue is most strongly influenced by these metabolic alterations, and severe brain damage and mental retardation result.

Fortunately, this condition is detectable shortly after birth, because it produces elevated levels of phenylalanine in the blood and phenylketone, a metabolic byproduct, in the blood and urine of the newborn infant. (During pregnancy, the normal mother metabolizes phenylalanine for the PKU fetus, so fetal levels are normal prior to delivery.) Treatment consists of controlling the amount of phenylalanine in the diet while plasma concentrations are monitored. This treatment is most important in infancy and childhood, when the nervous system is developing. Once the child has grown, dietary restriction of phenylalanine can be relaxed, except during pregnancy. A pregnant woman with PKU must protect the fetus from high levels of phenylalanine by following a strict diet that must actually begin before the pregnancy occurs.

Although the dietary restrictions are more relaxed for adults than for children, those with PKU must still monitor the ingredients used in the preparation of their meals. Because tyrosine cannot be synthesized from dietary phenylalanine, the diet of these patients must also contain adequate amounts of tyrosine. One popular artificial sweetener, *Nutrasweet*®, consists of phenylalanine and aspartic acid. The consumption of food or beverages that contain this sweetener can therefore cause problems for PKU sufferers.

In its most severe form, PKU affects approximately 1 infant in 20,000. Individuals who carry only a single gene for PKU will produce the affected enzyme but in lesser amounts. These individuals are asymptomatic but have slightly elevated phenylalanine levels in their blood. Statistical analysis of the incidence of fully developed PKU indicates that as many as 1 person in 70 may carry a gene for this condition.

25

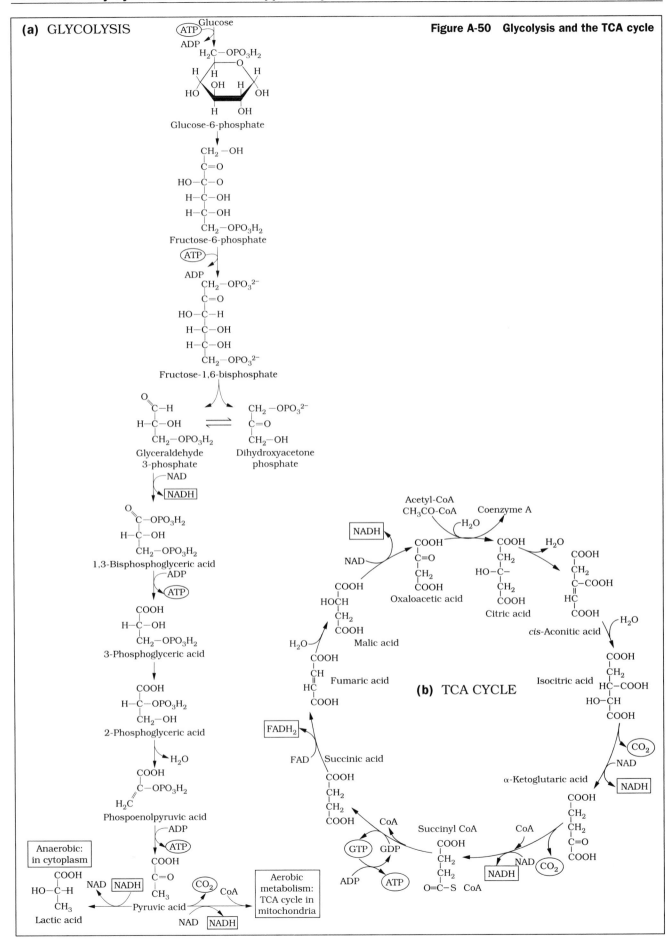

(a) GLYCOLYSIS

Figure A-50   Glycolysis and the TCA cycle

(b) TCA CYCLE

Protein Deficiency Diseases

FAP p. 936

Regardless of the energy content of the diet, if it is deficient in essential amino acids, the individual will be malnourished to some degree. In a **protein deficiency disease,** protein synthesis decreases throughout the body. As protein synthesis in the liver fails to keep pace with the breakdown of plasma proteins, plasma osmolarity falls. This reduced osmolarity results in a fluid shift as more water moves out of the capillaries and into interstitial spaces, the peritoneal cavity, or both. The longer the individual remains in this state, the more severe the ascites and edema that result.

This clinical scenario is relatively common in developing countries, where dietary protein is often scarce or prohibitively expensive. Growing infants suffer from **marasmus** (ma-RAZ-mus) when deprived of adequate proteins and calories. **Kwashiorkor** (kwash-ē-OR-kor) occurs in children whose protein intake is inadequate, even if the caloric intake is acceptable (Figure A-51). In each case, additional complications include damage to the developing brain. It is estimated that more than 100 million children worldwide suffer from protein deficiency diseases. War and civil unrest that disrupt local food production and distribution have been more instrumental than a shortage of food in itself in producing recent famines.

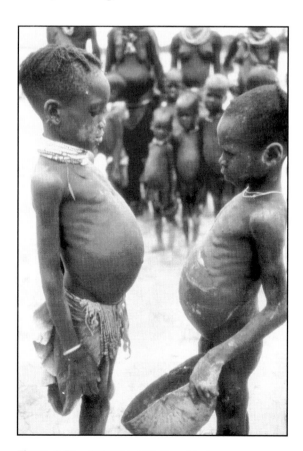

Figure A-51 Children with Kwashiorkor

Gout

FAP p. 936

At concentrations above 7.4 mg/dl, body fluids are supersaturated with uric acid. Although symptoms may not appear at once, uric acid crystals begin to precipitate in body fluids. The condition that results is called **gout.** The severity of the symptoms depends on the amount and location of the crystal deposits.

Initially, the joints of the extremities, especially the metatarsal/phalangeal joint of the great toe, are likely to be affected. This intensely painful condition, called *gouty arthritis,* may persist for several days and then disappear for a period of days to years. Recurrences often involve other joints and may produce generalized fevers. Precipitates may also form within cartilages, in synovial fluids, in tendons or other connective tissues, or in the kidneys. At serum concentrations of over 12–13 mg/dl, half the patients will develop kidney stones, and kidney function may be affected to the point of kidney failure.

The incidence of gout is much lower than that of hyperuricemia, with estimates ranging from 0.13 to 0.37 percent of the population. Only about 5 percent of the sufferers are women, and most affected males are over 50. Foods high in purines, such as meats or fats, may aggravate or initiate the onset of gout. These foods tend to cost more than carbohydrates, so "rich foods" have often been associated with this condition.

Obesity

FAP p. 939

Regulatory obesity results from a failure to regulate food intake so that appetite, diet, and activity are in balance. Most instances of obesity fall within this category. In most instances, there is no obvious organic cause, although in rare cases the problem may arise because some disorder, such as a tumor, affects the hypothalamic centers that deal with appetite and satiation. Typically, chronic overeating is thought to result either from psychological or sociological factors, such as stress, neurosis, long-term habits, family or ethnic traditions, or from inactivity. Genetic factors may also be involved, but because the psychological and social environment plays such an important role in human behavior, the exact connections have been difficult to assess. In short, individuals with regulatory obesity overeat for some reason and thereby extend the duration and magnitude of the absorptive state.

In **metabolic obesity,** the condition is secondary to some underlying organic malfunction that affects cell and tissue metabolism. For example, some cases of obesity have been linked to reduced insulin sensitivity due to a reduction in the number of insulin receptors in adipose tissue and in skeletal muscle. Metabolic obesity is commonly associated with chronic hypersecretion or

25

hyposecretion of metabolically active hormones, such as insulin or glucocorticoids.

Categorizing an obesity problem is less important in a clinical setting than is determining the degree of obesity and the number and severity of the related complications. The affected individuals are at a high risk of developing diabetes, hypertension, and coronary artery disease as well as gallstones, thromboemboli, hernias, arthritis, varicose veins, and some forms of cancer. A variety of treatments may be considered, ranging from behavior modification, nutritional counseling, psychotherapy, and exercise programs to gastric stapling or the bypass of a portion of the jejunum.

LIPOSUCTION
Liposuction is a surgical procedure for the removal of unwanted adipose tissue. Adipose tissue is flexible but not as elastic as loose connective tissue, and it tears relatively easily. In liposuction, a small incision is made through the skin, and a tube is inserted into the underlying adipose tissue. Suction is then applied. Because adipose tissue tears easily, chunks of tissue containing adipocytes, other cells, fibers, and ground substance can be vacuumed away. An estimated 115,000 liposuction procedures were performed in 1990.

This practice has received a lot of news coverage, and many advertisements praise the technique as easy, safe, and effective. In fact, it is not always easy, and it can be dangerous and have limited effectiveness. The density of adipose tissue varies from place to place in the body and from individual to individual, and it is not always easy to suck through a tube. An anesthetic must be used to control pain, and anesthesia always poses risks; blood vessels are stretched and torn, and extensive bleeding can occur. The most serious complication is probably the potential for infection after the liposuction treatment.

Finally, adipose tissue can repair itself, and adipocyte populations recover over time. The only way to ensure that fat lost through liposuction will not return is to adopt a lifestyle that includes a proper diet and adequate exercise. Over time, such a lifestyle can produce the same weight loss, *without liposuction,* eliminating the surgical expense and risk.

📖 Adaptations to Starvation FAP *p. 943*

Figure A-52 shows changes in the metabolic stores of a 70-kg individual during prolonged starvation. Carbohydrate utilization declines almost immediately as the stores are depleted. As blood glucose levels decline, gluconeogenesis accelerates, using glycerol, amino acids, and lactic acid. The glycerol is provided by adipocytes; the amino acids and lactic acid are provided primarily by skeletal muscle. At this point, the kidneys begin to assist the liver by deaminating amino acids and generating additional glucose molecules.

Gluconeogenesis is accompanied by an increase in circulating ketone bodies, some derived from ketogenic amino acids, others from the catabolism of fatty acids. As the starvation stress continues, peripheral tissues further restrict their glucose utilization, and the ketone bodies generated by fatty acid catabolism become the primary energy source.

The fasting individual gradually becomes weak and lethargic as peripheral systems are weakened by protein catabolism and stressed by pH changes. Buffer systems are challenged by the circulating amino acids, lactic acid, and ketone bodies, and ketoacidosis becomes a potential problem. Under these circumstances, most tissues begin catabolizing ketone bodies almost exclusively, and in extreme starvation more than 90 percent of the

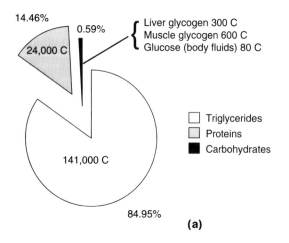

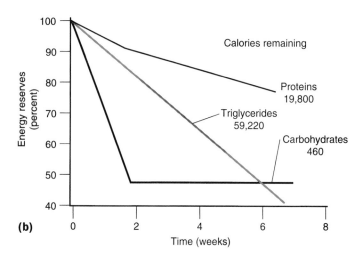

Figure A-52 Metabolic Reserves and the Effects of Starvation.
(a) Estimated metabolic reserves of a 70-kg individual. (b) Projected effects of prolonged starvation on the metabolic reserves of the same individual.

daily energy demands are met through the oxidation of ketone bodies. At this stage, even neural tissue relies on ketone bodies to supplement declining glucose supplies.

Structural proteins are the last to be mobilized, with other forms, such as the contractile proteins of skeletal muscle, more readily available. When peripheral tissues catabolize proteins, the amino acids are exported to the liver, where they can be safely deaminated. The carbon fragments are then catabolized to provide ATP, or they are used to manufacture glucose molecules or ketone bodies that can be broken down by peripheral tissues.

When lipid reserves are exhausted, crises soon follow. On a gram-for-gram basis, cells must catabolize almost twice as much protein as lipid to obtain the same energy benefits. Making matters worse, by this time most of the easily mobilized proteins have already been broken down. As structural proteins are disassembled, a variety of unpleasant effects may appear. Accelerated protein catabolism causes problems with fluid balance, because the nitrogenous and acidic wastes must be excreted in the urine. These waste products are excreted in solution, and the larger the amount of waste products eliminated, the greater the associated water loss. An increase in urinary water losses can lead to dehydration, and the combination of dehydration and acidosis can cause kidney damage.

When glucose concentrations can no longer be sustained above 40–50 mg/dl, the individual becomes disoriented and confused. The eventual cause of death may be kidney failure, ketoacidosis, protein deficiency, or hypoglycemia.

How long does it take to reach this critical state? That essentially depends on the size of the person's lipid reserves. Prolonged starvation for most people would last about 8 weeks, but the truly obese can hold out far longer. With adequate water and vitamin supplements, an 8-month fast has been used as a weight-loss technique. (This technique was an emergency treatment rather than a diet plan, because prolonged fasting can result in kidney damage or severe ketoacidosis.)

Metabolic adjustments during starvation are coordinated primarily by the endocrine system. The glucocorticoids produced by the adrenal cortex are the most important hormones, aided by growth hormone from the pituitary. The effects of these hormones on peripheral tissues are included in Table 25-1 of FAP, *p. 939.*

Nutrition and Nutritionists
FAP *p. 943*

Professional nutritionists are trained to provide daily dietary recommendations for specific individuals. There are many complexities involved in determining individual requirements, and a good nutritionist must be part biochemist, part pharmacist, and part clinician. He or she must also be able to work with people who have strong biases about what is acceptable or unacceptable in their diet—opinions that often stem from learned, social, or traditional values that have little correlation with nutritional facts. Good nutritionists are highly skilled, highly motivated individuals with extensive education. But unlike the situation for most other medically related professions, there are as yet no national standards to determine the use of the name *nutritionist* or qualifications for establishing a practice. You have doubtless encountered self-taught and self-confident "nutritionists" working in health-food stores or supermarkets or selling various products door to door. The differences between a balanced diet and one that produces avitaminosis, hypervitaminosis, or a degree of malnutrition are very slim indeed, so anyone who is seeking nutritional advice should find a qualified professional nutritionist.

Perspectives on Dieting   FAP *p. 948*

Dieting has become a U.S. national pastime, and magazines and bookstores are flooded with "how-to" guides for losing weight. Despite the unusual, astounding, and often preposterous claims these guides make, none of the claims has proved to be ideal, and many have turned out to be dangerous. Now that you have a basic familiarity with metabolic processes and interconversions, we can briefly review several dietary misconceptions. Analyzing the most popular fad diets is beyond the scope of this manual, but you might find the information summarized in Figure A-53 useful if you are considering an intensive diet program.

Identifying fad diets can be quite easy once you have learned the warning signs. These diets invariably promise almost immediate results, and most claim to be virtually effortless. The "immediate results" turn out to be temporary water losses, and if dieting in any form were easy, Americans would not have a weight problem in the first place.

Carried to extremes, some fad diets can produce fatal alterations in blood chemistry and physiological systems. Many trendy diets advocate the elimination of most or all of the members of one or more food groups, making it difficult to obtain dietary essentials. Several of the diets intentionally produce ketosis, a condition that can cause several unpleasant and even disastrous side effects. Equally significant, such restrictions are artificial and normally are intended to be followed only while dieting. Because the eating patterns that caused the original problem are not addressed, once the diet has ended, the individual immediately begins putting on weight. This weight gain leads to another dieting cycle 6 months later, commonly with a different fad diet. This cycle of dieting delights publishing companies, who thrive on a combination of diet books and cookbooks, but it can be very frustrating for the individual. Only about 5 percent of

25

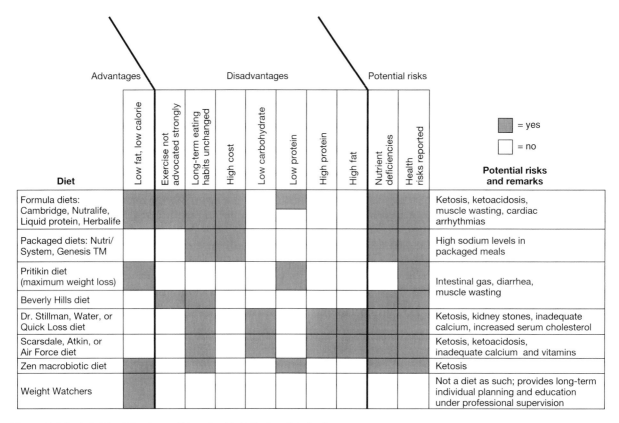

Figure A-53 A Critical Review of Popular Fad Dieting Methods.
Data Sources: "Fad Diet Summary," Kaiser Permanente; "Rating the Diets," *Consumer Reports*; "A Practical Guide to Fad Diets," *Clin. Sports Med.* 3 (3):723–729; "The Sense and Nonsense of Best-Selling Diet Books," *Can. Med. Assoc. Journal* 126 (6):F 696–701 (1982).

dieters sustain their weight loss for 1–5 years or more.

The advocacy of "secret ingredients" should also ring warning bells, and you should be on the alert for exaggerated claims about either the amount of weight loss or the potential fringe benefits of the program. Paid testimonials from successful dieters are highly suspect and hardly a suitable ground for decision making. Such testimonials typically take the place of scientific evidence, and pseudoscientific terms are often used to convince the public that scientific proof exists to support the inflated statements. In some cases, the "scientific justifications" are totally unrelated to the weight loss experienced by those who follow a particular diet. For example, eating fruits and vegetables, coupled with reduced quantities of beef, will lower caloric intake and bring the diet in close agreement with current dietary recommendations. Most people will lose weight on such a diet regardless of whether it is accompanied by complicated mumbo-jumbo about acupressure, cosmic contemplation, or toxic combinations in the stomach.

The most effective weight-loss program does not involve a crash diet, miracle drugs, elimination of all foods beginning with the letter B, or subscriptions to the diet-of-the-month club. The weight loss should be gradual rather than sudden. Dietary modifications should be moderate, not sweeping, and intended to alter long-term eating habits and to ensure that the weight loss will be permanent.

Given an opportunity, the human body will preserve homeostasis. By following the advice "everything in moderation," we can avoid testing our homeostatic limits, but many people tend to follow the "more must be better" philosophy when it comes to meeting nutritional needs. For example, it has been said that Americans have the most valuable urine in the world, thanks to our tendency to overdose ourselves on water-soluble vitamins. *Keep in mind that no amount of vitamins, minerals, dietary restrictions, or attention to nutritional recommendations can counteract the negative effects of even a single serious behavioral health risk, such as smoking, excessive alcohol consumption, or drug abuse.*

THERMOREGULATORY DISORDERS

Accidental Hypothermia   FAP *p. 954*

If body temperature declines significantly below normal levels, the thermoregulatory system begins to lose sensitivity and effectiveness. Cardiac output and respiratory rate decrease, and if the core temperature falls below 28°C (82°F), cardiac arrest is likely. The individual then has no heartbeat, no respiratory rate, and no response to external stimuli, even painful ones. Body temperature continues to decline, and the skin turns blue or pale and cold.

At this point, we would probably assume that the individual has died. But because metabolic activities have decreased systemwide, the victim may still be saved, even after several hours have elapsed. Treatment consists of cardiopulmonary support and gradual rewarming, both external and internal. The skin can be warmed up to 45°C (110°F) without damage; warm baths or blankets can be used. One effective method of raising internal temperatures involves the introduction of warm saline solution into the peritoneal cavity.

Hypothermia is a significant risk for those engaged in water sports, and it may complicate treatment of a drowning victim. Water absorbs heat roughly 27 times as fast as air does, and the body's heat-gain mechanisms are unable to keep pace over long periods or when faced with a large temperature gradient. But hypothermia in cold water does have a positive side. On several occasions, small children who have drowned in cold water have been successfully revived after periods of up to 4 hours. Children lose body heat quickly, and their systems stop functioning very quickly as body temperature declines. This rapid drop in temperature prevents the oxygen-starvation and tissue damage that would otherwise occur when breathing stops.

Resuscitation is not attempted if the individual has actually frozen. Water expands roughly 7 percent during ordinary freezing, and cell membranes throughout the body are destroyed in the process.

Very small organisms can be frozen and later thawed without ill effects, because their surface-to-volume ratio is enormous and the freezing process occurs so rapidly that ice crystals never form.

⚕ Fevers

FAP *p. 955*

When interleukin-1 increases the "thermostat setting" of the preoptic center of the hypothalamus, the heat-gain center is activated. The individual feels cold and may curl up in a blanket. Shivering may begin and continue until the temperature at the preoptic area corresponds to the new setting. The fever passes when the thermostat is reset to normal. The **crisis phase** then ensues as the heat-loss center is stimulated. The individual feels unbearably warm and discards the blanket; the skin is flushed, and the sweat glands work furiously to bring the temperature down. Repeated cycles of this type constitute the "chills and fever" pattern of many illnesses.

Fevers may be classified as *chronic* or *acute*. Chronic fevers may persist for weeks or months as the result of infections, cancers, or thermoregulatory disorders. In some cases, a discrete cause cannot be determined, leading to a classification as a *fever of unknown origin (FUO)*. Acute hyperthermia, as seen during heat stroke, in certain diseases, or in some marathon runners, is life-threatening. Immediate treatment may involve cooling the individual in an ice bath (to increase conduction) or giving alcohol rubs (to increase evaporation) in combination with the administration of **antipyretic drugs,** such as aspirin or acetaminophen.

25

CRITICAL-THINKING QUESTIONS

9-1. Tony is a chronic alcoholic who suffers from cirrhosis of the liver. Which of the following symptoms would you expect to observe in Tony?

   a. prolonged clotting time
   b. jaundice
   c. ascites
   d. portal hypertension
   e. all of the above

9-2. Paul, a 60-year-old man, complains of rectal bleeding for the last 24 hours. He has experienced four bloody stools in the last several hours and is beginning to weaken. Prior to this episode, he claims he was constipated. He is admitted to the hospital, where all the results of lab and diagnostic studies are within normal limits except for the following:

Hemoglobin: 9 g/dl

Hematocrit: 25

Barium enema: reveals narrowing of a small portion of the descending colon

Colonoscopy: mass detected in the descending colon

If the physician took a tissue biopsy during the colonoscopy, what results would you expect from the cytological analysis of the specimen?

9-3. Marva, a 25-year-old teacher, tells her physician she has been experiencing bouts of painful abdominal cramping, gas, and diarrhea during the last 6 months for unknown reasons. The episodes are short-lived and sporadic and are not accompanied by fever. When questioned about her diet, Marva cannot remember a specific food related to the episodes. A stool specimen is negative for bacteria and parasites. A barium enema is negative for any abnormalities. What might be Marva's problem? How would you check this diagnosis?

9-4. Barry is participating in a nutrition study in which he must follow a prescribed diet and have his blood tested regularly. On one occasion, his test results indicate low levels of insulin, elevated levels of glucagon and epinephrine, and moderate levels of cortisol and growth hormone. What would you expect the levels of glucose, ketones, and fatty acids to be?

Clinical Problems

1. Jose, 50 years old, is brought to the ER by his wife. Jose complains of severe shortness of breath on exertion and needs to rest after walking only a short distance. He recently almost fainted in the front yard after he checked the mailbox. Jose has been a heavy smoker for 25 years; he has a persistent cough with heavy mucus production. He has had several respiratory infections over the last few months.

Jose is placed on oxygen after he has gotten a chest X-ray and after the following blood test results were obtained:

RBC count: 6.2 million/mm^3

Hct: 55

Hb: 18 g/dl

MCV: 98 mm^3

MCH: 30 pg

WBC count: 10,000/mm^3

Platelets: 250,000/mm^3

Arterial blood gases:

   pH 7.25

   P_{CO_2} 48 mm Hg

   P_{O_2} 55 mm Hg

   HCO_3^- 30 mEq/l

Chest X-ray: normal heart and expanded lungs with small areas of probable fibrotic changes

What could cause Jose to suffer from shortness of breath? What are possible diagnoses?

2. Faith brings her 5-month-old infant to the pediatrician for treatment of a respiratory infection. The mucus is difficult to dislodge from the chest, and the infant has a persistent cough. This respiratory infection is the child's third in the last 2 months, and the infant has not gained weight during that period. The infant has diarrhea with bulky, foul-smelling, oily-looking stools. On the basis of the following test results, what is a preliminary diagnosis?

Aspirated duodenal contents after injection of secretin and pancreozymin: viscous fluid, trypsin absent, low amylase levels, low bicarbonate levels

Fecal fat test: 7 g/24 hr

Serum carotene: 25 mg/dl

Chest X-ray: bronchial wall thickening

The Urinary System

The urinary system consists of the kidneys, where urine production occurs, and the conducting system, which transports and stores urine prior to its elimination from the body. The conducting system includes the ureters, the urinary bladder, and the urethra. Although the kidneys perform all the vital functions of the urinary system, problems with the conducting system can have direct and immediate effects on renal function.

THE PHYSICAL EXAMINATION AND THE URINARY SYSTEM

The primary symptoms of urinary system disorders are pain and changes in the frequency of urination. The nature and location of the pain can provide clues to the source of the problem (see Figure 15-3, FAP *p. 495*). For example,

- Pain superior to the pubic region may be associated with urinary bladder disorders.

- Upper lumbar back pain radiating to the flank and the right upper quadrant or left upper quadrant can be caused by kidney infections such as *pyelonephritis* (FAP *p. 997*).

- *Dysuria* (painful or difficult urination) may occur with *cystitis* or *urethritis* (FAP *p. 997*) or with *urinary obstructions* (FAP *p. 995*). In males, prostatic enlargement can lead to compression of the urethra and dysuria.

Individuals with urinary system disorders may urinate more or less often than usual and may produce normal or abnormal amounts of urine:

- Irritation of the lining of the ureters or urinary bladder can lead to the desire to urinate with increased frequency, although the total amount of urine produced each day remains normal. When these problems exist, the individual feels the urge to urinate when the urinary bladder volume is very small. The irritation may result from trauma, urinary bladder infection (cystitis) or tumors, or increased acidity of the urine.

- *Incontinence*, an inability to control urination voluntarily, may involve periodic involuntary urination—a continual, slow trickle of urine from the urethra. Incontinence may result from urinary bladder or urethral problems, damage or weakening of the muscles of the pelvic floor, or interference with normal sensory or motor innervation in the region. Renal function and daily urinary volume are normal.

- In *urinary retention*, renal function is normal, at least initially, but urination does not occur. Urinary retention in males commonly results from prostatic enlargement and compression of the prostatic urethra. In members of both genders, urinary retention may result from obstruction of the outlet of the urinary bladder or from CNS damage, such as a stroke or damage to the spinal cord.

- Changes in the volume of urine produced by a normally hydrated person indicate that there are problems either at the kidneys or with the control of renal function. **Polyuria,** the production of excessive amounts of urine, may result from hormonal or metabolic problems, such as those associated with *diabetes* (pp. 97, 103), or from damage to the glomeruli, as in *glomerulonephritis* (FAP *p. 969*). **Oliguria** (a urine volume of 50–500 ml/day) and **anuria** (0–50 ml/day) are conditions that indicate serious kidney problems and potential renal failure. Renal failure can occur with *heart failure* (p. 128), renal ischemia, *circulatory shock* (FAP *p. 737*), burns (FAP *p. 168*), pyelonephritis (FAP *p. 997*), and a variety of other disorders.

Important clinical signs of urinary system disorders include the following:

- *Hematuria*, the presence of red blood cells in the urine, indicates bleeding at the kidneys or conducting system. Hematuria producing dark red urine typically indicates bleeding in the kidney, and hematuria producing bright red urine indicates bleeding in the lower urinary tract. Hematuria most commonly occurs with trauma to the kidneys, calculi (kidney stones), tumors, or urinary tract infections.

- *Hemoglobinuria* is the presence of hemoglobin in the urine. Hemoglobinuria indicates increased RBC hemolysis within the circulation due to cardiovascular or metabolic problems. Conditions that result in hemoglobinuria include the *thalassemias* (p. 110), *sickle cell anemia* (p. 110), *hypersplenism* (p. 138), and some autoimmune disorders.

- Changes in urine color may accompany some renal disorders. For example, the urine may become (1) cloudy due to the presence of bacteria, lipids, crystals, or epithelial cells; (2) red or brown from hemoglobin or myoglobin; (3) blue-green from bilirubin; or (4) brown-black from excessive concentration. Not all color changes are abnormal, however. Some foods and several prescription drugs can cause changes in urine color. A serving of beets can give urine a reddish color, whereas eating rhubarb can give the urine an orange tint, and B vitamins give urine a vivid yellow color.

- Renal disorders typically lead to a generalized edema in peripheral tissues. Facial swelling, especially around the eyes, is common.

- A fever commonly develops when the urinary system is infected by pathogens. Urinary blad-

26

der infections (cystitis) typically result in a low-grade fever; kidney infections, such as pyelonephritis, can produce very high fevers.

During the physical assessment, palpation can be used to check the status of the kidneys and urinary bladder. The kidneys lie in the *costovertebral area*, the region bounded by the lumbar spine and the twelfth rib on either side. To detect tenderness due to kidney inflammation, the examiner gently thumps a fist over each flank. This usually does not cause pain unless the underlying kidney is inflamed.

The urinary bladder can be palpated just superior to the pubic symphysis. On the basis of palpation alone, urinary bladder enlargement due to urine retention can be difficult to distinguish from the presence of an abdominal mass.

Many procedures and laboratory tests may be used in the diagnosis of urinary system disorders. The functional anatomy of the urinary system can be examined by using a variety of sophisticated procedures. For example, administering a radiopaque compound that will enter the urine permits the creation of an **intravenous pyelogram** (PĪ-el-ō-gram), or **IVP**, by taking an X-ray of the kidneys (Figure 26-18, FAP *p. 994*). This procedure, sometimes called an *EU (excretory urogram)* permits detection of unusual kidney, ureter, or urinary bladder structures and masses. Computerized tomography (CT) scans may also provide useful information about localized abnormalities; representative scans of the kidneys are shown in the Scanning Atlas, p. S-12. Other diagnostic procedures and important laboratory tests are detailed in Table A-28, and Figure A-54 (p. 188) outlines the major classes of disorders of the urinary system.

PAH and the Calculation of Renal Blood Flow
FAP *p. 970*

Although seldom used in clinical practice, *para-aminohippuric acid*, or *PAH*, can be administered to determine the rate of blood flow through the kidneys. PAH enters the filtrate through filtration at the glomerulus. As blood flows through the peritubular capillaries, any remaining PAH diffuses into the peritubular fluid, and the tubular cells actively secrete it into the filtrate. By the time blood leaves the kidney, virtually all the PAH has been removed from the circulation and filtered or secreted into the urine. You can therefore calculate renal blood flow if you know the PAH concentrations of the arterial plasma and urine. The calculation proceeds in a series of steps. The first step is to determine the plasma flow through the kidney by using the formula

$$P_f = \frac{PAH_u \times V_u}{PAH_p}$$

where

P_f is the plasma flow, also known as the *effective renal plasma flow,* or *ERPF;*

PAH_u is the concentration of PAH in the urine, usually expressed in milligrams per milliliter (mg/ml);

V_u is the volume of urine produced, usually in terms of milliliters per minute; and

PAH_p is the concentration of PAH in arterial plasma in milligrams per milliliter.

Consider the following example: A patient producing urine at a rate of 1 ml per minute has a urinary PAH concentration of 15 mg/ml with an arterial PAH concentration of 0.02 mg/ml. The patient's hematocrit is normal (Hct = 45). The plasma flow is as follows:

$$P_f = \frac{15 \text{ mg/ml} \times 1 \text{ ml/min} = 750 \text{ ml/min}}{0.02 \text{ mg/ml}}$$

This value is an estimate of the plasma flow through the glomeruli and around the kidney tubules each minute. However, plasma accounts for only part of the volume of whole blood; the rest consists of formed elements. To have a plasma flow of 750 ml/min, the blood flow must be considerably greater. The patient's hematocrit is 45, which means that plasma accounts for 55 percent of the whole blood volume. To calculate the renal blood flow, we must multiply the plasma volume by 1.8 (each 100 ml of blood has 55 ml of plasma, and 100/55 = 1.8):

$$750 \text{ ml/min} \times 1.8 = 1350 \text{ ml/min}$$

This value, 1350 ml/min, is the estimated tubular blood flow. The final step is to adjust this estimate to account for blood that enters the kidney but flows to the renal pelvis, the capsule, or other areas not involved with urine production. This value is usually estimated as 10 percent of the total blood flow. We can therefore complete the calculation for this example as follows:

$$1350 \text{ ml/min} = 90 \text{ percent of total blood flow}$$
$$10 \text{ percent} = 1350/9 = 150 \text{ ml/min}$$
$$\text{Total blood flow} = 1350 + 150 = 1500 \text{ ml/min}$$

Conditions Affecting Filtration
FAP *p. 975*

Changes in filtration pressure (P_f) can result in significant alterations in kidney function. Factors that can disrupt normal filtration rates include physical damage to the filtration apparatus and interference with normal filtrate or urine flow.

PHYSICAL DAMAGE TO THE FILTRATION APPARATUS

The lamina densa and podocytes can be injured by mechanical trauma, such as a blow to the kidneys,

Table A-28   Examples of Tests Used in the Diagnosis of Urinary System Disorders

Diagnostic Procedure	Method and Result	Representative Uses	Notes
Cystoscopy	A small tube (cystoscope) is inserted along the urethra into the urinary bladder to permit visualization of the lining of the urinary bladder	Procedure can be used to obtain a biopsy specimen or to remove stones (calculi) and small tumors as well as to provide direct visualization of the urinary bladder and urethral openings	
Retrograde pyelography	Radiopaque dye is injected into the ureters through a catheter in the cystoscope inserted into the urinary bladder; X-ray films are then taken	To detect obstructions of the ureter caused by tumors, calculi, or strictures	Does not require functional kidneys
Renal biopsy	Using ultrasound to guide the needle, the special biopsy needle is inserted through the back and into the kidney. The specimen is then removed for analysis.	To determine the cause of renal disease, to detect rejection of transplanted kidney, or to perform a tumor biopsy	
Intravenous pyelography (IVP)	Dye is injected intravenously; it is filtered at the kidney and excreted into the urinary tract; a series of X-rays is then taken to view the kidneys, ureters, and urinary bladder	To determine the presence of kidney disease, obstructions such as calculi or tumors, or anatomical abnormalities	Requires some level of kidney function; potential allergic response to the dye used
Cystography	Dye is inserted through a catheter placed in the urethra and threaded into the bladder. A series of X-ray films is then taken.	Identification of tumors of the bladder and rupture of the urinary bladder by trauma; if X-rays are taken as the patient voids, determination of reflux of urine from the bladder to the ureter can be achieved	Congenital urethral reflux can contribute to UTIs in children

Laboratory Test	Normal Values	Significance of Abnormal Values	Notes
URINALYSIS			
pH	4.6–8.0	Alkaline or acidic may indicate increased or decreased blood pH	Affected by diet
Color	Pale yellow-amber	Color may change with certain drugs and foods; dark red-brown urine may indicate bleeding from kidney; bright red blood comes from lower urinary tract; some bacteral infections cause a green tint	

26

Table A-29 *(continued)*

Laboratory Test	Normal Values in Blood Plasma or Serum	Significance of Abnormal Values	Notes
Estradiol (serum)	Follicular phase: 20–150 pg/ml Ovulation: 100–500 pg/ml Luteal phase: 60–260 pg/ml	Decreased in ovarian dysfunction and in amenorrhea	
FSH (serum)	Before and after ovulation: 4–20 mIU/ml Midcycle: 10–40 mIU/ml	Determination of cause of infertility and menstrual dysfunction; increased levels can be related to absence of estrogens; decreased levels can be caused by anorexia nervosa or hypopituitarism	Elevated in menopause
LH (serum)	Follicular: 3–30 mIU/ml Midcycle: 30–150 mIU/ml	Elevated in ovarian hypofunction; useful in determining if ovulation has occurred	Urine test available to recognize ovulation
Progesterone (serum)	Before ovulation: <70 ng/dl Midcycle: 250–2800 ng/dl	Useful in determining the timing of ovulation; levels are increased in early pregnancy	
Pregnanediol (urine)	Before ovulation: 0.5–1.5 mg/24 hr Midcycle: 2–7 mg/24 hr	Elevated with pregnancy or an ovarian cyst. Decreased with impending miscarriage, ovarian tumor, or preeclampsia.	
Prolactin (serum)	Nonlactating females: 0–23 ng/ml	Values >100 ng/ml in a nonlactating female indicate pituitary tumor	

MALES

Semen analysis		Decreased sperm count causes infertility; infertility could result if >40% of sperm are nonmotile or >30% of sperm are abnormal	Normal values vary widely among individuals and laboratories
Volume	2–5.0 ml		
Sperm count	60–150 million/ml		
Motility	60–80% are motile		
Sperm morphology	70–90% normal structure		
Testosterone (serum)	Adult males: 0.3–1.0 µg/dl	Decreased level could indicate testicular disorder, alcoholism, or pituitary hypofunction	
Acid phosphatase (ACP) (serum)	Adults: 0.0–0.8 U/l at 37°C	Elevated levels occur with carcinoma of the prostate, Paget's disease, and multiple myeloma of the bone marrow	Highest rise occurs in prostatic malignancy
Prostate-specific antigen (PSA)	<4 ng/ml	Elevated levels occur with prostatic cancer, benign prostatic hypertrophy, and increasing age	Upper limits of normal are higher in African-Americans

FEMALES AND MALES

Serologic test for syphilis	Negative	Presence of antibodies indicates past or present infection with syphilis	
Gonorrhea culture test	Negative	Positive test indicates gonorrheal infection	
Herpes simplex virus	Negative	Positive for presence of virus in culture	
Chlamydia smear	Negative	Positive for presence of *Chlamydia* upon culturing	Most common sexually transmitted disease

3. A *digital rectal examination* (DRE) is performed as a screening test for prostatitis or inflammation of the seminal vesicles. In this procedure, a gloved finger is inserted into the rectum and pressed against the anterior rectal wall to palpate the posterior walls of the prostate gland and seminal vesicles.

If urethral discharge is present or if discharge occurs in the course of any of these procedures, the fluid can be cultured to check for the presence of pathogenic microorganisms. Other potentially useful diagnostic procedures and laboratory tests are included in Table A-29.

Assessment of the Female Reproductive System

Important signs and symptoms of female reproductive disorders include the following:

- *Acute pelvic pain,* a symptom that may accompany disorders such as pelvic inflammatory disease (PID), ruptured tubal pregnancy, a ruptured ovarian cyst, or inflammation of the uterine tubes (*salpingitis*).

- *Bleeding outside menses,* which can result from oral contraceptive use, hormonal fluctuation, *pelvic inflammatory disease* (FAP *p. 1060*), or *endometriosis* (FAP *p. 1065*).

- *Amenorrhea* (FAP *p. 1065*), which may occur in women with *anorexia nervosa* (FAP *p. 950*), women who overexercise and are underweight, in extremely obese women, and in postmenopausal women.

- *Abnormal vaginal discharge,* which may be the result of a bacterial, fungal, or protozoan infection, including some STDs.

- *Dysuria.* Although the female reproductive and urinary tracts are distinct, dysuria may accompany an infection of the reproductive system due to migration of the pathogen to the urethral entrance.

- *Infertility,* which may be related to hormonal disturbances, a variety of ovarian disorders (see FAP *p. 1103*), or anatomical problems along the reproductive tract.

A physical examination generally includes the following steps:

1. Inspection of the external genitalia for skin lesions, trauma, or related abnormalities. Swelling of the labia majora may result from (a) regional cellulitis with lymphedema, (b) a *labioinguinal hernia,* (c) bleeding within the labia as the result of local trauma, or (d) *Bartholinitis,* an abscess within one of the greater vestibular glands *(Bartholin's glands).*

2. Inspection and/or palpation of the perineum, vaginal opening, labia, clitoris, urethral meatus, and vestibule to detect lesions, abnormal masses, or discharge from the vagina or urethra. Samples of any discharge present can be cultured to detect and identify any pathogens involved.

3. Inspection of the vagina and cervix by using a *speculum,* an instrument that retracts the vaginal walls to permit direct visual inspection. Changes in the color of the vaginal wall may be important diagnostic clues. For example:

- Cyanosis of the vaginal and cervical mucosa normally occurs during pregnancy (see below), but it may also occur when a pelvic tumor exists or in persons with congestive heart failure.

- Reddening of the vaginal walls occurs in *vaginitis* (p. 201), bacterial infections such as gonorrhea, protozoan infection by *Trichomonas vaginalis,* yeast infections, or postmenopausally in some women (a condition known as *atrophic vaginitis*).

The cervix is inspected to detect lacerations, ulceration, polyps, or cervical discharge. A spatula or brush is then used to collect cells from the cervical os and transfer them to a glass slide. After the sample is fixed by a chemical spray, cytological examination is performed. This technique is the best-known example of a *Papanicolaou (PAP) test* (Table A-2, p. 9), and the sampling process is commonly called a Pap smear. This test screens for the presence of cervical cancer. Ratings of the PAP smear are given in Table A-29.

4. *Bimanual examination* is a method for the palpation of the uterus, uterine tubes, and ovaries. The physician inserts two fingers vaginally and places the other hand against the lower abdomen to palpate the uterus and surrounding structures. The contour, shape, size, and location of the uterus can be determined, and any swellings or masses will be apparent. Abnormalities in other reproductive organs, such as ovarian cysts, endometrial growths, or tubal masses, can also be detected in this way.

Normal and Abnormal Signs Associated with Pregnancy

Pregnancy imposes a number of stresses on the maternal body systems. The major physiological changes are discussed in Chapter 29 (FAP *p. 1101*). Several clinical signs may be apparent in the course of a physical examination, including the following:

- *Chadwick's sign* is a normal cyanosis of the vaginal wall and cervix during pregnancy.

- There are drastic changes in the size of the uterus during pregnancy; the uterus at full-term extends almost to the level of the xiphoid process.

- Significant uterine bleeding, causing vaginal discharge of blood, most commonly occurs in *placenta previa* (p. 205), when the placenta forms near the cervix. Subsequent cervical stretching leads to tearing and bleeding of the vascular channels of the placenta. Vaginal bleeding may also occur prior to miscarriage.

- Nausea and vomiting tend to occur in pregnancy, especially during the first 3 months.

- Edema of the extremities, especially the legs, typically occurs because increased blood volume and weight of the uterus compress the inferior vena cava and its tributaries. As venous pressures rise in the lower limbs and inferior trunk, *varicose veins* and *hemorrhoids* (p. 126) may develop.

- Back pain due to increased stress on muscles of the lower back is common. These muscles balance the weight of the uterus over the lower limbs by accentuating the lumbar curvature.

- A weight gain of 10–12.5 kg (22–27.5 lbs) is now considered normal, although 20 years ago weight increases of 20–25 kg (44–55 lbs) were considered acceptable. Failure to gain adequate weight during a pregnancy can indicate serious problems.

- The combination of estrogen, progesterone, prolactin, human placental lactogen (hPL), and other hormones that are elevated during pregnancy appears to promote the development of insulin resistance. As a result, pregnant women are at greater risk for ketoacidosis. Women unable to increase insulin levels sufficiently to compensate for the increased insulin resistance can develop *gestational diabetes*. Glucose levels must be monitored and stabilized to prevent increased risk of fetal mortality and developmental defects. Gestational diabetes develops in 1–3 percent of pregnancies.

- In some cases, a dangerous combination of hypertension, proteinuria, and edema may occur. We will consider this condition, called *preeclampsia*, in a later section (p. 205).

DISORDERS OF THE REPRODUCTIVE SYSTEM

Disorders of the reproductive system can be organized by gender and region of the body, as in Figure A-57a. This arrangement has the advantage of linking disorders that primarily affect the reproductive system to their target organs. An alternative scheme groups the disorders as primary or secondary, depending on whether the reproductive system is the focus of the disorder. This arrangement is indicated in Figure A-57b.

⚕ The Diagnosis and Treatment of Ovarian Cancer   *FAP p. 1059*

A woman in the United States has a lifetime risk of 1 chance in 70 of developing ovarian cancer. In 1997, 26,800 ovarian cancers were diagnosed, and 14,200 women died from this condition. Ovarian cancer is the third most common reproductive cancer among women, and it is the most dangerous, because it is seldom diagnosed early in its early stages. The prognosis is relatively good for cancers that originate in the general ovarian tissues or from abnormal oocytes. These cancers respond well to some combination of chemotherapy, radiation, and surgery. However, 85 percent of ovarian cancers develop from epithelial cells, and sustained remission can be obtained in only about one-third of these patients. Early diagnosis would greatly improve the chances of successful treatment, but as yet there is no standardized screening procedure. (*Transvaginal sonography* can detect ovarian cancer at Stage I or Stage II, but there is a high incidence of false-positive results.)

The minimal treatment at Stage I or Stage II involves unilateral removal of an ovary and uterine tube (a *salpingo-oophorectomy*), or *bilateral salpingo-oophorectomy* (BSO) and *total hysterectomy* (removal of the uterus). Treatment of more-dangerous forms of early stage ovarian cancer includes radiation and chemotherapy in addition to surgery.

Treatment of Stage III or Stage IV ovarian cancer commonly involves removal of the omentum, in addition to a BSO and total hysterectomy and aggressive chemotherapy. Bone marrow transplantation may be required, because stem cells in the bone marrow are destroyed by these chemicals. Some chemotherapy agents may be introduced into the peritoneal cavity, because higher concentrations can be administered without the systemic effects that would accompany infusion of the drugs into the bloodstream. This procedure is called *intraperitoneal therapy*.

⚕ Uterine Tumors and Cancers   *FAP p. 1061*

Uterine tumors are the most common tumors in women. It has been estimated that 40 percent of women over age 50 have benign uterine tumors involving smooth muscle and connective tissue cells. If small, these leiomyomas (lē-ō-mī-Ō-mas), or *fibroids*, generally cause no problems. If stimulated by estrogens, they can grow quite large, reaching weights as great as 13.6 kg (30 lb). Occlusion of the uterine tubes, distortion of adjacent organs, and compression of blood vessels may then lead to a variety of complications. In symptomatic young women, observation or conservative treatment with drugs or restricted surgery may be utilized. In older women, a decision may be made to

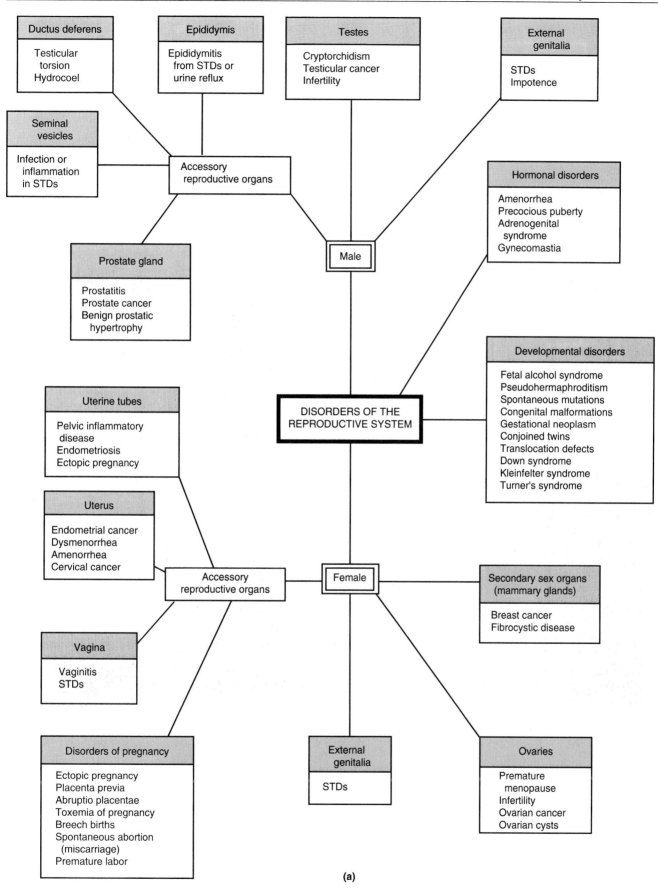

Figure A-57a   Disorders of the Reproductive System.
Arranged by gender and region of the body.

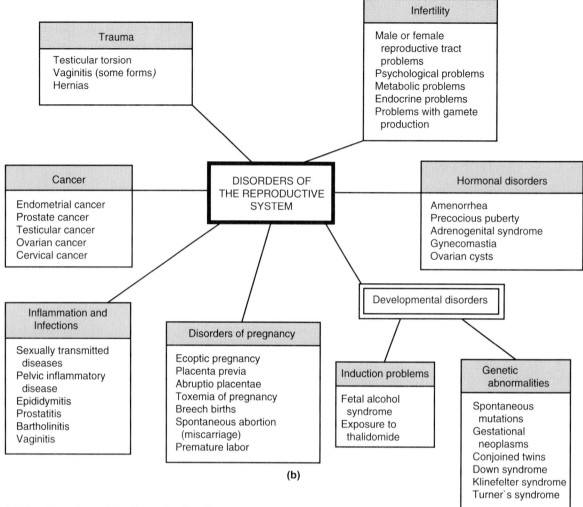

Figure A-57b   Disorders of the Reproductive System.
Arranged by type of disease.

remove the uterus, a procedure called a **hysterectomy** (his-ter-EK-to-mē).

Benign epithelial tumors in the uterus are called endometrial polyps. Roughly 10 percent of women probably have polyps, but because the polyps tend to be small and cause no symptoms, the condition passes unnoticed. If bleeding occurs, if the polyps become excessively enlarged, or if they protrude through the cervical os, they can be surgically removed.

Uterine cancers are less common, affecting approximately 11.9 per 100,000 women. In 1997, roughly 100,000 new cases were reported in the United States, and approximately 10,800 women died from the disease. There are two types of uterine cancers, (1) *endometrial* and (2) *cervical.*

Endometrial cancer is an invasive cancer of the endometrial lining. About 34,900 cases are reported each year in the United States, with approximately 6000 deaths. The condition most commonly affects women age 50–70. Estrogen therapy, used to treat osteoporosis in postmenopausal women, increases the risk of endometrial cancer by 2–10 times. Adding progesterone therapy to the estrogen therapy seems to reduce this risk.

There is no satisfactory screening test for endometrial cancer. The most common symptom is irregular bleeding, and diagnosis typically involves examination of a biopsy of the endometrial lining by suction or scraping. The prognosis varies with the degree of metastasis. Treatment of early-stage endometrial cancer involves a hysterectomy, perhaps followed by localized radiation therapy. In advanced stages, more aggressive radiation treatment is recommended. Chemotherapy has not proved to be very successful in treating endometrial cancers; only 30–40 percent of patients benefit from this approach.

Cervical cancer is the most common reproductive system cancer in women age 15–34. Roughly 14,500 new cases of invasive cervical cancer are diagnosed each year in the United States, and approximately 33 percent of them will eventually die of this condition. Another 33,500 patients are diagnosed with less-aggressive forms of cervical cancer.

Most women with cervical cancer fail to develop symptoms until late in the disease. At that stage, vaginal bleeding, especially after intercourse, pelvic pain, and vaginal discharge may appear. Early detection is the key to reducing the mortality rate for cervical cancer. The standard screening test is the *Pap smear*, named for Dr. George Papanicolaou, an anatomist and cytologist. The cervical epithelium normally sheds its superficial cells, and a sample of cells scraped or brushed from the epithelial surface can be examined for abnormal or cancerous cells. The American Cancer Society recommends yearly Pap tests at ages 20 and 21, followed by smears at 1-year to 3-year intervals until age 65.

The primary risk factor of cervical cancer is a history of multiple sexual partners. It appears likely that these cancers develop after viral infection by one of several different *human papilloma viruses* (HPV) that can be transmitted through sexual contact.

Early treatment of abnormal but not cancerous lesions detected by mildly abnormal Pap smears may prevent progression to cancer formation. Treatment of localized, noninvasive cervical cancer involves the removal of the affected portion of the cervix. Treatment of more-advanced cancers typically involves a combination of radiation therapy, hysterectomy, lymph node removal, and chemotherapy.

⚕ Vaginitis

FAP *p. 1066*

There are several forms of vaginitis, and minor cases are relatively common. **Candidiasis** (kan-di-DĪ-a-sis) results from a fungal (yeast) infection. The organism responsible appears to be a normal component of the vaginal ecosystem in 30–80 percent of healthy women. Antibiotic administration, immunosuppression, stress, pregnancy, and other factors that change the local environment can stimulate the unrestricted growth of the fungus. Symptoms include itching and burning, and a lumpy white discharge may also be produced. Topical antifungal medications are used to treat this condition.

Bacterial (nonspecific) vaginitis results from the combined action of several bacteria. The bacteria involved are normally present in about 30 percent of adult women. In this form of vaginitis, the vaginal discharge contains epithelial cells and large numbers of bacteria. The discharge has a homogeneous, sticky texture and a characteristic odor sometimes described as fishy or aminelike. Topical or oral antibiotics are typically effective in controlling this condition.

Trichomoniasis (trik-ō-mō-NĪ-a-sis) involves infection by a parasite, *Trichomonas vaginalis*, introduced by sexual contact with a carrier. Because it is a sexually transmitted disease, both partners must be treated to prevent reinfection.

A vaginal infection by *Staphylococcus* bacteria is responsible for **toxic shock syndrome** (TSS). Symptoms include high fever, sore throat, vomiting and diarrhea, and a generalized rash. As the condition progresses, shock, respiratory distress, and kidney or liver failure may develop, and 10–15 percent of all cases prove fatal. These symptoms result from the entry of bacterial toxins and even bacteria into the bloodstream. This disorder was first recognized in 1978, when it affected a group of children with skin infections. Since that time, roughly 3000 cases have been diagnosed, 95 percent of them adult women with vaginal staphylococcal infections. More than 90 percent of these women developed the condition while they were using a tampon, but the precise link between tampon use and TSS remains uncertain. The use of superabsorbent tampons was initially thought to be responsible, and the incidence of TSS did decline as those items were removed from the market. However, TSS continues to occur at a low but significant rate (6.2 per 100,000 menstruating women per year) in those who use ordinary tampons, and in men or women after abrasion or burn injuries that promote bacterial infection. Treatment for TSS involves fluid administration, removal of the focus of infection (such as removal of a tampon or cleansing of a wound), and antibiotic therapy.

⚕ Sexually Transmitted Diseases

FAP *p. 1079*

Sexually transmitted diseases, or **STDs,** are transferred from individual to individual, in most cases by sexual intercourse. A variety of bacterial, viral, and fungal infections are included in this category. At least two dozen STDs are currently recognized. All are unpleasant. *Chlamydia* can cause pelvic inflammatory disease and infertility. Other types of STDs are quite dangerous, and a few, including AIDS, are deadly. Here we will discuss five of the most common sexually transmitted diseases: *gonorrhea, syphilis, herpes, genital warts,* and *chancroid.*

GONORRHEA

The bacterium *Neisseria gonorrhoeae* is responsible for gonorrhea, one of the most common STDs in the United States. Nearly 2 million cases are reported each year. These bacteria normally invade epithelial cells that line the male or female reproductive tract. In relatively rare cases, they will also colonize the pharyngeal or rectal epithelium.

The symptoms of genital infection differ according to the gender of the infected individual. It has been estimated that up to 80 percent of women infected with gonorrhea experience no symptoms, or symptoms so minor that medical treatment is not sought. As a result, these women act as carriers, spreading the infection through their sexual contacts. An estimated 10–15 percent of women infected with gonorrhea experience more-acute symptoms because the bacteria invade the epithelia of the uterine tubes. This infection probably accounts for many of the cases of pelvic inflammatory disease (PID) in the U.S. population. As many as 80,000 women may become infertile each year as the result of scar tissue formation along the uterine tubes after gonorrheal infections.

Seventy to eighty percent of infected males develop symptoms painful enough to make them seek antibiotic treatment. The urethral invasion is accompanied by pain on urination (*dysuria*) and typically by a viscous urethral discharge. A sample of the discharge can be cultured to permit positive identification of the organism involved.

SYPHILIS

Syphilis (SIF-i-lis) results from infection by the bacterium *Treponema pallidum.* The first reported syphilis epidemics occurred in Europe during the sixteenth century, possibly introduced by early explorers returning from the New World. The death rate from the "Great Pox" was appalling, far greater than it is today, even after we take into account the absence of antibiotic therapies at that time. It appears likely that the syphilis organism has mutated during the interim. These changes have reduced the immediate mortality rate but have prolonged the period of chronic illness and have increased the likelihood of successful transmission. Syphilis still remains a life-threatening disease. Untreated syphilis can cause serious cardiovascular and neurological illness years after infection, or it can be spread to a fetus during pregnancy to produce congenital malformations. The annual reported incidence of this disease peaked in 1994, and has since declined to roughly 6 cases per 100,000 population, the lowest rate since 1960.

Primary syphilis begins as the bacteria cross the mucous epithelium and enter the lymphatics and bloodstream. At the invasion site, the bacteria multiply; after an incubation period ranging from 1.5 to 6 weeks, their activities produce a painless raised lesion, or **chancre** (SHANG-ker). This lesion remains for several weeks before fading away, even without treatment. In heterosexual men, the chancre tends to appear on the penis; in women, it may develop on the labia, vagina, or cervix. Lymph nodes in the region often enlarge and remain swollen even after the chancre has disappeared.

Symptoms of *secondary syphilis* appear roughly 6 weeks later. Secondary syphilis generally involves a diffuse, reddish skin rash. Like the chancre, the rash fades over a period of 2–6 weeks. These symptoms may be accompanied by fever, headaches, and uneasiness. The combination is so vague that the disease may easily be overlooked or diagnosed as something else. In a few instances, more-serious complications such as *meningitis* (p. 74), *hepatitis* (p. 172), or *arthritis* (p. 55) may develop.

The individual then enters the *latent phase.* The duration of the latent phase varies widely. Fifty to seventy percent of untreated individuals with latent syphilis fail to develop the symptoms of *tertiary syphilis,* or *late syphilis,* although the bacterial pathogens remain within their tissues. Those destined to develop tertiary syphilis may do so 10 or more years after infection.

The most severe symptoms of tertiary syphilis involve the CNS and the cardiovascular system. **Neurosyphilis** may result from bacterial infection of the meninges or the tissues of the brain or spinal cord. **Tabes dorsalis** (TĀ-bēz dor-SAL-is) results from the invasion and demyelination of the posterior columns of the spinal cord and the sensory ganglia and nerves. In the cardiovascular system, the disease affects the major vessels, leading to *aortic stenosis* (p. 121), *aneurysms* (p. 125), or *focal calcification* (FAP *p. 713*).

Equally disturbing are the effects of transmission from mother to fetus across the placenta. These cases of congenital syphilis are marked by infections of the developing bones and cartilages of the skeleton and progressive damage to the spleen, liver, bone marrow, and kidneys. The risk of transmission may be as high as 95 percent, so maternal blood testing is recommended early in pregnancy. Treatment of syphilis involves the administration of *penicillin* or other antibiotics.

HERPES VIRUS

Genital herpes results from infection by herpes viruses. Two different viruses are involved. Eighty to ninety percent of genital herpes cases are caused by the virus known as HSV-2 (herpes simplex virus Type 2), which is usually associated with the external genitalia. The remaining cases are caused by HSV-1, the virus that is also responsible for cold sores on the mouth. Typically, within a week of the initial infection, the individual develops painful, ulcerated lesions on the external genitalia. In women, ulcerations may also appear on the cervix. These ulcerations gradually heal over the next 2–3 weeks. Recurring lesions are common, although subsequent incidents are less severe.

During delivery, infection of the newborn infant with herpes viruses present in the mother's vagina can lead to serious illness, because the infant has few immunological defenses. The antiviral agent *acyclovir* has helped treatment of initial infections.

GENITAL WARTS

Genital warts, or *condyloma acuminata,* result from infection with one of a number of different strains of *human papillomavirus* (HPV). Several of these strains (notably types 16, 18, and 31) are thought to be responsible for cases of cervical, anal, vaginal, and penile cancer. Roughly 1.2 million cases of genital warts are diagnosed each year in the U.S.. There is no satisfactory treatment for this problem. The traditional treatments have included cryosurgery, erosion by caustic chemicals, surgical removal, and laser surgery to remove the warts. These treatments remove the visible signs of infection, but the virus remains within the epidermis. Alpha-interferons have been tried with limited success.

CHANCROID

Chancroid is an STD caused by the bacterium *Haemophilus ducreyi.* Chancroid cases were rarely seen in the United States before 1984, but since then the number of cases has risen dramatically, reaching 4000–5000 cases per year. The primary sign of this disease is the development of *soft chancres,* soft lesions otherwise resembling those of syphilis. The majority of chancroid patients also develop prominent inguinal lymphadenopathy.

Experimental Contraceptive Methods   FAP *p. 1076*

A number of experimental contraceptive methods are being investigated. For example, researchers are attempting to determine whether low doses of inhibin will suppress GnRH release and prevent ovulation. Another approach is to develop a method of blocking human chorionic gonadotropin (hCG) receptors at the corpus luteum. Produced by the placenta, hCG maintains the corpus luteum for the first 3 months of pregnancy. If the corpus luteum were unable to respond to hCG, normal menses would occur despite implantation of a blastocyst.

Several male contraceptives are also under development:

- *Gossypol,* a yellow pigment extracted from cottonseed oil, produces a dramatic decline in sperm count and sperm motility after 2 months. It can be administered topically, as it is readily absorbed through the skin. Fertility returns within a year after treatment is discontinued. Unfortunately, gossypol has not been approved as yet, because it has about a 10 percent risk of permanent sterility and the potential for development of hypokalemia.

- Weekly doses of testosterone suppress GnRH secretion over a period of 5 months. The result is a drastic reduction in the sperm count. The combination of a testosterone implant, comparable to that used in the Norplant system, with a GnRH antagonist, *cetrorelix,* effectively suppresses spermatogenesis. A new synthetic form of testosterone, *alpha-methyl-nortestosterone (MENT),* appears to be even more effective than testosterone in suppressing GnRH production.

- A drug used to control blood pressure appears to cause a temporary, reversible sterility in males. This drug is now being evaluated to see if low dosages will affect fertility in normal males without affecting blood pressure.

If contraceptive methods fail, options exist to either prevent implantation or terminate the pregnancy. The "morning-after pills" contain estrogens or progestins. They must be taken within 72 hours of intercourse, and they appear to alter the transport of the zygote or to prevent its attachment to the uterine wall. The drug known as *RU-486 (Mifepristone)* blocks the action of progesterone at the endometrial lining. The result is a normal menses and the degeneration of the endometrium whether or not a pregnancy has occurred.

DISORDERS OF DEVELOPMENT

Teratogens and Abnormal Development   FAP *p. 1090*

Teratogens (TER-a-tō-jenz) are stimuli that disrupt normal development by damaging cells, altering chromosome structure, or acting as abnormal inducers. **Teratology** (ter-a-TOL-o-jē—literally, the "study of monsters"—deals with extensive departures from the pathways of normal development. Teratogens that affect the embryo in the first trimester will potentially disrupt cleavage, gastrulation, or neurulation. The embryonic survival rate will be low, and most survivors will have severe anatomical and physiological defects that affect all the major organ systems. Errors introduced into the developmental process during the second or third trimester will be likely to affect specific organs or organ systems, for the major organizational patterns are already established. Nevertheless, the alterations reduce the chances for long-term survival.

Many powerful teratogens are encountered in everyday life. The location and severity of the resulting defects vary with the nature of the stimulus and the time of exposure. Radiation is a powerful teratogen that can affect all living cells. Even the X-rays used in diagnostic procedures can break chromosomes and produce developmental errors; thus nonionizing procedures such as ultrasound are used to track embryonic and fetal development. Fetal exposure to the microorganisms responsible for *syphilis* (p. 202) or *rubella* ("German measles") can also produce serious developmental abnormalities, including congenital heart defects, mental retardation, or deafness.

Some chemical agents, especially those acting as abnormal inducers, will be teratogenic only if present at a time when embryonic or fetal targets have the competence to respond to them. Thousands of critical inductions are underway during the first trimester, initiating developmental sequences that will produce the major organs and organ systems of the body. In almost every case, the nature of the inducing agent remains unknown, and the effects of unusual compounds within the maternal circulation cannot be predicted. As a result, virtually any unusual chemical that reaches an embryo has the potential for producing developmental abnormalities. For example, during the 1960s, the European market was strong for **thalidomide,** a drug effective in promoting sleep and preventing nausea. Thalidomide was commonly prescribed for women in early pregnancy, with disastrous results. The drug crossed the placenta and entered the fetal circulation, where (among other effects) it interfered with the induction process responsible for limb development. As a result, many infants were born without limbs or with drastically reduced ones. Thalidomide had not been approved by the U.S. Food and Drug Administration (FDA), and it could not be sold legally in the United States. Although today the FDA is often criticized for the slow pace of its approval process, in this case the combination of rigorous testing standards and complex bureaucratic procedures protected the public.

However, even when extensive testing is performed with laboratory animals, uncertainties still remain, because the chemical nature of the inducer responsible for a specific process may vary from one species to another. For example, thalidomide produces abnormalities in humans and monkeys, but developing mice, rats, and rabbits are unaffected by the drug.

More-powerful teratogenic agents will have an effect regardless of the time of exposure. Pesticides, herbicides, and heavy metals are common around agricultural and industrial environments, and these substances can contaminate the drinking water in the area. A number of prescription drugs, including certain antibiotics, tranquilizers, sedatives, steroid hormones, diuretics, anesthetics, and analgesics also have teratogenic effects. Pregnant women should read the "Caution" label before using any drug without the advice of a physician.

Fetal alcohol syndrome (FAS) occurs when maternal alcohol consumption produces developmental defects, such as skeletal deformation, cardiovascular defects, and neurological disorders. Mortality rates can be as high as 17 percent, and the survivors are plagued by problems in later development. The most severe cases involve mothers who consume the alcohol content of at least 7 ounces of hard liquor, 10 beers, or several bottles of wine each day. But because the effects produced are directly related to the degree of exposure, there is probably no level of alcohol consumption that can be considered completely safe. Fetal alcohol syndrome is the number one cause of mental retardation in the United States today, affecting roughly 7500 infants each year.

Smoking presents another major risk to the developing fetus. In addition to introducing potentially harmful chemicals, such as nicotine, smoking lowers the P_{O_2} of maternal blood and reduces the amount of oxygen that arrives at the placenta. A fetus carried by a smoking mother will not grow as rapidly as one carried by a nonsmoker, and smoking increases the risks of spontaneous abortion, prematurity, and fetal death. There is also a higher rate of infant mortality after delivery, and postnatal development can be adversely affected.

INDUCTION AND SEXUAL DIFFERENTIATION

The physical (phenotypic) gender of a newborn infant depends on the hormonal cues received during development not on the genetically determined gender of the individual. If something disrupts the normal inductive processes, the individual's genetic and anatomical genders may be different. Such a person is called a **pseudohermaphrodite** (soo-dō-her-MAF-ro-dīt). For example, if a female embryo becomes exposed to male hormones, it will develop the sexual characteristics of a male. Cases of this sort are relatively rare. The most common cause is the hypertrophy of the fetal adrenals and their production of androgens in high concentrations; in some cases this condition has been linked to genetic abnormalities. Maternal exposure to androgens, in the form of anabolic steroids or as a result of an endocrine tumor, can also produce a female pseudohermaphrodite.

Male pseudohermaphrodites may result from an inability to produce adequate concentrations of androgens due to some enzymatic defect. In **testicular feminization syndrome,** the infant appears to be a normal female at birth. Typical physical changes occur at puberty, and the individual has the overt physical and behavioral characteristics of an adult woman. Menstrual cycles do not appear, however, for the vagina ends in a blind pocket, and there is no uterus. Biopsies performed on the gonads reveal normal testicular structure, and the interstitial cells are busily secreting testosterone. The problem apparently involves a defect in the cellular receptors sensitive to circulating androgens. Neither the embryo nor the adult tissues can respond to the testosterone produced by the gonads, so the person develops and remains physically a female.

If detected in infancy, many cases of pseudohermaphroditism can be treated with hormones and surgery to produce males or females of normal appearance. Depending on the arrangement of the internal organs and gonads, normal reproductive function may be more difficult to achieve.

Pseudohermaphroditism is one example of a developmental problem caused by hormonal miscues or by an inability to respond appropriately to hormonal instructions. Another example is provided by male infertility associated with maternal exposure to *diethylstilbestrol* (DES), a synthetic steroid prescribed in the 1950s to prevent miscarriages. An estimated 28 percent of male offspring produced abnormally small amounts of semen with marginal sperm counts at maturity. Daughters also have higher than normal infertility rates, due to uterine, vaginal, and uterine tube abnormalities, and they have an increased risk of developing vaginal cancer.

Ectopic Pregnancies   FAP *p. 1091*

Implantation normally occurs at the endometrial surface that lines the uterine cavity. The precise location within the uterus varies, although in most cases implantation occurs in the uterine body. In an ectopic pregnancy, implantation occurs somewhere other than within the uterus.

The incidence of ectopic pregnancies is approximately 0.6 percent of all pregnancies. Women who douche regularly have a 4.4 times higher risk of experiencing an ectopic pregnancy, presumably because the flushing action pushes the zygote away from the uterus. If the uterine tube has been scarred by a previous episode of pelvic inflammatory disease, there is also an increased risk of an ectopic pregnancy. Although implantation may occur within the peritoneal cavity, in the ovarian wall, or in the cervix, 95 percent of ectopic pregnancies involve implantation within a uterine tube. Because it cannot expand enough to accommodate the developing embryo, the tube normally ruptures during the first trimester. At that time, the bleeding that occurs in the peritoneal cavity may be severe enough to pose a threat to the woman's life.

In a few instances, the ruptured uterine tube releases the embryo with an intact umbilical cord, and further development can occur. About 5 percent of these abdominal pregnancies actually complete full-term development; normal birth cannot occur, but the infant can be surgically removed from the abdominopelvic cavity. Because abdominal pregnancies are possible, it has been suggested that men as well as women could act as surrogate mothers if a zygote were surgically implanted into the peritoneal wall. It is not clear how the endocrine, cardiovascular, nervous, and other systems of a man would respond to the stresses of pregnancy. However, the procedure has been tried successfully in mice, and experiments continue.

Problems with Placentation

   FAP *p. 1097*

In a **placenta previa** (PRĒ-vē-uh; "in the way"), implantation occurs in or near the cervix. This condition causes problems as the growing placenta approaches the internal os (internal cervical orifice). In a **total placenta previa,** the placenta extends across the internal os, whereas a partial placenta previa only partially blocks the os. The placenta is characterized by a rich fetal blood supply and the erosion of maternal blood vessels within the endometrium. Where the placenta passes across the internal os, the delicate complex hangs like an unsupported water balloon. As the pregnancy advances, even minor mechanical stresses

can be enough to tear the placental tissues, leading to massive fetal and maternal bleeding.

Most cases are not diagnosed until the seventh month of pregnancy, as the placenta reaches its full size. At that time, the cervical canal is dilated and the uterine contents are pushing against the placenta where it bridges the internal os. Minor, painless bleeding normally appears as the first sign of the condition. The diagnosis can be confirmed by ultrasound scanning. Treatment of total placenta previa involves bed rest for the mother until the fetus reaches a size at which cesarean delivery can be performed with a reasonable chance of neonatal (newborn) survival.

In an **abruptio placentae** (ab-RUP-shē-ō pla-SEN-tē), part or all of the placenta tears away from the uterine wall sometime after the fifth month of gestation. The bleeding into the uterine cavity and the pain that follows are normally noted and reported, although in some cases the shifting placenta may block the passage of blood through the cervical canal. In severe cases, the bleeding leads to maternal anemia, shock, and kidney failure. Although maternal mortality is low, the fetal mortality rate from this condition ranges from 30 to 100 percent, depending on the severity of the fetal blood loss.

Problems with the Maintenance of a Pregnancy   FAP *p. 1105*

The rate of maternal complications during pregnancy is relatively high. Pregnancy stresses maternal systems, and the stresses can overwhelm homeostatic mechanisms. The term **toxemia** (tok-SĒ-mē-uh) **of pregnancy** refers to disorders that affect the maternal cardiovascular system. Chronic hypertension is the most characteristic symptom, but problems with fluid balance and CNS disturbances, leading to coma or convulsions, may also occur. Some degree of toxemia occurs in 6–7 percent of third-trimester pregnancies. Severe cases account for 20 percent of maternal deaths and contribute to an estimated 25,000 neonatal deaths each year.

Toxemia of pregnancy includes **preeclampsia** (prē-ē-KLAMP-sē-uh) and **eclampsia** (ē-KLAMP-sē-uh). Preeclampsia is most common during a woman's first pregnancy. The mother's systolic and diastolic pressures become elevated, reaching levels as high as 180/110. Other symptoms include fluid retention and edema, along with CNS disturbances and alterations in kidney function. Roughly 4 percent of individuals with preeclampsia develop eclampsia.

Eclampsia is heralded by the onset of severe convulsions lasting 1–2 minutes followed by a variable period of coma. Other symptoms resemble those of preeclampsia, with additional evidence of liver and kidney damage. The mortality rate from eclampsia is approximately 5 percent; the mother can be saved only if the fetus is delivered immediately. Once the fetus and placenta have been removed from the uterus, symptoms of eclampsia disappear over a period of hours to days.

Complexity and Perfection

FAP *p. 1106*

The expectation of prospective parents that every pregnancy will be idyllic and every baby will be perfect reflects deep-seated misconceptions about the nature of the developmental process. These misconceptions lead to the belief that when serious developmental errors occur, someone or something is at fault and that blame might be assigned to maternal habits, such as smoking, alcohol consumption, or improper diet, maternal exposure to toxins or prescription drugs, or the presence of other disruptive stimuli in the environment. The prosecution of women who give birth to severely impaired infants for "fetal abuse" (exposing a fetus to known or suspected risk factors) is an extreme example of this philosophy.

Although environmental stimuli may indeed lead to developmental problems, such factors are only one component of a complex system normally subject to considerable variation. Even if every pregnant woman were packed in cotton and confined to bed from conception to delivery, developmental accidents and errors would continue to appear with regularity.

Spontaneous mutations are the result of random errors in the replication process; such incidents are relatively common. At least 10 percent of fertilizations produce zygotes with abnormal chromosomes. Because most spontaneous mutations fail to produce visible defects, the actual number of mutations must be far larger. Most of the affected zygotes die before completing development, and only about 0.5 percent of newborn infants show chromosomal abnormalities that result from spontaneous mutations.

Due to the nature of the regulatory mechanisms, prenatal development does not follow precise, predetermined pathways. For example, much variation exists in the pathways of blood vessels and nerves, because it does not matter how the blood or neural impulses get to their destination, as long as they do get there. If the variations fall outside acceptable limits, however, the embryo or fetus fails to complete development. Very minor changes in heart structure may result in the death of a fetus, whereas large variations in venous distribution are extremely common and relatively harmless. Virtually everyone can be considered abnormal to some degree, because no one has characteristics that are statistically average in every respect. An estimated 20 percent of your genes are subtly different from those found in the majority of the population, and minor defects such as extra nipples or birthmarks are quite common.

Current evidence suggests that as many as half of all conceptions produce zygotes that do not survive the cleavage stage. These zygotes disintegrate within the uterine tubes or uterine cavity; because implantation never occurs, there are no obvious signs of pregnancy. These instances of preimplantation mortality are commonly associated with chromosomal abnormalities. Of those embryos that implant, roughly 20 percent fail to complete 5 months of development, with an average survival time of 8 weeks. In most cases, severe problems affecting early embryogenesis or placenta formation are responsible. Figure A-58 graphically shows the relation of prenatal mortality to gestational age.

Prenatal mortality tends to eliminate the most severely affected fetuses. Those with less-extensive defects may survive, completing full-term gestation or arriving via premature delivery.

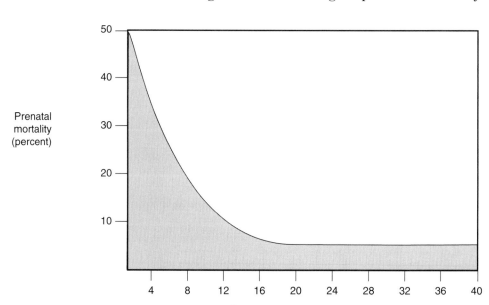

Figure A-58 Prenatal Mortality

Prenatal mortality (percent)

Gestational age (weeks)

Congenital malformations are structural abnormalities, present at birth, that affect major systems. Spina bifida, hydrocephaly, anencephaly, cleft lip, and Down syndrome are among the most common congenital malformations; we described those conditions in earlier chapters of the text. The incidence of congenital malformations at birth averages about 6 percent, but only 2 percent are categorized as severe. Of these congenital problems, only 10 percent can be attributed to environmental factors in the absence of chromosomal abnormalities or genetic factors, including a family history of similar or related defects.

Medical technology continues to improve our abilities to understand and manipulate physiological processes. Genetic analysis of potential parents may now provide estimates about the likelihood of specific problems, although the problems themselves remain outside our control. But even with a better understanding of the genetic mechanisms involved, we will probably never be able to control every aspect of development and thereby prevent spontaneous abortions and congenital malformations. There are simply too many complex, interdependent steps involved in prenatal development, and malfunctions of some kind are statistically inevitable.

Monitoring Postnatal Development
FAP *p. 1108*

Each newborn infant is closely scrutinized after delivery. The maturity of the newborn may also be determined prior to delivery by means of ultrasound or amniocentesis (see Table A-30, p. 208). Potential complications of premature delivery include neonatal *respiratory distress syndrome* (p. 158), the absence of some reflexes, hyperbilirubinemia, and a predisposition to infection.

A physical examination of the newborn focuses on the status of vital systems. Inspection of the infant normally includes the following:

- The head of a newborn infant may be misshapen after vaginal delivery, but it generally assumes its normal shape within the next few days. However, the size of the head must be checked to detect *hydrocephalus* (FAP *p. 452*), and the cranial vault is checked to ensure that *anencephaly* (FAP *p. 503*) does not exist.

- The abdomen is palpated to detect abnormalities of internal organs.

- The external genitalia are inspected. The scrotum of a male infant is checked to see if the testes have descended.

- Cyanosis of the hands and feet is normal in the newborn, but the rest of the body should be pink. A generalized cyanosis may indicate congenital circulatory disorders, such as *erythro-*

blastosis fetalis (FAP *p. 654*), a patent foramen ovale or ductus arteriosus, tetralogy of Fallot (FAP *p. 759*), or other problems.

Measurements of body length, head circumference, and body weight are taken. A weight loss in the first 48 hours is normal, because fluid shifts occur as the infant adapts to the change from weightlessness (floating in amniotic fluid) to normal gravity. (Comparable fluid shifts occur in astronauts returning to Earth after extended periods in space.)

The excretory systems of the newborn infant are assessed by examination of the urine and feces. The first urination may be pink, owing to the presence of uric acid derivatives. The first bowel movement consists of a mixture of epithelial cells and mucus. This *meconium* is greenish-black.

The **Apgar rating** is a measurement that takes into account heart rate, respiratory rate, muscle tone, response to stimulation, and color at 1 and 5 minutes after birth. In each category, the infant receives a score ranging from 0 (poor) to 2 (excellent), and the scores are then totaled. An infant's Apgar rating (0–10) has been shown to be an accurate predictor of newborn survival and the presence of neurological damage. For example, newborn infants with *cerebral palsy* (FAP *p. 499*) tend to have a low Apgar rating.

In the course of this examination, the breath sounds, the depth and rate of respirations, and the heart rate are noted. Both the respiratory rate and the heart rate are considerably higher in the infant than in the adult (see Table A-1, p. 7). The nervous and muscular systems are assessed for normal reflexes and muscle tone. Reflexes commonly tested include the following:

- The *Moro reflex* is a reflex triggered when support for the head of a supine infant is suddenly removed. The reflex response consists of trunk extension and a rapid cycle of extension–abduction and flexion–adduction of the limbs. This reflex normally disappears at an age of about 3 months.

- The *stepping reflex* consists of walking movements triggered by holding the infant upright, with a forward slant, and placing the soles of the feet against the ground. This reflex normally disappears at an age of about 6 weeks.

- The *placing reflex* can be triggered by holding the infant upright and drawing the top of one foot across the bottom edge of a table. The reflex response is to flex and then extend the leg on that side. This reflex also disappears at an age of about 6 weeks.

- The *sucking reflex* is triggered by stroking the lips. The associated *rooting reflex* is initiated by stroking the cheek, and the response is to turn the mouth toward the site of stimulation. These reflexes persist until age 4–7 months.

Table A-30 **Tests Performed During Pregnancy and on the Neonate**

Diagnostic Procedure	Method and Result	Representative Uses	Notes
Amniocentesis	A needle is inserted through the abdominal wall into the uterine cavity; ultrasonography is used to guide the needle as amniotic fluid is removed for analysis	Detection of chromosomal abnormalities; level of hemolysis in erythroblastosis fetalis; determination of fetal maturity; detection of birth defects such as spina bifida; to evaluate fetal distress	Procedure carries a small risk of miscarriage
Pelvic ultrasonography	Standard ultrasound	Detection of multiple fetuses, fetal abnormalities, and placenta previa; estimation of fetal age and growth	
External fetal monitoring	Monitoring devices on the external abdominal surface measure fetal heart rate and force of uterine contraction	Detection of irregular heart rate or fetal stress	Interpretation can be difficult; may increase incidence of cesarean sections
Internal fetal monitoring	Electrode is attached to fetal scalp to monitor heart rate; catheter is placed in uterus to monitor uterine contractions	As above	As above
Chorionic villi biopsy	Test performed during weeks 8–10 of gestation; a cannula is placed in the vagina and led through the cervix to the placenta; ultrasound is used to help in guiding the cannula; small pieces of villi are suctioned into a syringe	Detection of chromosomal abnormalities and biochemical disorders	Procedure has a higher risk of miscarriage than does amniocentesis

Laboratory Test	Normal Values	Significance of Abnormal Values	Notes
Amniotic fluid analysis			
Karyotyping	Normal chromosomes	Detects chromosomal defects such as those in Down syndrome	
Bilirubin	Traces only	Increased values indicate amount of hemolysis of fetal RBCs by mother's Rh antibodies	
Meconium	Not present	Present in fetal distress	
Lecithin/sphingomyelin ratio (L/S ratio)	$\geq$2:1 ratio	Indicates maturity of fetal lungs and maturity of fetus	
Creatinine	$\geq$2 mg/dL of amniotic fluid indicates week 36 of gestation	Assessed with L/D ratio for fetal maturity (not as accurate)	
Alpha-fetoprotein (AFP)	Week 16 of gestation: 5.7–31.5 ng/ml (lowers with increasing gestational age)	Increased values indicate possible neural tube defect such as spina bifida	
Human chorionic gonadotropin (hCG) (serum or urine)	Nonpregnant females (serum): <0.005 IU/ml Nonpregnant females (urine): negative Pregnant females (urine): <500,000 IU/24 hr	Useful in determination of pregnancy; used in home pregnancy tests	

Table A-30 (*continued*)

Laboratory Test	Normal Values	Significance of Abnormal Values	Notes
TORCH Toxoplasmosis Other (syphilis, group B beta-hemolytic strep and *Varicella*) Rubella Cytomegalovirus Herpes simplex Type II virus	Pregnant females: negative for IgM antibodies to these pathogens Neonates: negative for IgM antibodies to these pathogens	Pathogens causing these disorders can cross the placenta and infect the fetus; these fetal infections cause mild to severe problems, such as stillbirth; mother must be free of active herpetic lesions to deliver vaginally	Pregnant women are normally tested when infection is suspected to determine danger to fetus; most-common congenital infections are cytomegalovirus and rubella
Alpha-fetoprotein AFP (serum)	Adults: <40 ng/ml	>500 ng/ml occur in liver tumors; in pregnancy the levels peak at week 16–18; elevated levels occur with Down syndrome, anencephaly, and spina bifida	
Blood type, Rh factor	Rh$^+$ or Rh$^-$	A sensitized Rh$^-$ mother carrying an Rh$^+$ baby can result in erythroblastosis fetalis.	*See* FAP *p. 653*
2-hour postprandial glucose test	Adults: 70–140 mg/dl (2 hr after glucose administration)	Increased level indicates possible diabetic state.	
Human placental lactogen (hPL)	>4 mg/dl	<4 mg/dl may result from fetal distress; may indicate possible miscarriage; useful in evaluation of placental function	
NEONATES			
Blood from umbilical cord	Blood typing	Detection of maternal Rh antibody titer	
Bilirubin	<12 mg/dl	Increased levels occur in jaundice due to immaturity of newborn's liver	
Phenylalanine (serum)	1–3 mg/dl	>4 mg/dl occurs in phenylketonuria	
Galactose-1-phosphate	18.5–28.5 U/g hemoglobin	Decreased value indicates galactosemia	

- The *Babinski reflex* is positive, with fanning of the toes in response to stroking of the side of the sole of the foot (FAP *p. 440*). This reflex disappears at about age 3 as descending motor pathways become established.

These procedures check for the presence of anatomical and physiological abnormalities. They also provide baseline information useful in assessing postnatal development. In addition, newborn infants are typically screened for genetic or metabolic disorders, such as *phenylketonuria* (PKU) (p. 175), congenital *hypothyroidism* (p. 97), *galactosemia* (p. 17), and *sickle cell anemia* (p. 110).

Pediatrics is a medical specialty focusing on postnatal development from infancy through adolescence. Infants and young children cannot clearly describe the problems they are experiencing, so pediatricians and parents must be skilled observers. Standardized testing procedures are also used to assess an individual's developmental progress. In the **Denver Developmental Screening Test** (DDST), infants and children are checked repeatedly during their first 5 years. The test checks gross motor skills, such as sitting up or rolling over, language skills, fine motor coordination, and social interactions. The results are compared with normal values determined for individuals of similar age. These screening procedures assist in identifying children who may need special teaching and attention.

Too often parents tend to focus on a single ability or physical attribute, such as the age at first step or the rate of growth. This kind of one-track analysis has little practical value, and the parents may become overly concerned with how their infant compares with the norm. *Normal values are statistical averages,* not absolute realities. For example, most infants begin walking between 11 and 14 months of age. But about 25 percent start before then, and another 10 percent have not started walking by the fourteenth month. Walking early does not indicate true genius, and walking late does not mean that the infant will need physical therapy. The questions on screening tests such as the DDST are intended to identify any *patterns* of developmental deficits. Such patterns appear only when a broad range of abilities and characteristics are considered.

Death and Dying   FAP *p. 1111*

Despite exaggerated claims, there have been few substantiated cases of individuals who have reached an age of 120 years. Estimates for the life span of individuals born in the United States during 1996 are 77 years for males and 84 years for females. Interestingly enough, the causes of death vary with the age group under discussion. Consider the graphs shown in Figure A-59, which indicate the mortality statistics for various age groups. The major cause of death in young people is accidents, and in those over 40–45, it is cardiovascular disease. More-specific information about the major causes of death is given in Table A-31. Many of the characteristic differences in mortality values result from changes in the functional capabilities of the individuals linked to development or senescence. These values would differ significantly if tabulated for countries and cultures with different genetic and environmental pressures.

The differences in mortality values for males and females are related to differences in the accident rates among young people and in the rates of heart disease and cancer among older individuals. The upswing in cancer rates in females reflects a rising breast cancer incidence for those over age 34, whereas lung cancer is the primary cancer killer of older men. Among women, the incidence of lung cancers and related killers, including pulmonary disease, heart disease, and pneumonia, has been steadily increasing as the number of women smokers has increased. This change has narrowed the difference between male and female life expectancies.

Experimental evidence and calculations suggest that the human life span has an upper limit of about 150 years. As medical advances continue, research must focus on two related issues: (1) extending the average life span toward that maximum and (2) improving the functional capabilities of long-lived individuals. The first objective may be the easiest from a technical and moral standpoint. It is already possible to reduce the number of deaths attributed to specific causes. For example, new treatments promote remission in a variety of cancer cases, and anticoagulant therapies may reduce the risks of death or permanent damage after a stroke or heart attack. Many defective organs can be replaced with functional transplants, and the use of controlled immunosuppressive

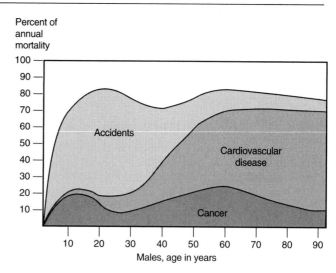

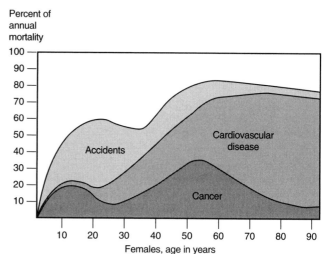

Figure A-59   Major Causes of Postnatal Mortality

drugs will increase the success rates of these operations. Artificial hearts have been used with limited success, and artificial kidneys and endocrine pancreases are under development.

The second objective poses more of a problem. Few people past their mid-nineties lead active, stimulating lives, and most would find the prospect of living another 50 years rather horrifying unless the quality of their lives could be significantly improved. Our abilities to prolong life now involve making stopgap corrections in systems on the brink of complete failure. Reversing the process of senescence would entail manipulating the biochemical operations and genetic programming of virtually every organ system. Although investigations continue, breakthroughs cannot be expected in the immediate future.

In the interim, we are left with some serious ethical and moral questions. Now that we can postpone the moment of death almost indefinitely, how do we decide when it is appropriate to do so? How can medical and financial resources be fairly allo-

Table A-31   The Five Major Causes of Death in the U.S. Population

Rank	Ages 1–14		Ages 15–34		Ages 35–54		Ages 55–74		Ages 75+	
	Males	Females	Males	Females	Males	Females	Males	Females	Males	Females
1	Accidents		Accidents		Heart disease	Cancer	Heart disease		Heart disease	
2	Cancer		Homicide	Cancer	Cancer	Heart disease	Cancer		Cancer	CVD
3	Congenital anomalies		Suicide		Accidents		CVD		CVD	Cancer
4	Homicide		Cancer	Homicide	Cirrhosis of the liver	CVD	Accidents	Diabetes	Pneumonia, influenza	
5	Pneumonia, influenza		Heart disease		Suicide	Cirrhosis of the liver	COPD	Accidents	COPD	Arterio-sclerosis

Note: CVD = cerebrovascular disease; COPD = chronic obstructive pulmonary disease.

cated? Who gets the limited number of hearts, livers, kidneys, and corneas available for transplant? Who should be selected for experimental therapies of potential significance? Should we take into account that care of an infant or child may add decades to a life span, whereas the costly insertion of an artificial heart in a 60-year-old may add only months to years? How shall we allocate the costs of sophisticated procedures that may reach hundreds of thousands of dollars over the long run? Are these individual or family responsibilities? Will only the rich be able to survive? Should the funds be provided by the government? If yes, what will happen to tax rates as the baby boomers become elderly citizens? And what about the role of the individual involved? If you decline treatment, are you mentally and legally competent? Could your survivors bring suit if you were forced to survive or if you were allowed to die? These and other difficult questions will not go away, and in the years to come we will have to find answers we are content to live and die with.

CRITICAL-THINKING QUESTIONS

11-1. Sally is an avid runner, and she trains incessantly. She has slimmed down so much that she is now underweight for her height and has very little fat tissue. Some of her laboratory values are as follows:

FSH: early cycle, 2 mIU/ml; midcycle, 3 mIU/ml

LH: early cycle, 2 mIU/ml; midcycle, 15 mIU/ml

Estrogen (serum): midcycle, 60 pg/ml

From this information, you would expect Sally to
a.   have heavy menstrual flows
b.   double ovulate
c.   be amenorrheic
d.   have painful menstrual cramps

11-2. Many male athletes using anabolic steroids show low sperm counts and some feminizing characteristics, such as enlarged breasts. Why do these symptoms occur?

11-3. Diane has peritonitis, which she is told resulted from a gonorrheal infection. Why does this secondary problem occur in women but not in men?

11-4. David is brought to the hospital after he received a severe blow to the scrotal region during a baseball game. The trauma has caused a breakdown in the blood–testis barrier. What effect will this have on David's immune system?

Clinical Problems

1. A 20-year-old woman lost consciousness while in a shopping mall. She had no identification with her. In the emergency room, the physician assessed her comatose state and ordered blood and urine studies. The physician obtained the following laboratory test results:

Blood Studies

Complete blood count (CBC): within normal limits

Serum electrolytes:

Sodium: 135 mEq/l

Potassium: 4.1 mEq/l

Chloride: 100 mEq/l

Blood glucose: 800 mg/dl

Blood pH: 7.23

P_{CO_2}: 30 mm Hg

HCO_3^-: 12 mEq/l

Blood urea nitrogen (BUN): 9 mg/dl

Urinalysis

pH: 4.6

Glucose: (+)

Protein: (–)

Blood: (–)

Ketones: (+)

Without further assessment, the physician ordered intravenous fluids and another intravenous substance. The patient soon regained consciousness. What was the diagnosis, and what did the physician give intravenously?

2. Gayle is in her sixth month of pregnancy and notices that her face and hands are swelling. At her next checkup, she mentions this swelling to her obstetrician. Gayle's blood pressure is now 150/110. The obstetrician examines the results of Gayle's urinalysis:

pH: 5

Color: amber

Appearance: cloudy

Specific gravity: 1.040

Protein: 2+ (160 mg/day or 16 mg/dl)

Glucose: trace

Ketones: none

Gayle is immediately hospitalized. What is the likely initial diagnosis?
a. placenta previa
b. preeclampsia
c. ectopic pregnancy
d. erythroblastosis fetalis
e. gestational diabetes

3. Moira and Sam have been attempting to conceive for the last year. They have consulted a fertility specialist, who conducted lab and diagnostic tests. The lab results are as follows:

Moira

LH (serum)

follicular = 30 mIU/ml

midcycle = 80 mIU/ml

Progesterone (serum)

before ovulation = 140 ng/dl

after ovulation = 400 ng/dl (within normal range)

FSH (serum)

pre- and post-ovulation = 25 mU/ml

midcycle = 80 mU/ml

What do these tests indicate about Moira's reproductive status? What aspects of her reproductive status remain uncertain?

Sam

Semen Analysis

Volume: 3.0 ml

Sperm count: 10 million/ml

Sperm motility: 65%

Abnormal sperm: 20%

What do these tests indicate about Sam's reproductive status? What else remains uncertain about his reproductive status? What treatment options are available to this couple?

4. John, a 66-year-old man, reports to his physician that he has had trouble urinating for the last several months, with narrowed stream and difficulty initiating urination. For the last 2 days, urination has become painful and produces only small amounts of urine. Urinalysis results are within normal limits, with the exception of the presence of bacteria on culturing. The following observations and lab tests are recorded in John's chart:

Digital rectal exam: enlarged prostate and prostatic nodule in one lobe

Transrectal ultrasonography: abnormal mass present

Urine culture: bacteria present

Prostate-specific antigen: 12 ng/ml

What is the likely diagnosis? Why is it difficult for John to urinate?

Case Studies

❏ CASE STUDY 1—George's Demise

George P., a 65-year-old man, is brought to the emergency room by his family because he almost fainted when he stood up at home. George reports that he has felt tired and weak for the last few days. He is short of breath but attributes this symptom to his chronic bronchitis (he has smoked for 25 years).

When George arrives at the emergency room (ER), his blood pressure is 100/70 while he is lying down, and his pulse is 110 bpm. While he is sitting, his blood pressure is 90/70 and his pulse is 115 bpm. On palpation, he has mild epigastric tenderness, and his stool guaiac test is positive for blood.

A stat Hct/Hb is performed, and the results are Hct = 33, Hb = 10.0 g/dl. He is immediately given IV fluids and oxygen, and his blood is sent for typing and cross-match testing.

A nasogastric tube is inserted into George's stomach; suction reveals that the stomach contains black-brown material resembling coffee grounds. A guaiac test on this material is positive for blood. The stomach is flushed with ice water introduced through the tube, and the suction fluid gradually clears. Acid inhibitors are administered intravenously, and George is given antacids orally.

During the administration of IV fluids, George's blood pressure stabilizes at 120/70, and his pulse slows to 90 bpm. A second blood test reveals a hematocrit of 29 and a hemoglobin concentration of 9.0 g/dl. George now feels more comfortable and is breathing easier; he is transferred from the emergency room to the intensive care unit (ICU) for further observation. An hour later, his Hct = 30 and his Hb = 9.1.

While talking with his family, George suddenly experiences anterior chest pain that radiates into his neck and left shoulder, with diaphoresis (cold sweats) and shortness of breath. Nitroglycerin and antacids have no effect. His blood pressure falls, and IV fluids, blood transfusion, and further medication cannot stop the trend. Four hours after the first chest pain, George dies, and resuscitation attempts prove futile.

Discussion Questions

1. What is the likely explanation for fainting on standing? What is this phenomenon called?

2. Disorders affecting which systems may be responsible for sensations of tiredness and weakness?

3. What are the long-term effects of smoking on systems other than those discussed in Question 2?

4. What is the significance of these postural changes in blood pressure and heart rate?

5. What might account for the combination of epigastric tenderness and a positive guaiac test?

6. How do these values compare with normal values?

7. Why is George treated with IV fluids and oxygen? Why does the hospital staff perform typing and cross-match testing?

8. What is a nasogastric tube? Why did the physician decide to perform this test?

9. Why use ice water? Is it significant that the fluid clears?

10. What is the preliminary diagnosis at this point? Why are acid inhibitors and antacids administered?

11. What disorders may account for these symptoms?
12. Why are nitroglycerin and antacids administered? What does their lack of effect mean?
13. What can account for the decline in blood pressure despite the therapies administered?
14. What is the final diagnosis? Discuss the mechanisms responsible for the progression of this case.

❏ CASE STUDY 2—Christine's Workout

On Monday, Christine K., a 35-year-old woman, complains to her physician of chest and arm pain. The pain started after Christine worked hard in the backyard the previous Saturday, lifting and carrying plants, fertilizer, and concrete blocks. (She had continued working despite arm pain that developed about midday.) Christine has no problem breathing, but she cannot raise her arms above her shoulders, and her shoulder and chest muscles are sore to the touch. On questioning, Christine reports that her hands are not numb, and her urine is not dark.

On physical examination, muscles around the elbows, arms, shoulders, and anterior chest are swollen and tender. Christine can move her fingers and wrists normally, but she has some pain in the proximal forearm muscles. The ranges of motion at the elbow and shoulder joints are reduced. No bruises are evident, and Christine has normal sensation in her fingertips.

On inspection, her urine is cloudy and light brown; on dipstick testing, it tests positive for protein and hemoglobin (or myoglobin). Microscopically, there are no RBCs in the urine.

Blood tests are ordered to check for CPK and LDH, and kidney function tests are performed to assay BUN and creatinine. Treatment initially consists of pain medication, ice applied to the sore muscles, and a high fluid intake (at least 4 quarts every 24 hours).

Christine is asked to call the physician if her urine remains cloudy, if the volume decreases, or if her hands or forearms feel worse. Christine returns the next day to have her urine rechecked and to get the results of her lab tests. Her arms and chest are still sore and swollen, her hands still have normal sensation, and her urine is clear and dilute, but tests still show traces of protein and hemoglobin. Her lab tests from the previous day show CPK and LDH levels 10 times normal values, but kidney tests were normal.

Christine continues a high fluid intake for the next week, and her sore muscles gradually became less swollen and tender. During this period, her urine and blood enzyme tests returned to normal.

Discussion Questions

1. On the basis of the case history, discuss the possible causes of these symptoms, and rank them from high probability to low probability.

2. What is significant about breathing and arm elevation?

3. Why ask about her hands and her urine?

4. Discuss the potential importance of these observations.

5. What do these findings suggest?

6. Why are these tests selected? What is the physician trying to determine?

7. What is the preliminary diagnosis?

8. What problems might be present if Christine's urine remains cloudy or declines in volume or if her peripheral sensation changes?

9. What is the significance of these lab results?

10. What is the final diagnosis?

❏ CASE STUDY 3—Charlenes's Pain

Charlene, a 65-year-old woman, arrives at her physician's office reporting pain in her right back, thigh, and leg. This pain, which is continuous, was initially dull and aching but has gotten progressively more intense, especially over the last few days. The pain seems unrelated to a position change or to eating, and Charlene has had no problems with micturition or defecation. Her general

Discussion Questions

1. What are the most likely sources of pain that is distributed in this way?

2. What do these observations tell you about the nature of the problem?

medical health is poor, with severe COPD that requires continuous low-flow oxygen administration. Despite the oxygen, she is dyspneic, with labored respirations at 20–30 breaths per minute, and she can sleep only in a semisitting position. When she removes her oxygen, she can walk only a short distance. (Charlene has to do that frequently to sneak a quick cigarette. She has been smoking for 50 years and is unable or unwilling to quit.) She is currently on bronchodilators and has been taking prednisone for 18 months.

The physician notes that the physical examination reveals a thin, pale, weak elderly woman with a round, moon-shaped face, breathing with pursed lips and getting oxygen from nasal prongs. Her vital signs are normal except for the increased respiratory rate. She is not tender over her back, right thigh, or leg and has a normal range of motion in her right hip, knee, and ankle. She can walk normally. Her skin is clear, and her abdominal exam is normal. Her breath sounds are decreased, but she has no increased cough or sputum production. A urinalysis is performed, with normal results.

X-rays of her lower back and hip show osteoporosis with DJD, a possible compression fracture at L_5 (of uncertain age), but no acute problems. The physician reaches a tentative diagnosis of musculoskeletal pain possibly associated with the compression fracture. Blood samples are sent to the lab for analysis, and pain medications are prescribed.

The blood test results are normal. However, over the next 3 days, the pain intensifies. There are no new signs or symptoms, and Charlene continues her pain medications. A bone scan is ordered for the following day, but that night her friends bring Charlene to the emergency room with severe mental confusion.

The consulting internist wonders about (1) a toxic reaction to the pain medications, (2) a stroke, or (3) an infection masked by chronic steroid use. Laboratory tests are then performed; the results are normal except the chest X-ray and the hypoxia consistent with the preexisting COPD. A lumbar puncture is performed, and the results suggest meningitis. Charlene is held for observation and started on antibiotics pending the results of bacterial culture of blood, urine, sputum, and CSF samples.

The next morning, Charlene's family physician reexamines her in the hospital and notes the appearance of a red, papular rash along the side of Charlene's right ankle and the top of her right foot. Red papules are also distributed in a linear fashion across her trunk and arms. During the next day, the areas involved develop distinctive clear vesicles typical of a herpes infection.

3. Discuss the pattern of symptoms characteristic of COPD.

4. On the basis of this history, what else might be responsible for Charlene's pain?

5. How do bronchodilators work? What are the systemic effects of prednisone?

6. Relate these observations to what you now know about Charlene's medical history.

7. What do these observations reveal about the nature of the problem?

8. Why is it significant that there has been no change in coughing or sputum production and that the urinalysis was normal?

9. What was not found on the X-rays that could have accounted for the reported symptoms?

10. Explain the proposed linkage between the compression fracture and musculoskeletal pain.

11. Would you expect this, on the basis of the tentative diagnosis?

12. Why is the bone scan ordered?

13. Discuss each of these possibilities. What tests might be performed to determine which of these factors (if any) is responsible for Charlene's condition?

14. Does the rash affect the diagnosis, or is it an interesting but relatively insignificant secondary problem?

Charlene is started on IV acyclovir, and during the next 2 weeks, her rash fades and her mental status improves. The pain, diagnosed as postherpetic neuralgia, persists for 3 months but then subsides. There is no recurrence.

15. What is acyclovir used for?
16. What was the final diagnosis? Explain the linkage between (a) the pattern and timing of pain development, (b) the meningitis and its effects on Charlene's mental state, and (c) the distribution of the rash.

❏ CASE STUDY 4—Hal's Athletic Adventure

Hal, a 50-year-old politician, was a college basketball player and stayed fairly active with sports and weekend pickup games through his thirties and early forties. Having a members-only gym near his office facilitated this practice. However, intense campaigning and successful election to the U.S. Congress meant extra travel, fried chicken dinners, hotel rooms, and prolonged sitting.

After years of relative inactivity, with only sporadic visits to the gym, Hal participates in an intense basketball game. Midway through the game, he collapses suddenly with a sharp pain behind the right ankle. While he is on the floor, the pain eases, and he is able to move his ankle and foot reasonably well. But when he tries to stand, his ankle is very weak and painful. He proceeds (hopping) to a physician.

On examination, there is normal and painless range of motion in his right hip, knee, and toes. With his right leg dangling, he can dorsiflex and plantar flex the foot with only minimal pain. The region posterior to the ankle and proximal to the heel is swollen and tender. When he lies prone, with the right knee bent, there is a palpable depression posterior to the ankle, extending toward the knee. Grasping and compressing the calf muscles on this leg does not produce plantar flexion, whereas it produces plantar flexion when performed on the left leg.

Discussion Questions

1. What are the effects of (a) aging and (b) reduced levels of activity on the systems critical to the performance of athletic activities? How do these effects interact?

2. List at least three possible causes, and relate them to relevant factors detailed in your answer to Question 1.
3. What is the significance of the reduction in pain and Hal's ability to move the ankle and foot when he is not standing?

4. What does this tell the physician about the nature of the injury?

5. What structures are located in this region?

6. What does this mean, and what was the final diagnosis? Why was Hal still able to plantar flex the foot when it was not bearing weight?

❏ CASE STUDY 5—Jim and That Darned Cat

Jim, 65 years old, visits his new physician on Friday afternoon. Jim has a headache that will not go away. He and his wife retired and moved into a new home about 3 months ago. A week ago he was looking for one of their cats, which he found under the steps to the porch. When backing out from beneath the porch, he banged his head. Jim thinks that the headache started either at that time or shortly thereafter during an argument with his wife. Jim wants to get rid of the cats, because he thinks they are too much trouble and because they make him sneeze, but his wife won't consider it. Their arguments have been heated, and he has often developed mild headaches afterward.

Discussion Questions

1. Discuss the various possible causes of headaches.

2. List the possible causes for Jim's headache that match this history.

Jim describes this headache as continuous, all over his head, and extending into his posterior neck muscles. His neck muscles are tight. His vision is normal, as is his hearing, and he has no nausea. Tylenol and ibuprofen have helped slightly. He hasn't been sleeping well and remains upset about the cats.

On physical examination, Jim is alert and oriented, and his vital signs are normal. His neck is supple, with normal range of motion. The skull is normal on palpation, and there are no signs of abnormal cranial nerve function. His tympanic membranes are normal, as are his optic reflexes and retinae. His gait and deep tendon reflexes are also normal. The physician reaches a preliminary diagnosis of a muscle tension headache and gives Jim a mild sedative/muscle relaxant/pain medication. Jim has instructions to call or return if the headache does not improve.

On Monday, Jim calls the physician's office to report that the headache is worse. He is seen that morning at 11 A.M. His appearance and manner have changed markedly over the weekend. He could not drive, so his wife brought him to the office. He seems drowsy (which he attributes to the medication), and his headache is worse. It is a throbbing pain that increases if he lies down, presses against his head, or flexes his neck. His vision is still normal. Although he is not vomiting, he has not wanted to eat today.

On physical examination, Jim still has normal vital signs, and he has no fever. However, his blood pressure is 140/90, up from 120/80 on Friday. His neck motion is normal, although flexion increases the severity of the headache. His cranial nerve, optic, and gait/balance tests are normal and symmetrical. The skull remains tender to the touch.

Because there has been no improvement and the symptoms have worsened, Jim is sent for a CT scan of the head and a series of blood tests. The radiologist reports to the family physician that the CT scan showed a large, bilateral subdural hematoma of uncertain age. There is no sign of a skull fracture, and the blood tests (CBC, WBC, diff. count) are normal.

Jim is referred to a neurosurgeon. After discussion, Jim signs informed consent papers and is taken to the operating room. The neurosurgeon draining the subdural hematoma reports that there is an initial high-pressure release of blood when the skull is pierced.

After recovering from anesthesia, Jim finds that his headache is virtually gone. He jokes with his family physician that he is glad that he didn't have to get his head shaved (he is already bald) and says he has decided to buy a big dog.

3. Discuss the significance of each of these symptoms.

4. Does the lack of substantial pain relief with these drugs suggest anything about the nature of the problem?

5. Discuss the significance of each of these findings. What possible problems were ruled out in the course of this examination?

6. How does this situation match the preliminary diagnosis?

7. Discuss the significance of these symptoms.

8. Why is it important that the results are symmetrical? What would be the implications if they had been asymmetrical?

9. Is it significant that the blood tests are normal? How does this result affect the diagnosis? If results had been abnormal, what else could be involved?

10. What does the high-pressure release tell you about the status of the hematoma?

❑ CASE STUDY 6—Robin's Abdominal Pains

Robin, a 70-year-old man, goes to the hospital with severe upper abdominal pain and vomiting that began after his last meal. An X-ray and blood analysis show elevated liver function tests and indications of pancreatic and gallbladder problems. The initial diagnosis is acute cholelithiasis.

Robin is sent to surgery, and his inflamed gallbladder is removed. After surgery, he remains hospitalized for 6 weeks with elevated serum levels of amylase. Over this period, he is given nutrients through an intravenous infusion.

During surgery, the surgeon noted that Robin's liver had a lumpy, scarred appearance, and a liver biopsy was taken. The pathologist reports signs of chronic inflammation and scarring. The attending physician visits Robin and obtains more information about his history. Robin had vascular surgery for femoral artery occlusion 10 years ago and surgery to repair an aortic aneurysm 5 years ago. During each of these procedures, he received blood transfusions. Robin was a "social drinker" (two or three drinks a day for 35 years), often at business lunches or dinners. He has never used injected drugs and has been heterosexual and monogamous with his wife for the last 40 years. Robin had been a POW in a German concentration camp during World War II, where unknown substances were injected into his chest as part of an experiment. (The experimental goals and methods are not known.) He has had hypertension and occasional symptoms of post-traumatic stress disorder since that time, but until his vascular surgery, his health was otherwise good. Blood tests on admission showed no signs of hepatitis A or B, but his blood contained antibodies to hepatitis C. He is told to stop drinking and to restrict the use of drugs, such as tranquilizers.

In the 3 years since his release from the hospital, Robin develops ascites, peripheral edema, splenomegaly, lowered serum protein levels, a reduced platelet count, and elevated blood ammonia levels. He becomes somewhat disoriented and has difficulty speaking.

After another year, Robin reenters the hospital by ambulance. He is unresponsive, his hematocrit is 25, and he has vomited dark blood. His blood ammonia levels are above 500. A gastroscopic examination reveals ruptured blood vessels along the esophagus, with bleeding into the stomach. Epinephrine is injected into the broken vessels, and the bleeding stops. The gastric contents are aspirated, and the digestive tract evacuated. His hematocrit rises, his ammonia levels decline, and he awakens confused but relatively alert. He is discharged from the hospital after 2 weeks and placed on a low-protein diet.

Discussion Questions

1. What disorders may produce these symptoms?

2. What was probably seen on the X-ray? What blood tests were likely to have been abnormal?
3. Discuss the origin, symptoms, and treatment options for cholelithiasis.

4. What condition does Robin have? Is it serious? Why does he receive nutrients by infusion rather than by mouth?
5. What are the technical terms for liver inflammation and liver scarring? What are the most likely causes of these conditions?
6. What factors are apparent in this history that may have affected liver function?
7. Discuss (a) the condition responsible for femoral artery occlusion and (b) the goal of the surgery.
8. Describe the repair of an aortic aneurysm.

9. Which forms of hepatitis are unlikely, given this history?

10. What were the most likely routes of infection?
11. Why was Robin given these instructions?

12. Are these symptoms related to Robin's hepatitis? Discuss the linkage, if any.

13. Why is Robin experiencing disorientation and difficulty in speaking?

14. Discuss his status on admission and the implications of his symptoms.

15. Why was epinephrine injected at the bleeding site?
16. Why aspirate the gastric contents and clean the digestive tract?
17. Discuss the factors that accounted for his hospitalization and the reason for the dietary restrictions.
18. What other treatments might be used to relieve Robin's symptoms? Are any of these options practical in this case?

❏ CASE STUDY 7—Mrs. M.'s Summer Problems

Mrs. M. is a 47-year-old woman who spends 3 long, warm days in a wet swimming suit while she participates in canoe races. Afterward, she develops vulvar pruritis and notices a white, curdy ("cottage cheese") vaginal discharge. She consults her doctor for the first time in 3 years. While evaluating and treating her vaginitis, the physician notices that Mrs. M. has not had a Pap test or breast exam since her tubal ligation 6 years ago and that there is no history of diabetes. Because Mrs. M. is usually healthy, active, and has trouble scheduling regular checkups, her physician convinces her to have those tests performed while she is still in the office.

The pelvic exam is normal except for inflammation of the vulva and vagina. A KOH prep of the vaginal discharge shows a yeast infection. On breast exam, a 2 cm, hard, painless lump that dimpled the skin is found superior to the left nipple. There is no other breast abnormality or palpable axillary lymphadenopathy. Mrs. M. had felt the lump for several months, but she did not want to mention (or think about) it, because her aunt recently died of breast cancer.

A mammogram shows changes consistent with breast cancer. During the next few weeks, Mrs. M. and her husband consult surgeons, library books, and nutritionists. A lumpectomy is recommended, with axillary node biopsy followed by irradiation and/or chemotherapy to be determined by the extent of lymph node involvement. Mrs. M. decides to have the lumpectomy with no radiation. Because the lymph node biopsies are negative and the tumor cells are estrogen-sensitive, she agrees to have tamoxifen therapy.

The surgeon removes a large area around the tumor to avoid possible recurrence, as the area will not be irradiated. The surgery leaves a loss of tissue, including the nipple, but Mrs. M. declined breast augmentation therapy and prosthesis. Six years later, there is no evidence of disease, and she remains on tamoxifen.

Discussion Questions

1. Discuss the likely cause of Mrs. M.'s initial problem.
2. What organisms may be responsible?

3. Many people believe that if they are healthy, regular checkups and screening tests are unnecessary. Discuss the relationship between normal good health and diseases such as cancer.

4. What is the likely treatment for this condition?

5. What conditions may produce such symptoms?

6. Is it significant that her aunt died of breast cancer?

7. What stage is her breast cancer? Discuss the pros and cons of aggressive treatment of breast cancer and other cancers.

8. Why is estrogen sensitivity important?

❏ CASE STUDY 8—Cynthia's Bad Back

Cynthia, 28 years old, barely manages to make it to her physician's office. She has terrible right lower back pain. She has had a few twinges over the last few days that disappeared in seconds, but today she woke up at 3 A.M. with steadily increasing pain in her right back at about waist level. She at first attributed it to physical activity, but she remembered no injury and was able to move and bend her back, arms, and legs without any change in the level of pain. Cynthia has no previous history of back pain or chronic health problems. She has no cough or

Discussion Questions

1. Discuss the range of potential causes of acute back pain.

2. Is the history consistent with musculoskeletal injury?

3. What does the stability of the pain imply about the condition?

respiratory symptoms, no dysuria or dark urine, and no nausea, vomiting, or bowel problems. Her last menses, 3 weeks ago, was normal.

On entering the examination room, the physician finds Cynthia kneeling on the floor with her head resting on a chair seat. On request she is able to stand, walk, and sit on the exam table without limitation of movement. She does not have a fever, but her blood pressure and pulse are elevated. Her head, neck, and chest exams are normal. She is not tender over her kidneys, spinal vertebrae, or muscles. The neurological exam is normal. On abdominal examination, she is tender in her right lower abdomen but has no masses, guarding, or rebound. Neither the liver nor the spleen is enlarged, and bowel sounds are normal. A urinalysis is performed, and a pregnancy test is negative.

On pelvic examination, Cynthia's vulva, vagina, and cervix are normal in appearance, but on palpation the pain on the right side increases with movement of the cervix. Her uterus is small, firm, and nontender, but her right uterine tube and ovary are very tender, with a 3-cm mass detected on palpation. Cynthia feels pain both in her abdomen and in her back when this area is palpated. Rectal examination confirms the pelvic findings; the stool guaiac test is negative.

After Cynthia is admitted to the hospital, blood tests, X-rays, and ultrasound exams rule out pneumonia, cholecystitis, ectopic pregnancy, and appendicitis but show an enlarged cystic right ovary. Abdominal surgery is performed, and the ovary is found to be twisted on its supporting ligaments. The twisting is reduced, the cyst removed, and the ovary stabilized to prevent recurrence of the problem. Cynthia recovers from the surgery and resumes a normal life.

4. Discuss the implications of these findings; what would have been likely diagnoses if there had been a cough, dysuria, dark urine, nausea, vomiting, bowel problems, or abnormal menstrual periods?

5. Why might her bp and pulse be elevated?

6. Discuss the relevance of each of these findings. What potential causes of acute back pain are being eliminated?

7. What results would you expect to have seen, if these conditions were not ruled out?

8. Why was the stool guaiac test performed?

9. What is a comparable condition that affects males?

Answer Key

CRITICAL-THINKING QUESTIONS

THE INTEGUMENTARY SYSTEM (p. 44)

1-1. (d).

1-2. (b).

1-3. Because Carrie is a vegetarian, her diet includes an abundant amount of carrots, carrot juice, oranges, apricots, and green vegetables. These food sources contain carotene, a yellow-orange pigment that is stored in lipids, notably those in the skin. The skin discoloration is not dangerous, but Carrie may elect to cut back a bit on carrot juice for aesthetic reasons.

1-4. The organic solvents that Sam uses for cleaning dissolve the lipids that help provide a water barrier in his skin and help keep the skin flexible. Without the lipids, the skin becomes dry and loses more water than normal to the surroundings.

1-5. The hormonal changes that accompany pregnancy are responsible for the increased pigmentation, because hormones can affect melanin production by melanocytes. The hormone progesterone is responsible for the changes noted. In fact, these skin changes are so consistent that they are referred to as the "mask of pregnancy."

THE SKELETAL SYSTEM (p. 58)

2-1. (c).

2-2. Trauma. Arthroscopy revealed torn ligaments, which caused bleeding in the synovial cavity, and a damaged medial meniscus, which restricts movement.

2-3. (b). Tooth decay occurs when the acid produced by normal oral bacteria erodes the harder tooth structure. This particular tooth infection had spread through the tooth and into the root. The bony structure surrounding the tooth root became infected as well, producing osteomyelitis and an erosion of the adjacent bone. Antibiotics were given, and the decay was removed from the tooth, followed by a root canal procedure.

2-4. Since osteopenia generally begins between the ages of 30 and 40, these individuals were probably older than 40 when they died. The characteristics of the arm bones indicate that the men used their arms a great deal—perhaps in lifting, pulling, chopping, or shoveling. But because the leg bones do not indicate large or well-developed muscles, these men probably did not do a great deal of walking or running.

2-5. The first fracture probably involved the greater trochanter or the shaft of the femur inferior to the neck. The second fracture was probably a fracture of the femoral neck, which disrupted the circulation and led to avascular necrosis at the joint. The most likely treatment that would restore normal function is the insertion of an artificial hip joint.

THE MUSCULAR SYSTEM (p. 68)

3-1. (c).

3-2. (a). CeCe is concerned about the paralysis of her respiratory muscles after the injection of this drug. Dr. R. reassures her that she will be on a respirator until either the drug wears off or the drug's effects are counteracted by the administration of an antagonistic drug, *neostigmine methylsulfate.*

3-3. (b).

3-4. As a result of training, we would expect the tennis player to have larger reserves of glycogen and creatine phosphate than does the nonathlete. These materials would help supply energy during short bursts of strenuous activity. Thus we would expect to find higher levels of the enzymes required for the breakdown of glycogen and for the conversion of creatine phosphate to creatine in the muscle of the tennis player. In addition, training increases the level of enzymes that hydrolyze ATP during contraction, so we would also expect to see more of this enzyme in the tennis player than in the nonathlete.

3-5. Calvin has probably injured the lateral ligament, which stabilizes the lateral aspect of the ankle; this ligament tenses during inversion. He may also have torn the tibialis posterior muscle, which is the prime mover for inversion of the foot.

THE NERVOUS SYSTEM (p. 95)

4-1. (e).

4-2. (b).

4-3. The expected response would be plantar flexion. This ankle jerk reflex is an example of a stretch reflex: (1) Tapping on the tendon stretches intrafusal and extrafusal fibers in the gastrocnemius and soleus muscles; (2) the sensory processes at the muscle spindles are distorted; (3) the rate of action potentials along afferent fibers from the muscle spindles increases; (4) motor neurons in the spinal cord are stimulated; and (5) the extrafusal fibers in the stretched muscles contract. Stretch reflexes include important postural reflexes and reflexes involved with the automatic control of muscle tension during a contraction. Other examples include the patellar reflex, the biceps reflex, and the triceps reflex. Testing stretch reflexes provides information about the status of simple reflex arcs whose components are clearly identifiable. These are monosynaptic reflexes, and each reflex involves a limited number of spinal nerves and spinal cord segments. As a result, testing stretch reflexes is one way to check for damage to specific spinal nerves or to the related segments of the spinal cord.

4-4. It appears that Mrs. Glenn is suffering from Parkinson's disease. The motor neurons are more excitable than they should normally be, so we would expect to see exaggerated spinal reflexes due to increased muscle tone.

4-5. Chelsea probably suffered a subdural or epidural hemorrhage, which can result from a blow or other injury to the head that ruptures the blood vessels in the meninges. As blood accumulates in the subdural or epidural space, it puts pressure on the brain, leading to impaired neural function and (at high enough pressures) cessation of function. The pattern of sensory and motor impairment is consistent with a lateral hemorrhage in the area of the temporal lobe that gradually spreads downward to the brain stem, ultimately causing Chelsea's death.

4-6. If Molly's condition is the result of taking a sympathomimetic drug, you might observe the following symptoms: increased heart rate (pulse), increased blood pressure, rapid breathing, sweating even though the room is a comfortable temperature, and dilated pupils.

4-7. Fifty percent of the fibers decussate (cross over) at the optic chiasm. As a result, Hido will lose vision from the temporal field of the right eye and the nasal field of the left eye. After a period of time, some accommodation will occur, helping fill in the missing parts of the visual field.

4-8. Bone conduction tests can be used to discriminate between conductive and nerve deafness. In this type of test, the physician places a vibrating tuning fork against the patient's skull. If the patient hears the tuning fork when it touches the skull but not when it is held next to the ear, the problem must lie within the external or middle ear. If the patient does not respond to either stimulus, the problem must be at the receptors or the auditory pathway.

THE ENDOCRINE SYSTEM (p. 104)

5-1. (e).

5-2. (b).

5-3. (c).

5-4. John should not take GH (human growth hormone), because there is no medical reason for him to receive the hormone and the long-term effects of such treatments are not well known. GH affects many different tissues and can have widespread metabolic effects, such as a decline in body fat content. This decline can in turn affect the metabolism of many organs.

5-5. Angie's diabetes was due to excess GH (growth hormone) from the anterior pituitary. Chronic hypersecretion of GH produces chronic hyperglycemia (high blood sugar) and the same problems that are associated with insulin-dependent diabetes. Removal of the anterior pituitary takes away the source of the problem, allowing blood sugar levels to return to normal.

5-6. You could inject a large dose of TRH into the patient and then monitor the blood for elevated TSH. If there was an increase in TSH after the injection, the problem would involve the hypothalamus. But if there was no increase in TSH in response to the exogenous TRH, then the problem would be with the pituitary cells.

THE CARDIOVASCULAR SYSTEM (p. 130)

6-1. (b).

6-2. (c).

6-3. The most obvious possibility is patent ductus arteriosus, which occurs when the ductus arteriosus fails to close. When the baby is not being stressed (bathing creates heat loss and thermal stress) or eating (less air is entering the lungs), she appears normal. Because some of the blood flow is shunted to the aorta during stress and eating, not enough blood is being oxygenated, so the infant becomes cyanotic.

THE LYMPHATIC SYSTEM AND IMMUNITY (p. 149)

7-1. This type of pulmonary infection most commonly occurs in patients with AIDS. Matthew could be tested for AIDS by an ELISA test, which would detect antibodies to HIV in his blood. The Western blot test is another test for HIV and is more sensitive.

7-2. A key characteristic of cancer cells is their ability to break free from a tumor and migrate to other tissues of the body, forming new tumors. This process is called metastasis. The primary route for the spread of cancer cells is the lymphatic system, and cancer cells invade lymph nodes on their way to other tissues. Examination of regional lymph nodes for the presence of cancer cells can help the physician determine if the cancer was caught in the early stage or whether it has started to spread to other tissues. It can also give the physician an idea of what other tissues may be affected by the cancer to help choose the proper treatment.

7-3. Anne probably has infectious mononucleosis. The blood work reveals values within normal limits with the exception of the lymphocyte count. This diagnosis could be confirmed by testing for the presence of EBV in the blood.

THE RESPIRATORY SYSTEM (p. 162)

8-1. (c).

8-2. In not finishing the antibiotics, the patient has allowed the bacteria to propagate and enter the pleural cavity through the lymphatic system. The patient has pleurisy accompanied by an increase of fluid in the pleural space. This increase in fluid is called a *pleural effusion.*

8-3. (c). The tuberculin skin test is a confirming test for the presence of antibodies to the bacterium that causes tuberculosis. Ann will have a positive reaction if she has been exposed to the TB bacterium at some time in her life. The sputum culture may take up to 6 weeks; it will indicate whether she is currently infected with *Mycobacterium tuberculosis.*

THE DIGESTIVE SYSTEM (p. 182)

9-1. (e).

9-2. Cytological examination of the tissue specimen would most likely reveal malignant cells. The mass growing in Paul's colon is cancerous, and protrudes into the colon lumen. The rectal bleeding originates at the tumor.

9-3. Marva could have lactose intolerance due to the deficiency of the enzyme lactase. A diet analysis over a period of several weeks might reveal that Marva's episodes of diarrhea followed the ingestion of dairy products. The physician could order a lactose tolerance test. Marva may instead have irritable bowel syndrome, for which a diagnosis would be made only after all other, more dangerous disorders had been ruled out.

9-4. You would expect low blood glucose levels and elevated ketone bodies and fatty acids.

THE URINARY SYSTEM (p. 193)

10-1. (a).

10-2. (b). Use the following three steps to answer this type of question:

1. Determine if the pH is within normal limits (7.35–7.45):

pH > 7.45 is an alkalosis.

pH < 7.35 is an acidosis.

In our example, Fred has a pH of 7.30, which indicates an acidosis.

2. Determine the primary cause of the acidosis/alkalosis by checking the values for P_{CO_2} and bicarbonate against the corresponding normal values:

P_{CO_2} = 35–45 mm Hg

bicarbonate = 22–28 mEq/l

For the respiratory system to be the primary cause of an acidosis, the P_{CO_2} value must be above 45 mm Hg.

For the respiratory system to be the primary cause of an alkalosis, the P_{CO_2} value must be below 35 mm Hg.

For there to be a metabolic primary cause of an acidic pH, the bicarbonate value must be less than 22 mEq/l.

For there to be a metabolic primary cause of an alkaline pH, the bicarbonate value must be greater than 28 mEq/l.

In Fred's case, the primary cause of his acidic pH cannot be of a metabolic nature, because the bicarbonate level is greater than 28 mEq/l. The primary cause of Fred's acidic pH is the respiratory system, as indicated by the increased P_{CO_2} above the normal range.

3. Determine if the system that is not the primary cause is compensating for the imbalance in pH. If so, the value should be as follows:

P_{CO_2}: <35 mm Hg is a compensation for metabolic acidosis.

P_{CO_2}: >45 mm Hg is a compensation for metabolic alkalosis.

bicarbonate: <22 mEq/l is a compensation for respiratory alkalosis.

bicarbonate: >28 mEq/l is a compensation for respiratory acidosis.

In Fred's case, the increased level of bicarbonate is compensating for his respiratory acidosis. This process of compensation involves the kidneys, where some of the excess carbon dioxide is converted into carbonic acid and the carbonic acid is allowed to dissociate. The hydrogen ions are secreted, and the newly formed bicarbonate is conserved, to maintain a proper buffering capacity.

10-3. Narrowing of the right renal artery is related to atherosclerosis. This narrowing restricts blood flow to the kidney and produces renal hypotension and ischemia. Decreased blood flow and ischemia trigger the juxtaglomerular apparatus to produce more renin, which leads to elevated levels of angiotensin II and aldosterone. Both substances can increase the blood pressure by vasoconstriction (angiotensin II) and by increasing the blood volume (aldosterone).

10-4. Her GFR would be reduced, because (a) she has been hemorrhaging and her blood volume is reduced, and (b) she is probably in shock, and activation of the sympathetic nervous system will have reduced blood flow through the afferent arterioles.

10-5. Strenuous exercise causes sympathetic activation, which produces (1) powerful vasoconstriction of the afferent arterioles that deliver blood to the glomeruli and (2) a dilation of peripheral blood vessels, so blood is shunted away from the renal vessels. As a result, the GFR is reduced. Potentially dangerous conditions develop as the circulating concentration of metabolic wastes increases and peripheral water losses mount.

10-6. Because the inulin is neither reabsorbed nor secreted, the rate at which the inulin is excreted is equal to the glomerular filtration rate (GFR). Thus, if you find the rate at which the inulin is cleared from the plasma, you will have the GFR.

$$GFR = \frac{0.1 \text{ mg/ml} \times 1.0 \text{ ml/min}}{0.75 \text{ mg/dl}} = 133 \text{ ml/min}$$

10-7. Because Mr. Smith is a diabetic, he is probably suffering from acidosis resulting from ketosis. Increased production of ketone bodies lowers the blood pH. The increased concentration of hydrogen ions causes more hydrogen ions to move into the cells and more potassium ions to leave the cells, so the level of potassium in the ECF is elevated. The urine pH should be quite acidic, as the kidneys work to eliminate the excess hydrogen ions. The decreased pH would lead to increased rate and depth of breathing, so we would expect to see declining levels of CO_2, especially because the cause of the acidosis is not respiratory.

THE REPRODUCTIVE SYSTEM (p. 211)

11-1. (c).

11-2. Anabolic steroids are chemically related to the hormone testosterone. Elevated levels of such hormones suppress the release of GnRH from the hypothalamus and FSH and LH from the pituitary. This suppression leads to low endogenous levels of testosterone, decreased spermatogenesis, and a low sperm count. Changes in male secondary sex characteristics may result from the conversion of anabolic steroids into estrogens.

11-3. In males, an infectious organism can move up the urethra to the urinary bladder or into the ejaculatory duct to the ductus deferens. There is no direct connection between the urinary or reproductive tract and the abdominopelvic cavity in the male. In females, the urethral opening is close to the vaginal orifice, so the same infectious organism that causes the UTI could proceed from the vagina to the uterus, into the uterine tubes, and into the abdominopelvic cavity via the infundibulum. This process can lead to infection and inflammation of the peritoneal lining.

11-4. The blood–testis barrier is maintained by tight junctions between adjacent sustentacular cells. This barrier prevents sperm antigens from triggering an immune response. A breakdown in the blood–testis barrier caused by a traumatic blow to the scrotum would allow sperm antigens to escape into the interstitial spaces. Their presence there could cause an autoimmune response that would result in sterility.

Clinical Problems

INTEGUMENTARY/SKELETAL/ MUSCULAR SYSTEMS (p. 68)

1. The patient is most likely suffering from rheumatoid arthritis (RA), although several other joint disorders could have been present. Osteoarthritis normally occurs in older individuals. Uric acid crystals in the joints cause gout, a form of arthritis (but note the absence of uric acid in the synovial fluid). Bacterial arthritis is ruled out because there are no bacteria in the synovial fluid. Systemic lupus erythematosus (SLE) also causes joint problems, but the patient did not have the characteristic skin lesions found in SLE.

2. The weakness of the pelvic girdle muscles in a young boy suggests a possible muscle disorder, such as muscular dystrophy (MD). Aldolase is an enzyme present in high concentrations in muscle fibers, and its appearance in serum indicates muscle damage. A muscle biopsy revealed histological changes characteristic of MD.

NERVOUS/ENDOCRINE SYSTEMS (p. 105)

1. The electromyography (a) and serum aldolase (d) gave abnormal results. Electromyography is a measure of the electrical activity of the resting and contracting muscle. Abnormal readings occur in disorders such as muscular dystrophy (MD). Aldolase is an enzyme found in many cells, but it is in high concentration in skeletal muscle cells. Muscle damage causes the aldolase to be released from the cell and enter into the blood circulation. The serum aldolase levels are elevated early in the onset of MD. This disease is a genetic disorder common in males. Another early sign of the disorder is the weakness of the pelvic girdle muscles, which this young boy was exhibiting by frequent falls and difficulty in rising to a standing position.

2. The occurrence of optic neuritis in a young adult, the plaques, and abnormal CSF values point to multiple sclerosis (MS). Clinically, a history of *multiple* neurological deficits at *multiple* times and involving *multiple* sites on the body is the classic pattern for MS. Because MS is an autoimmune problem, IgG levels increase in the CSF.

3. (a). The pituitary gland is anatomically seated in the sella turcica. Erosion or enlargement of the sella turcica indicates a possible pituitary gland tumor. The MRI confirmed the presence of an abnormal mass of the pituitary gland. Hypersecretion of ACTH, as shown by the increased plasma levels, was the result of the pituitary tumor. The increased level of ACTH would hyperstimulate the adrenal gland to produce increased adrenalcorticoids. This increased level of adrenalcorticoids would be the cause of Cushing's disease.

CARDIOVASCULAR/ LYMPHATIC SYSTEMS (p. 150)

1. Melanie should have informed her dentist about her MVP condition so that she could have taken antibiotics prior to the procedure. The antibiotics would have prevented the accumulation of bacteria in the blood after a procedure such as a dental cleaning. Cleaning the teeth with instruments forces bacteria, present in tartar and plaque into the pocket of gum tissue that surrounds the teeth. The bacteria enter the blood through these areas, which are highly vascular, and cause bacteremia. Moving through the heart, the bacteria are capable of lodging on the endocardium if a heart defect such as a prolapsed valve is present. The bacteria can then propagate there and cause multiple problems. Bacterial endocarditis can be fatal without treatment.

2. CPK, LDH, and AST are enzymes found in cardiac muscle cells. Because these are intracellular enzymes, elevated levels in the blood would indicate that the cells died as a result of the lack of oxygen. As the membranes of these cells break down, intracellular enzymes are released into the tissue fluid. Levels of CPK isoenzymes begin to rise 3–6 hours after the myocardial infarction (MI). AST levels begin to increase 6–10 hours after an MI, and LDH isoenzymes rise 12–24 hours after an MI. CPK-MB is the earliest predictor of an MI.

3. Teresa is suffering from hypovolemic shock secondary to acute hemorrhage. A ruptured spleen is the most likely bleeding site. The accumulation of blood in the peritoneal cavity causes irritation of the diaphragm and phrenic nerves, which in turn causes pain, and this leads to abdominal rigidity.

RESPIRATORY/DIGESTIVE SYSTEMS (p. 182)

1. The primary problem is chronic bronchitis, a form of COPD. It has led to chronic hypoxemia and hypercapnia. The low P_{O_2} has led to erythrocytosis, and the increased viscosity, added to pulmonary vasoconstriction shunting blood away from fibrotic areas, causes pulmonary hypertension. Any exertion can cause acute hypoxia; Joe may need home oxygen administration.

2. The infant is diagnosed with cystic fibrosis. The major effect is an alteration of exocrine gland function due to an abnormal chloride transport mechanism. Thick mucus predisposes the infant to respiratory infections. Mucus plugs the pancreatic ducts, leading to problems with enzyme release and signs of malabsorption syndrome.

URINARY/REPRODUCTIVE SYSTEMS (p. 212)

1. The patient had an undiagnosed case of diabetes mellitus, which created the emergency situation. The markedly elevated blood glucose and the presence of glucose in the urine confirmed this diagnosis. Blood pH was low and indicated acidosis. The bicarbonate levels were decreased, indicating a metabolic cause. The patient was suffering from a ketoacidosis, which is common in uncontrolled diabetes.

Ketoacidosis is the consequence of excessive fat catabolism for energy purposes. In the absence of insulin, the body's cells lack accessibility to the glucose in the blood, so the body utilizes fats instead of glucose for energy. Excessive catabolism of fats causes the buildup of keto acids in the blood, decreasing the blood pH. Respiratory compensation occurred, as shown by the decreased P_{CO_2}. Regardless of the compensation, the nervous system was adversely affected by the decreased pH, and loss of consciousness occurred.

The patient was given IV fluids and IV insulin. The pH balance was restored as the glucose entered her cells for energy use.

2. (b).

3. Moira's test results show that she is ovulating normally and has regular hormonal cycles. However, problems with her uterine tubes or uterine lining cannot be discounted unless further tests, such as a hysterosalpingogram and endometrial biopsy, are performed. Sam's sperm count was low, and this is the most likely cause of the couple's infertility problem. If low sperm count is the only factor involved, his sperm can be collected, concentrated, and artificially introduced into the uterus immediately after ovulation. The sperm may also be incapable of normal fertilization; if this is the case, a sperm donor can be used.

4. John has two problems. First, he has a urinary tract infection that caused the dysuria and increased frequency of urination. This condition is probably related to chronic partial obstruction of the urethra by an enlarged prostate; the reduction in urine force and volume during urination increases the likelihood of urinary tract infections. This problem can be treated with antibiotics. The second problem is the enlarged prostate, which appears to be the result of a malignancy in the prostate. Treatment will depend on the stage of the cancer and the type of prostate cells involved.

Index

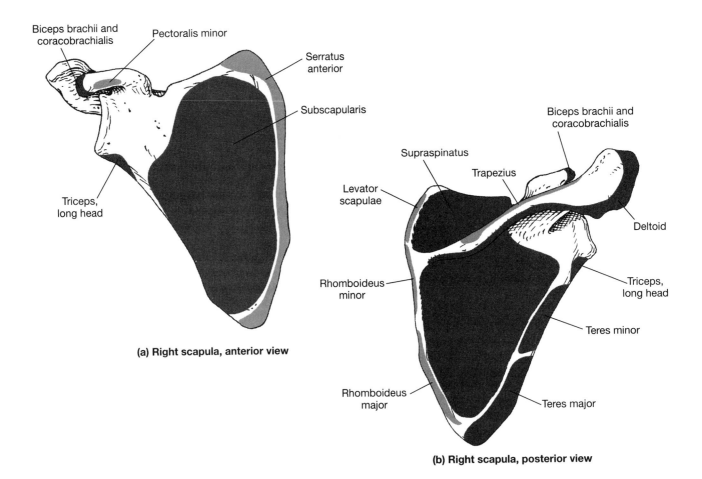

(a) Right scapula, anterior view

(b) Right scapula, posterior view

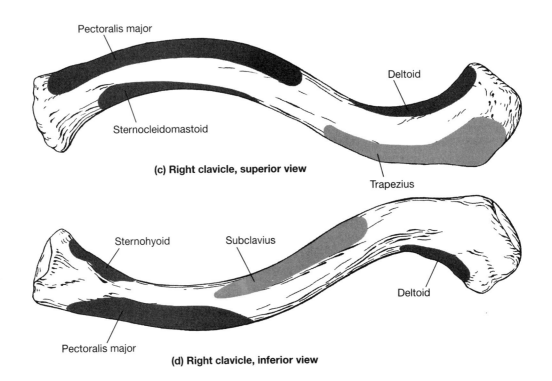

(c) Right clavicle, superior view

(d) Right clavicle, inferior view

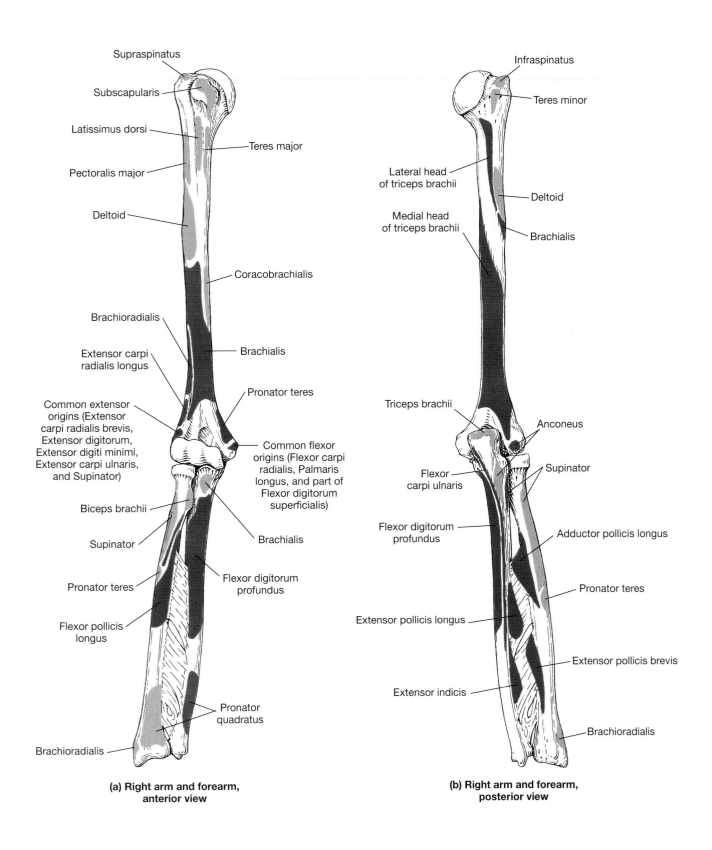

Supraspinatus

Subscapularis

Latissimus dorsi

Pectoralis major

Deltoid

Teres major

Coracobrachialis

Brachioradialis

Brachialis

Extensor carpi
radialis longus

Pronator teres

Common extensor
origins (Extensor
carpi radialis brevis,
Extensor digitorum,
Extensor digiti minimi,
Extensor carpi ulnaris,
and Supinator)

Common flexor
origins (Flexor carpi
radialis, Palmaris
longus, and part of
Flexor digitorum
superficialis)

Biceps brachii

Brachialis

Supinator

Pronator teres

Flexor digitorum
profundus

Flexor pollicis
longus

Pronator
quadratus

Brachioradialis

**(a) Right arm and forearm,
anterior view**

Infraspinatus

Teres minor

Lateral head
of triceps brachii

Deltoid

Medial head
of triceps brachii

Brachialis

Triceps brachii

Anconeus

Flexor
carpi ulnaris

Supinator

Flexor digitorum
profundus

Adductor pollicis longus

Pronator teres

Extensor pollicis longus

Extensor pollicis brevis

Extensor indicis

Brachioradialis

**(b) Right arm and forearm,
posterior view**

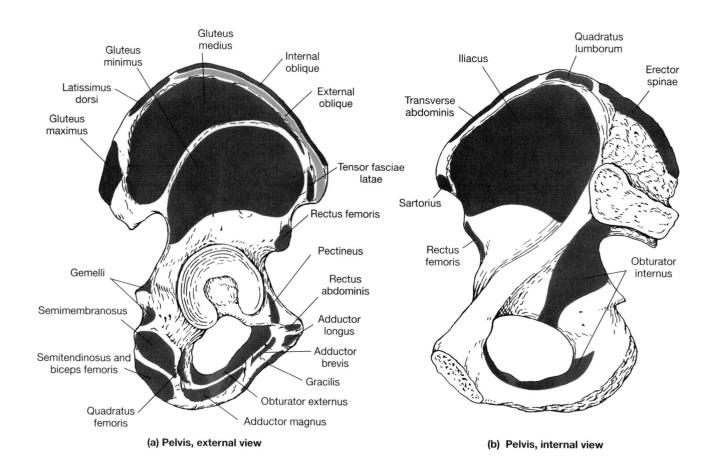

(a) Pelvis, external view

(b) Pelvis, internal view

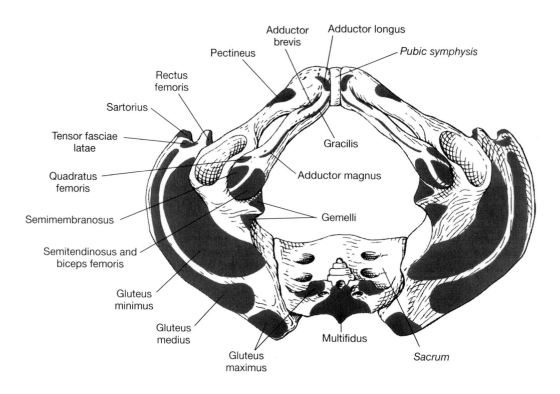

(c) Pelvis, inferior view

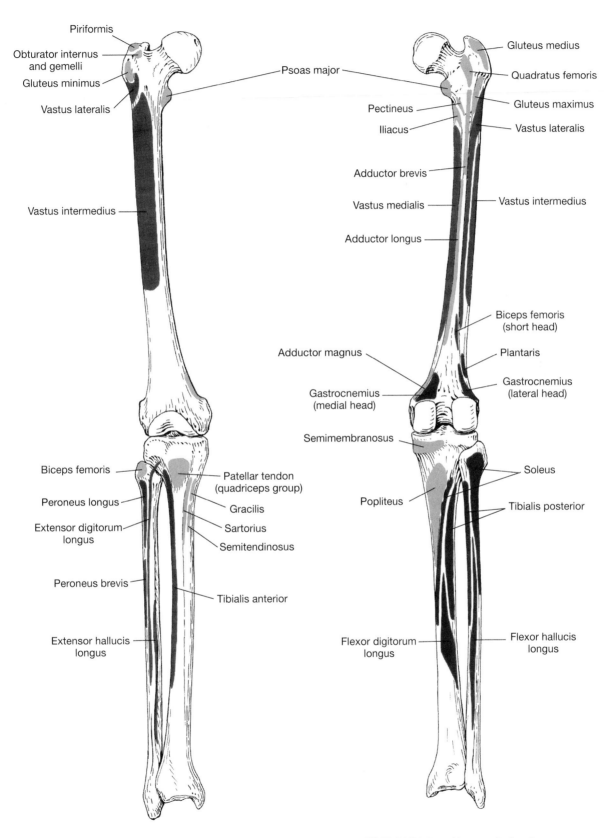

Piriformis

Obturator internus
and gemelli

Gluteus minimus

Vastus lateralis

Vastus intermedius

Psoas major

Gluteus medius

Quadratus femoris

Gluteus maximus

Pectineus

Iliacus

Vastus lateralis

Adductor brevis

Vastus medialis

Vastus intermedius

Adductor longus

Biceps femoris
(short head)

Adductor magnus

Plantaris

Gastrocnemius
(medial head)

Gastrocnemius
(lateral head)

Semimembranosus

Biceps femoris

Soleus

Peroneus longus

Popliteus

Patellar tendon
(quadriceps group)

Tibialis posterior

Extensor digitorum
longus

Gracilis

Sartorius

Semitendinosus

Peroneus brevis

Tibialis anterior

Extensor hallucis
longus

Flexor digitorum
longus

Flexor hallucis
longus

(a) Right thigh and leg, anterior view

(b) Right thigh and leg, posterior view

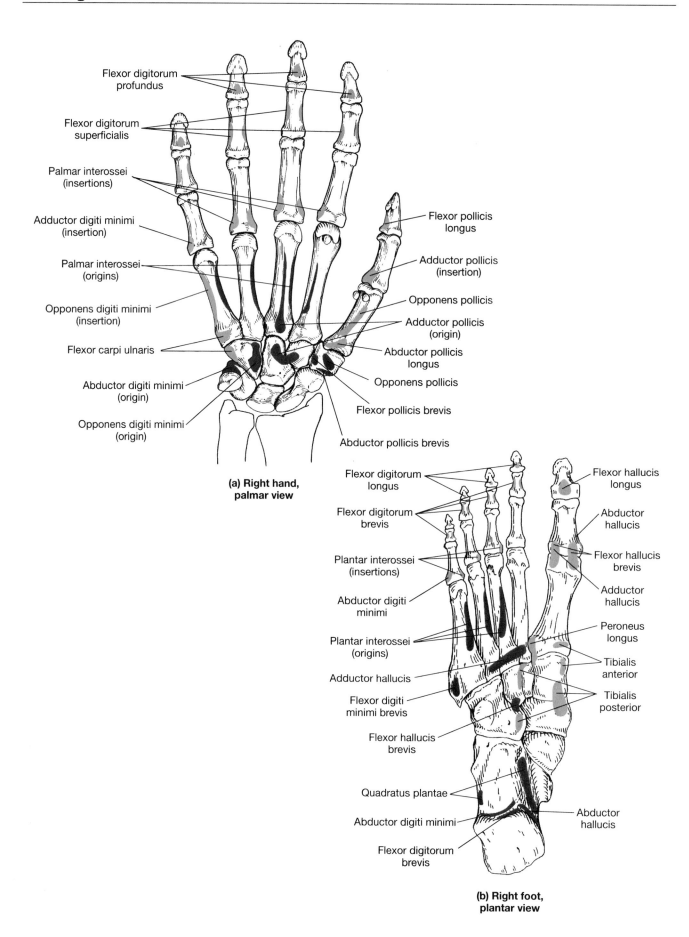

Flexor digitorum
profundus

Flexor digitorum
superficialis

Palmar interossei
(insertions)

Adductor digiti minimi
(insertion)

Palmar interossei
(origins)

Opponens digiti minimi
(insertion)

Flexor carpi ulnaris

Abductor digiti minimi
(origin)

Opponens digiti minimi
(origin)

Flexor pollicis
longus

Adductor pollicis
(insertion)

Opponens pollicis

Adductor pollicis
(origin)

Abductor pollicis
longus

Opponens pollicis

Flexor pollicis brevis

Abductor pollicis brevis

**(a) Right hand,
palmar view**

Flexor digitorum
longus

Flexor digitorum
brevis

Plantar interossei
(insertions)

Abductor digiti
minimi

Plantar interossei
(origins)

Adductor hallucis

Flexor digiti
minimi brevis

Flexor hallucis
brevis

Quadratus plantae

Abductor digiti minimi

Flexor digitorum
brevis

Flexor hallucis
longus

Abductor
hallucis

Flexor hallucis
brevis

Adductor
hallucis

Peroneus
longus

Tibialis
anterior

Tibialis
posterior

Abductor
hallucis

**(b) Right foot,
plantar view**

SCANNING ATLAS

Note to the Reader: Try to use the sequential images in Figures 1–9 to enhance your understanding of these three-dimensional structures. To test yourself, try to label images 9c, 9e, 9g, and 9h, using your mental image and the labeled scans in that series as references.

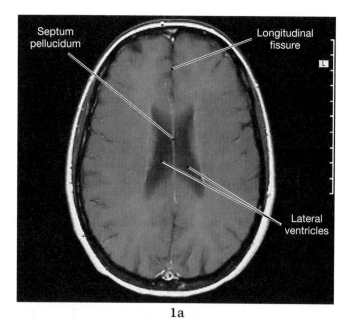

1a

Septum pellucidum

Longitudinal fissure

Lateral ventricles

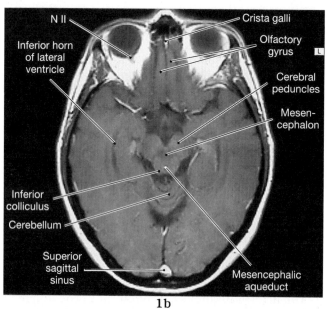

1b

N II

Crista galli

Olfactory gyrus

Inferior horn of lateral ventricle

Cerebral peduncles

Mesencephalon

Inferior colliculus

Cerebellum

Superior sagittal sinus

Mesencephalic aqueduct

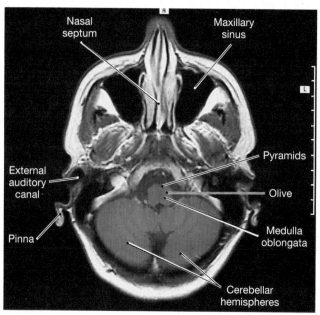

1c

Nasal septum

Maxillary sinus

External auditory canal

Pyramids

Olive

Pinna

Medulla oblongata

Cerebellar hemispheres

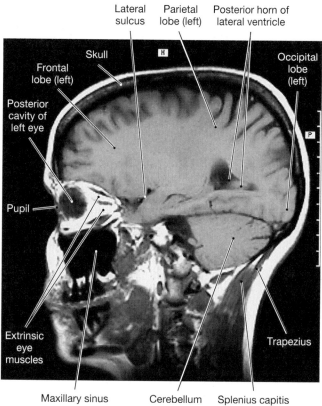

1d

Lateral sulcus

Parietal lobe (left)

Posterior horn of lateral ventricle

Skull

Occipital lobe (left)

Frontal lobe (left)

Posterior cavity of left eye

Pupil

Extrinsic eye muscles

Maxillary sinus

Cerebellum

Splenius capitis

Trapezius

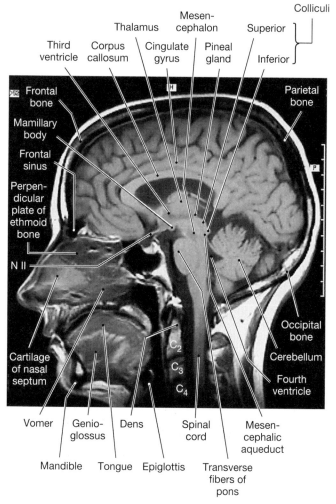

1e

Colliculi

Thalamus

Mesencephalon

Superior

Third ventricle

Corpus callosum

Cingulate gyrus

Pineal gland

Inferior

Frontal bone

Parietal bone

Mamillary body

Frontal sinus

Perpendicular plate of ethmoid bone

N II

Occipital bone

Cerebellum

Cartilage of nasal septum

C_2

C_3

C_4

Fourth ventricle

Vomer

Genioglossus

Dens

Spinal cord

Mesencephalic aqueduct

Mandible

Tongue

Epiglottis

Transverse fibers of pons

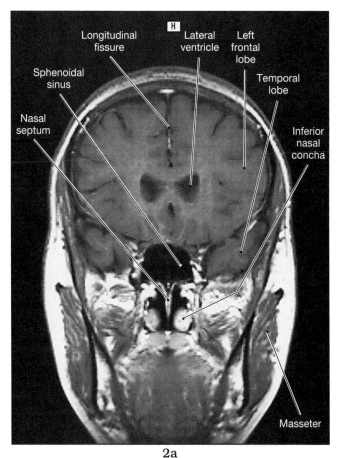

Longitudinal fissure
Lateral ventricle
Left frontal lobe
Temporal lobe
Sphenoidal sinus
Inferior nasal concha
Nasal septum
Masseter

2a

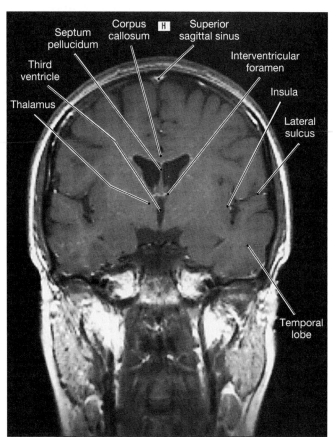

Septum pellucidum
Corpus callosum
Superior sagittal sinus
Interventricular foramen
Third ventricle
Insula
Thalamus
Lateral sulcus
Temporal lobe

2b

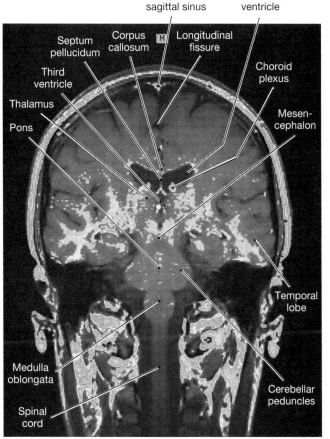

Superior sagittal sinus
Lateral ventricle
Septum pellucidum
Corpus callosum
Longitudinal fissure
Third ventricle
Choroid plexus
Thalamus
Mesen-cephalon
Pons
Temporal lobe
Medulla oblongata
Cerebellar peduncles
Spinal cord

2c

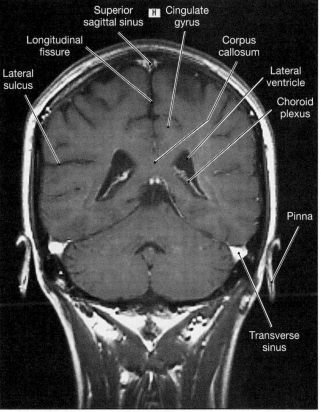

Superior sagittal sinus
Cingulate gyrus
Longitudinal fissure
Corpus callosum
Lateral sulcus
Lateral ventricle
Choroid plexus
Pinna
Transverse sinus

2d

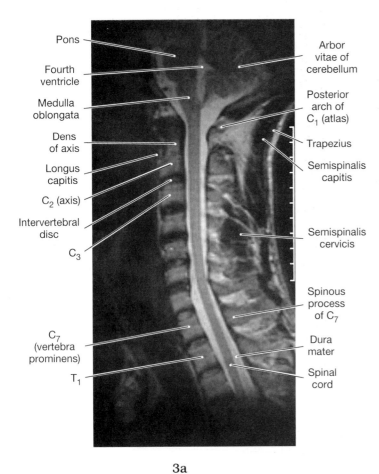

Pons

Fourth ventricle

Medulla oblongata

Dens of axis

Longus capitis

C_2 (axis)

Intervertebral disc

C_3

C_7 (vertebra prominens)

T_1

Arbor vitae of cerebellum

Posterior arch of C_1 (atlas)

Trapezius

Semispinalis capitis

Semispinalis cervicis

Spinous process of C_7

Dura mater

Spinal cord

3a

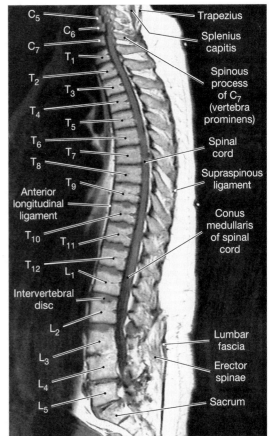

C_5

C_6

C_7

T_1

T_2

T_3

T_4

T_5

T_6

T_7

T_8

T_9

Anterior longitudinal ligament

T_{10}

T_{11}

T_{12}

L_1

Intervertebral disc

L_2

L_3

L_4

L_5

Trapezius

Splenius capitis

Spinous process of C_7 (vertebra prominens)

Spinal cord

Supraspinous ligament

Conus medullaris of spinal cord

Lumbar fascia

Erector spinae

Sacrum

3b

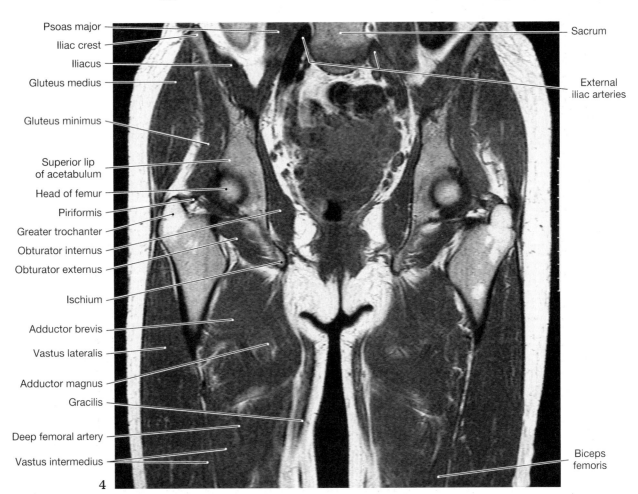

Psoas major

Iliac crest

Iliacus

Gluteus medius

Gluteus minimus

Superior lip of acetabulum

Head of femur

Piriformis

Greater trochanter

Obturator internus

Obturator externus

Ischium

Adductor brevis

Vastus lateralis

Adductor magnus

Gracilis

Deep femoral artery

Vastus intermedius

Sacrum

External iliac arteries

Biceps femoris

4

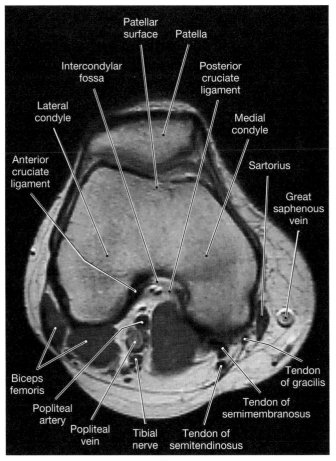

Patellar surface — Patella
Intercondylar fossa
Posterior cruciate ligament
Lateral condyle
Medial condyle
Anterior cruciate ligament
Sartorius
Great saphenous vein
Biceps femoris
Popliteal artery
Popliteal vein
Tibial nerve
Tendon of semitendinosus
Tendon of semimembranosus
Tendon of gracilis

5a

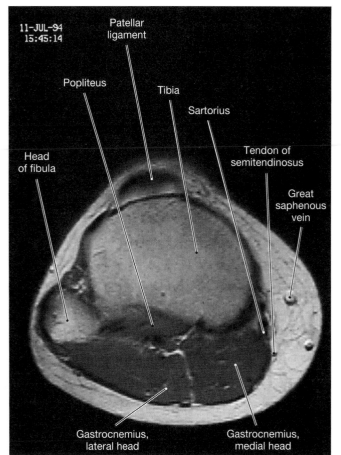

11-JUL-94 15:45:14
Patellar ligament
Popliteus
Tibia
Sartorius
Head of fibula
Tendon of semitendinosus
Great saphenous vein
Gastrocnemius, lateral head
Gastrocnemius, medial head

5b

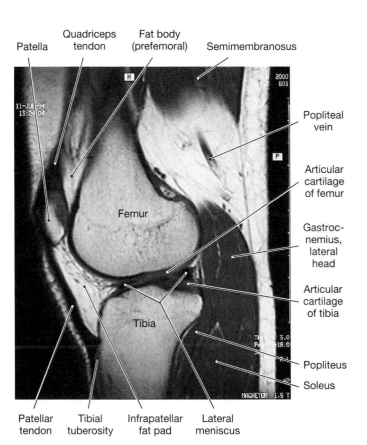

Patella
Quadriceps tendon
Fat body (prefemoral)
Semimembranosus
Popliteal vein
Articular cartilage of femur
Gastrocnemius, lateral head
Articular cartilage of tibia
Femur
Tibia
Popliteus
Soleus
Patellar tendon
Tibial tuberosity
Infrapatellar fat pad
Lateral meniscus

6a

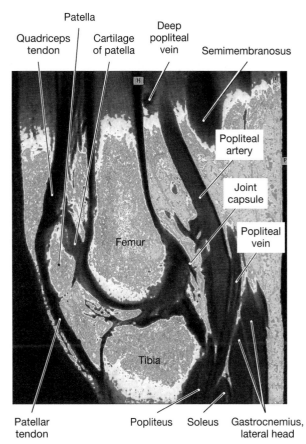

Patella
Quadriceps tendon
Cartilage of patella
Deep popliteal vein
Semimembranosus
Popliteal artery
Joint capsule
Popliteal vein
Femur
Tibia
Patellar tendon
Popliteus
Soleus
Gastrocnemius, lateral head

6b

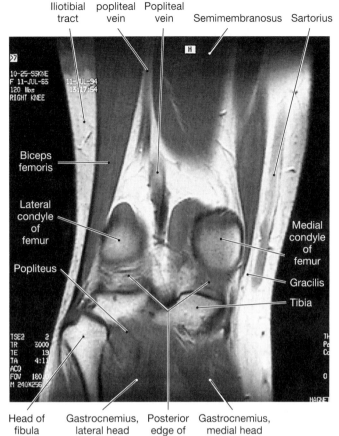

Iliotibial tract — Deep popliteal vein — Popliteal vein — Semimembranosus — Sartorius

Biceps femoris

Lateral condyle of femur

Popliteus

Medial condyle of femur

Gracilis

Tibia

Head of fibula — Gastrocnemius, lateral head — Posterior edge of joint capsule — Gastrocnemius, medial head

7a

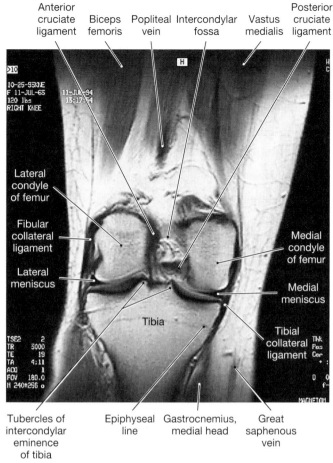

Anterior cruciate ligament — Biceps femoris — Popliteal vein — Intercondylar fossa — Vastus medialis — Posterior cruciate ligament

Lateral condyle of femur

Fibular collateral ligament

Lateral meniscus

Medial condyle of femur

Medial meniscus

Tibia

Tibial collateral ligament

Tubercles of intercondylar eminence of tibia — Epiphyseal line — Gastrocnemius, medial head — Great saphenous vein

7b

Flexor hallucis longus

Soleus

Tendon of tibialis anterior

Calcaneal tendon

First cuneiform

Talus

Tibia

Navicular

Head of first metatarsal

Flexor hallucis brevis — Flexor digitorum brevis — Quadratus plantae — Talocalcaneal ligament — Calcaneus

8a

Extensor digitorum longus and peroneus tertius

Tibia

Medial malleolus of tibia

Lateral malleolus of fibula

Tendon of tibialis posterior

Deltoid ligament

Talus

Tendon of flexor digitorum longus

Calcaneus

Tendon of flexor hallucis longus

Tendon of peroneus longus

Plantar artery

Abductor digiti minimi

Abductor hallucis

Quadratus plantae

Flexor digitorum brevis

8b

S-6

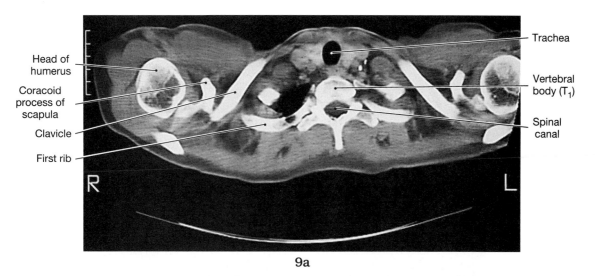

Head of humerus

Coracoid process of scapula

Clavicle

First rib

Trachea

Vertebral body (T$_1$)

Spinal canal

R

L

9a

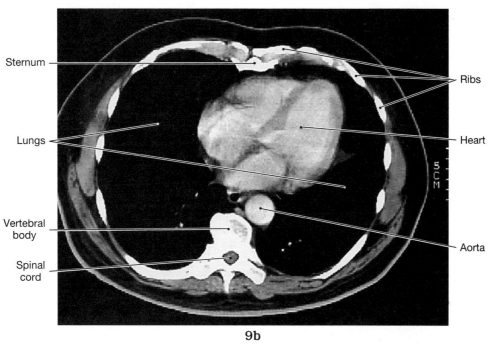

Sternum

Lungs

Vertebral body

Spinal cord

Ribs

Heart

Aorta

9b

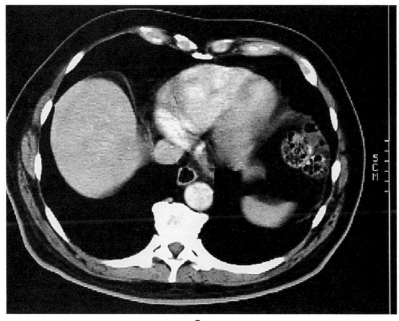

9c

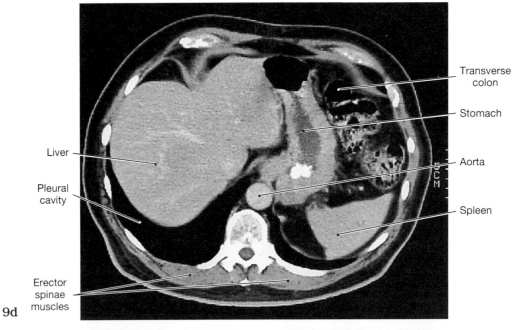

Transverse colon

Stomach

Aorta

Spleen

Liver

Pleural cavity

Erector spinae muscles

9d

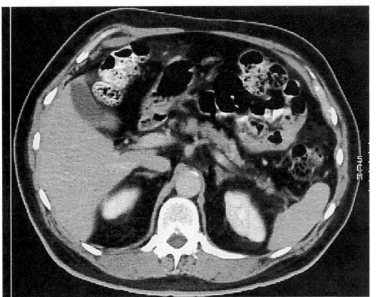

9e

Superior mesenteric vein

Superior mesenteric artery

Liver

Pancreas

Inferior vena cava

Kidneys

Erector spinae muscles

9f

Transverse colon

Small intestine

Colon

Renal vein

Aorta

Renal artery

Renal pelvis

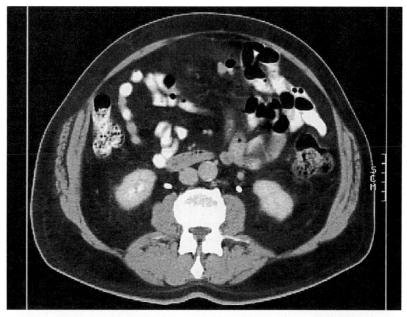

9g

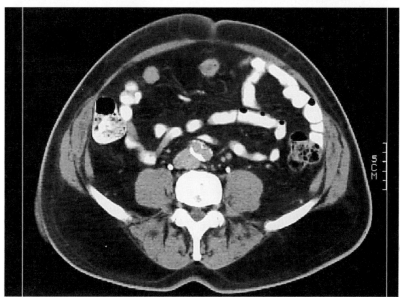

9h

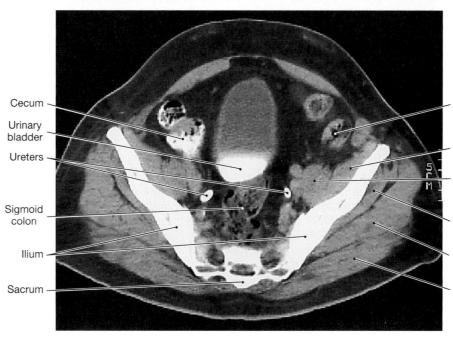

Cecum

Urinary
bladder

Ureters

Sigmoid
colon

Ilium

Sacrum

Descending
colon

Iliacus

Psoas
major

Gluteus
minimus

Gluteus
medius

Gluteus
maximus

9i

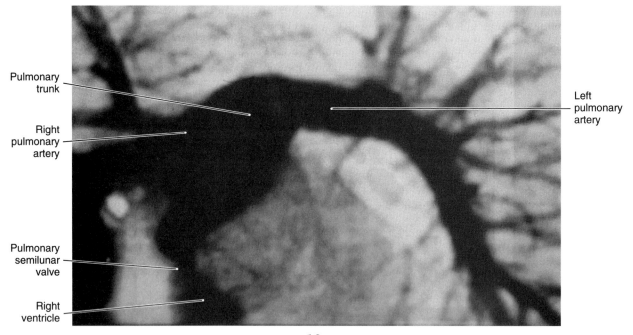

Pulmonary trunk

Right pulmonary artery

Pulmonary semilunar valve

Right ventricle

Left pulmonary artery

10

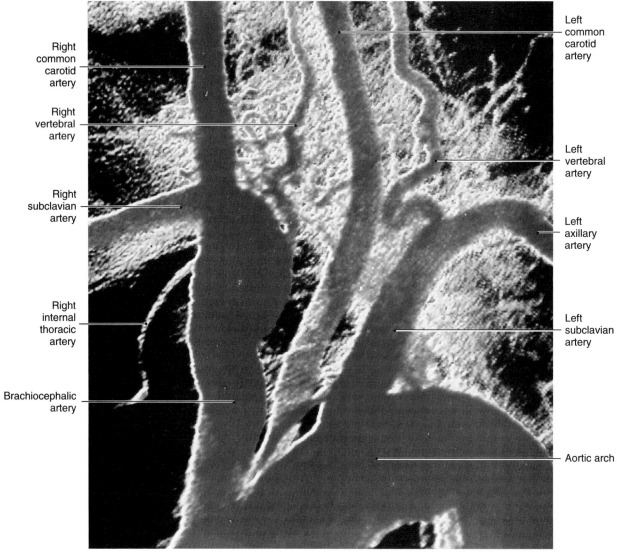

Right common carotid artery

Right vertebral artery

Right subclavian artery

Right internal thoracic artery

Brachiocephalic artery

Left common carotid artery

Left vertebral artery

Left axillary artery

Left subclavian artery

Aortic arch

11

S-10

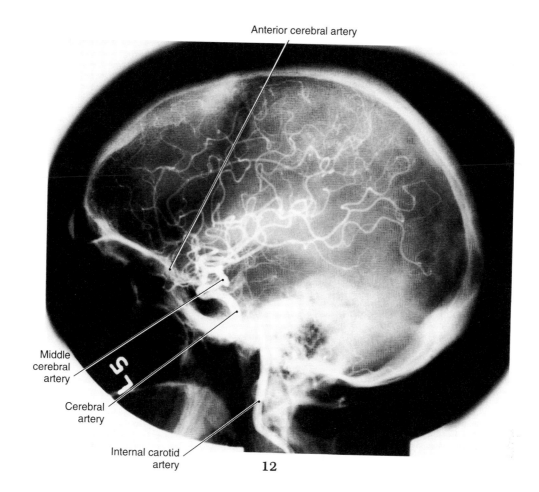

Anterior cerebral artery

Middle
cerebral
artery

Cerebral
artery

Internal carotid
artery

12

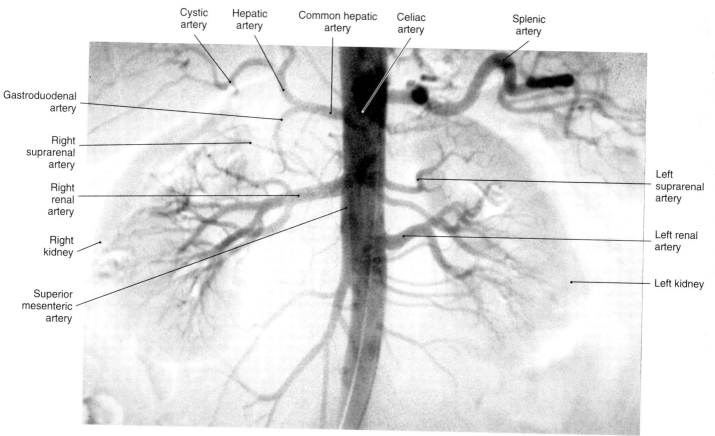

Cystic
artery

Hepatic
artery

Common hepatic
artery

Celiac
artery

Splenic
artery

Gastroduodenal
artery

Right
suprarenal
artery

Right
renal
artery

Right
kidney

Superior
mesenteric
artery

Left
suprarenal
artery

Left renal
artery

Left kidney

13

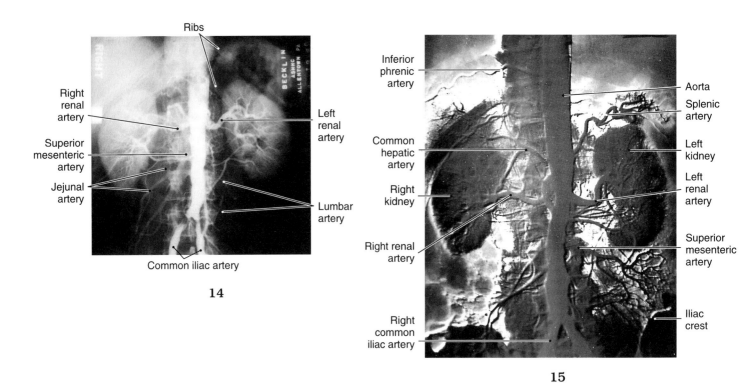

Ribs

Right renal artery

Superior mesenteric artery

Jejunal artery

Left renal artery

Lumbar artery

Common iliac artery

14

Inferior phrenic artery

Common hepatic artery

Right kidney

Right renal artery

Right common iliac artery

Aorta

Splenic artery

Left kidney

Left renal artery

Superior mesenteric artery

Iliac crest

15

Left hepatic duct

Right hepatic duct

Gallbladder { Neck

Body

Fundus

Common bile duct

Common hepatic duct

Cystic duct

16

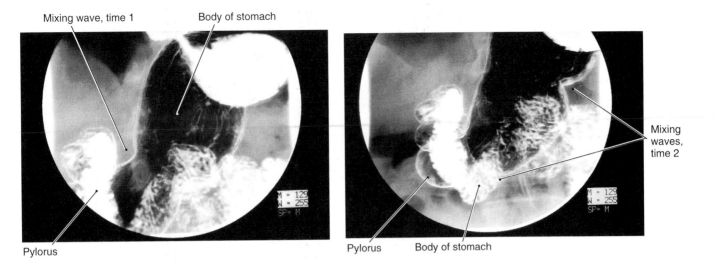

Mixing wave, time 1   Body of stomach

Pylorus

17

Pylorus   Body of stomach

Mixing waves, time 2

18

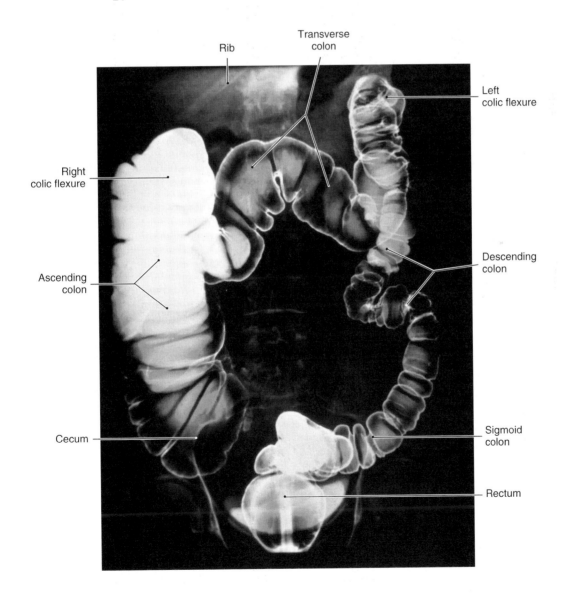

Rib   Transverse colon

Left colic flexure

Right colic flexure

Ascending colon

Descending colon

Cecum

Sigmoid colon

Rectum

19

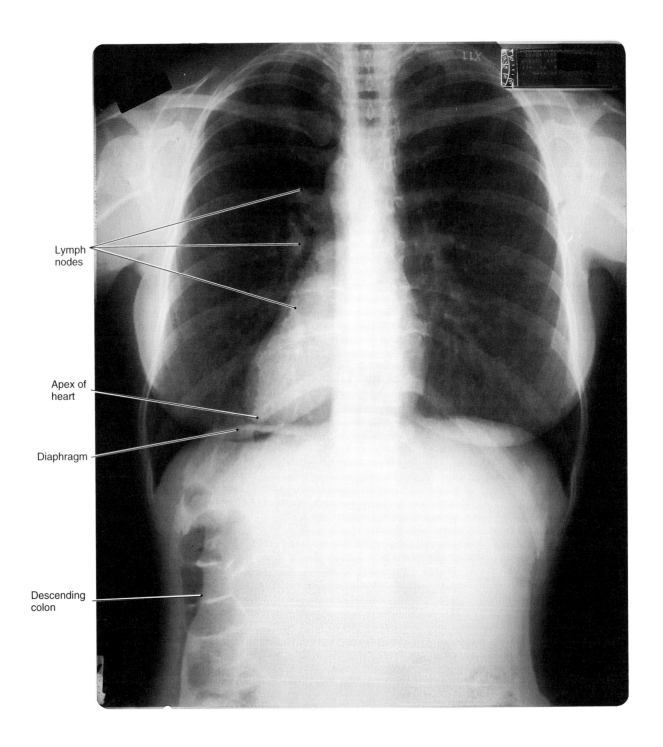

Lymph
nodes

Apex of
heart

Diaphragm

Descending
colon

20

Surface Anatomy and Cadaver Atlas

PLATE 7 THE LOWER LIMBS

PLATE 1   THE HUMAN SKELETON

1.1 Anterior view

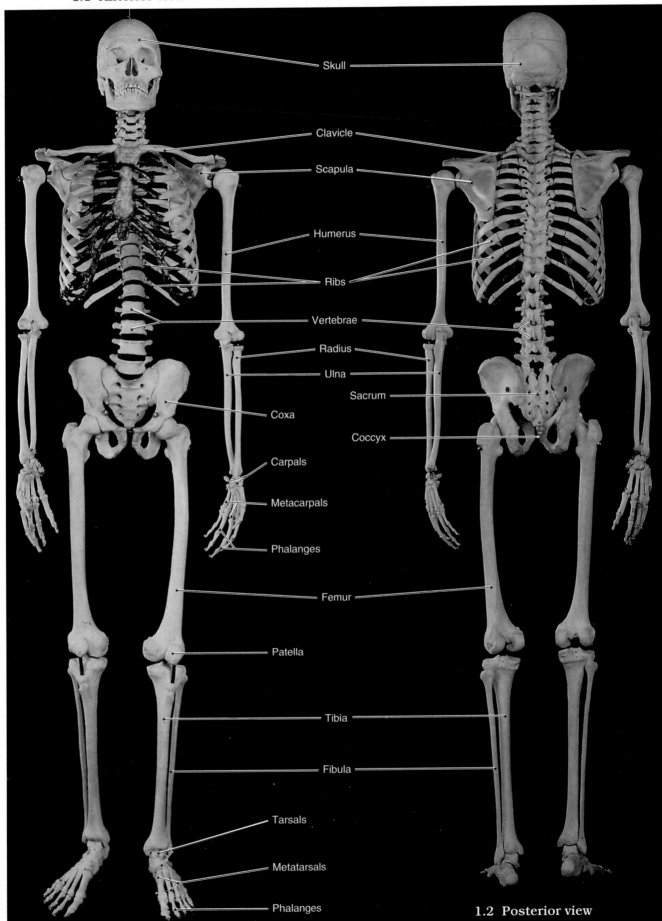

Skull

Clavicle

Scapula

Humerus

Ribs

Vertebrae

Radius

Ulna

Sacrum

Coxa

Coccyx

Carpals

Metacarpals

Phalanges

Femur

Patella

Tibia

Fibula

Tarsals

Metatarsals

Phalanges

1.2 Posterior view

PLATE 2   THE SKULL
Painted Skull Series

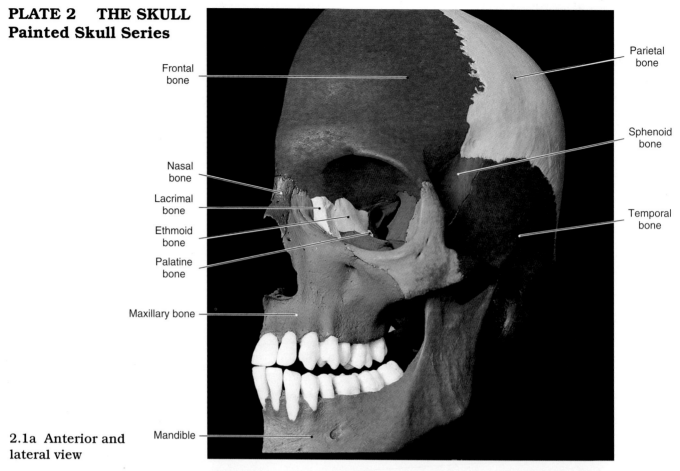

Frontal bone

Parietal bone

Nasal bone

Lacrimal bone

Ethmoid bone

Palatine bone

Sphenoid bone

Temporal bone

Maxillary bone

Mandible

2.1a Anterior and lateral view

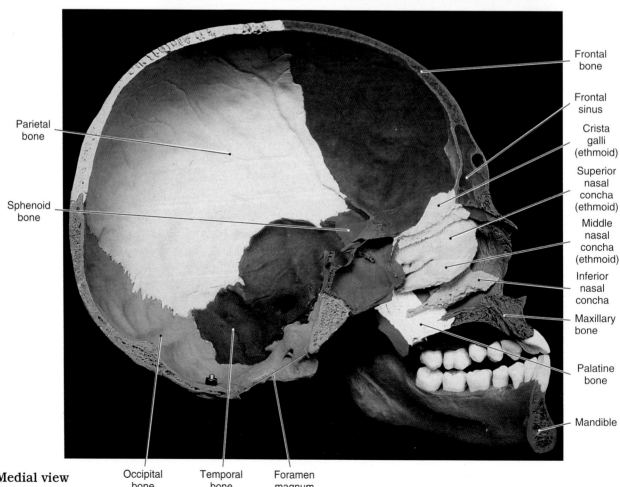

Parietal bone

Sphenoid bone

Frontal bone

Frontal sinus

Crista galli (ethmoid)

Superior nasal concha (ethmoid)

Middle nasal concha (ethmoid)

Inferior nasal concha

Maxillary bone

Palatine bone

Mandible

Occipital bone

Temporal bone

Foramen magnum

2.1b Medial view
C-2

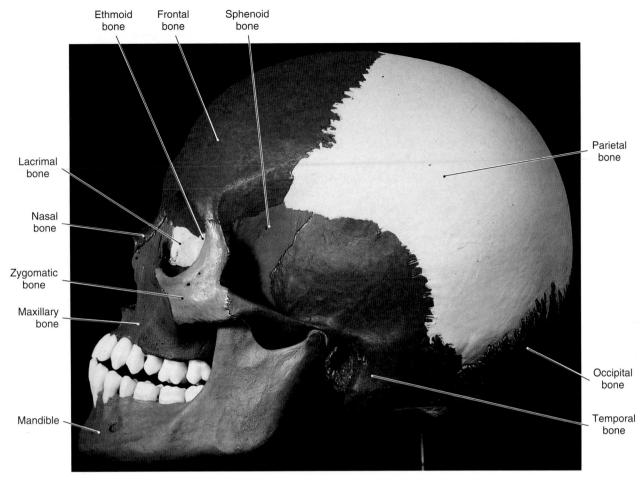

Ethmoid bone · Frontal bone · Sphenoid bone

Lacrimal bone

Nasal bone

Zygomatic bone

Maxillary bone

Mandible

Parietal bone

Occipital bone

Temporal bone

2.1c Lateral view

The Adult Skull

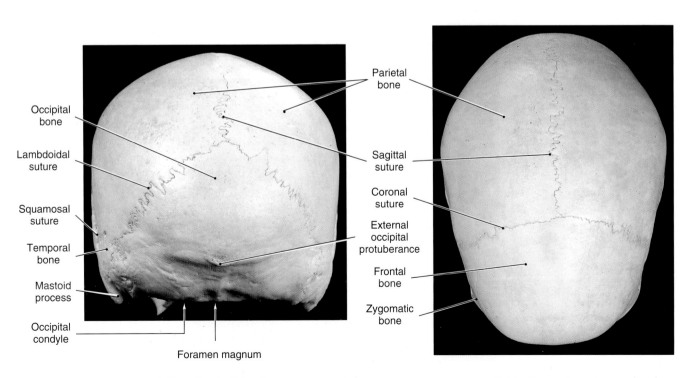

Occipital bone

Lambdoidal suture

Squamosal suture

Temporal bone

Mastoid process

Occipital condyle

Parietal bone

Sagittal suture

Foramen magnum

2.2a Posterior view

Parietal bone

Coronal suture

External occipital protuberance

Frontal bone

Zygomatic bone

2.2b Superior view

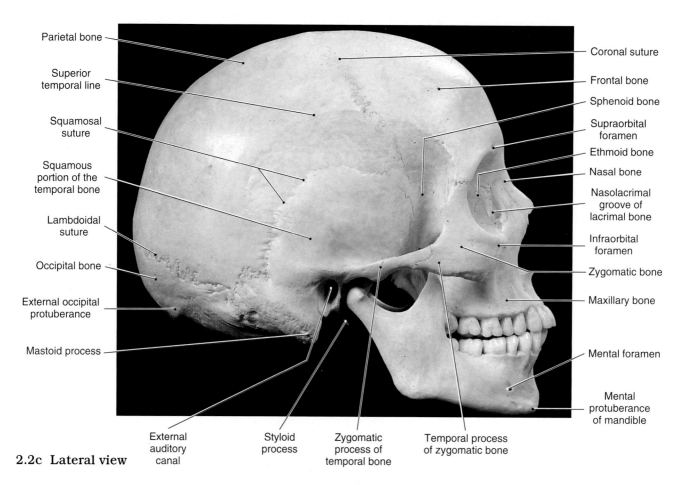

Parietal bone

Superior temporal line

Squamosal suture

Squamous portion of the temporal bone

Lambdoidal suture

Occipital bone

External occipital protuberance

Mastoid process

Coronal suture

Frontal bone

Sphenoid bone

Supraorbital foramen

Ethmoid bone

Nasal bone

Nasolacrimal groove of lacrimal bone

Infraorbital foramen

Zygomatic bone

Maxillary bone

Mental foramen

Mental protuberance of mandible

External auditory canal

Styloid process

Zygomatic process of temporal bone

Temporal process of zygomatic bone

2.2c Lateral view

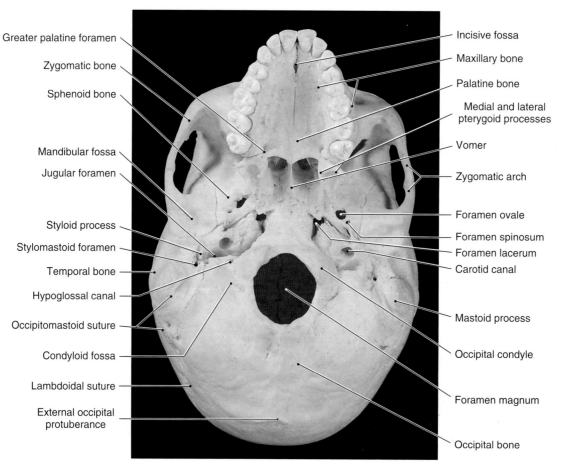

Greater palatine foramen

Zygomatic bone

Sphenoid bone

Mandibular fossa

Jugular foramen

Styloid process

Stylomastoid foramen

Temporal bone

Hypoglossal canal

Occipitomastoid suture

Condyloid fossa

Lambdoidal suture

External occipital protuberance

Incisive fossa

Maxillary bone

Palatine bone

Medial and lateral pterygoid processes

Vomer

Zygomatic arch

Foramen ovale

Foramen spinosum

Foramen lacerum

Carotid canal

Mastoid process

Occipital condyle

Foramen magnum

Occipital bone

2.2d Inferior view

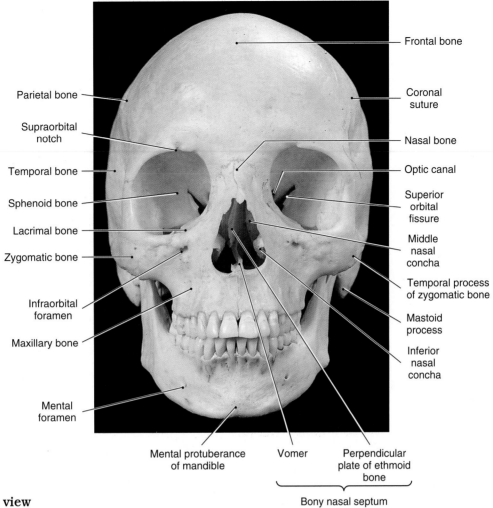

Frontal bone

Parietal bone

Supraorbital notch

Temporal bone

Sphenoid bone

Lacrimal bone

Zygomatic bone

Infraorbital foramen

Maxillary bone

Mental foramen

Coronal suture

Nasal bone

Optic canal

Superior orbital fissure

Middle nasal concha

Temporal process of zygomatic bone

Mastoid process

Inferior nasal concha

Mental protuberance of mandible

Vomer

Perpendicular plate of ethmoid bone

Bony nasal septum

2.2e Anterior view

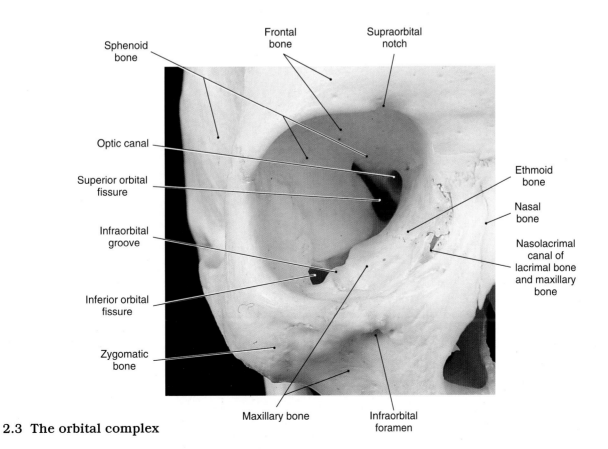

Sphenoid bone

Frontal bone

Supraorbital notch

Optic canal

Superior orbital fissure

Infraorbital groove

Inferior orbital fissure

Zygomatic bone

Ethmoid bone

Nasal bone

Nasolacrimal canal of lacrimal bone and maxillary bone

Maxillary bone

Infraorbital foramen

2.3 The orbital complex

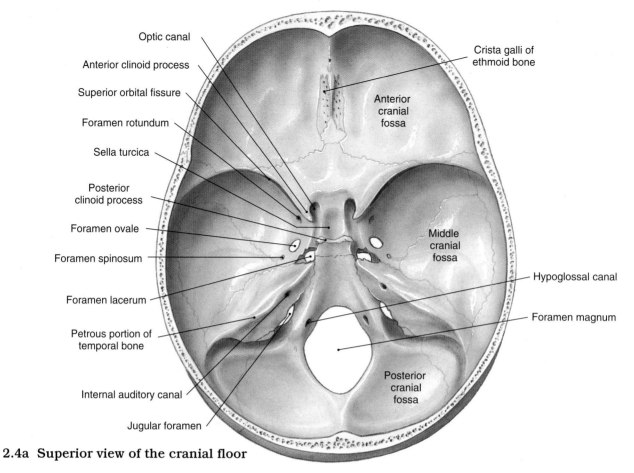

Optic canal

Anterior clinoid process

Superior orbital fissure

Foramen rotundum

Sella turcica

Posterior
clinoid process

Foramen ovale

Foramen spinosum

Foramen lacerum

Petrous portion of
temporal bone

Internal auditory canal

Jugular foramen

Crista galli of
ethmoid bone

Anterior
cranial
fossa

Middle
cranial
fossa

Hypoglossal canal

Foramen magnum

Posterior
cranial
fossa

2.4a Superior view of the cranial floor

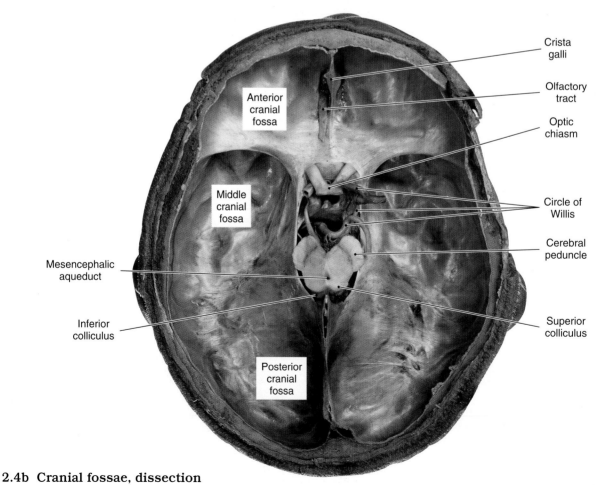

Crista
galli

Olfactory
tract

Optic
chiasm

Circle of
Willis

Cerebral
peduncle

Superior
colliculus

Anterior
cranial
fossa

Middle
cranial
fossa

Mesencephalic
aqueduct

Inferior
colliculus

Posterior
cranial
fossa

2.4b Cranial fossae, dissection
C-6

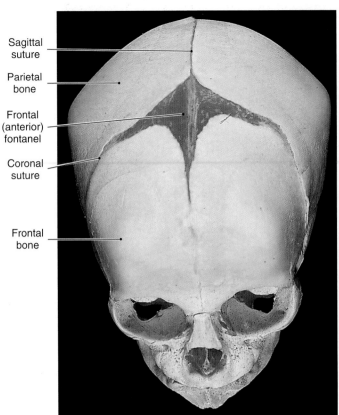

Sagittal suture

Parietal bone

Frontal (anterior) fontanel

Coronal suture

Frontal bone

2.5a The skull of an infant, superior view

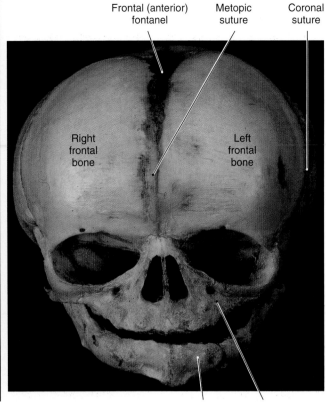

Frontal (anterior) fontanel

Metopic suture

Coronal suture

Right frontal bone

Left frontal bone

Mandible   Maxillary bone

2.5b Fetal skull, anterior view

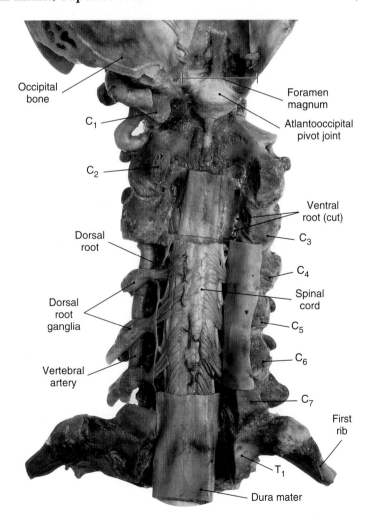

Occipital bone

Foramen magnum

Atlantooccipital pivot joint

C_1

C_2

Ventral root (cut)

C_3

Dorsal root

C_4

Spinal cord

Dorsal root ganglia

C_5

C_6

Vertebral artery

C_7

First rib

T_1

Dura mater

2.6 Cervical spine

PLATE 3   THE BRAIN AND SPINAL CORD
The Head

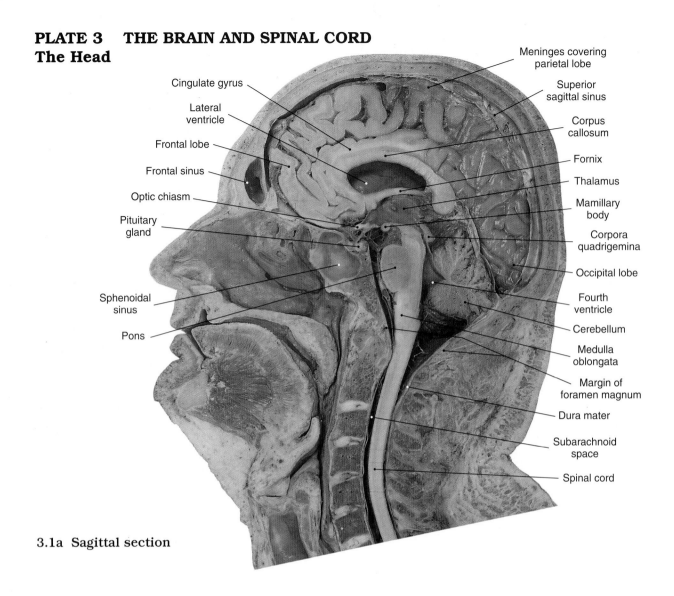

Cingulate gyrus

Lateral ventricle

Frontal lobe

Frontal sinus

Optic chiasm

Pituitary gland

Sphenoidal sinus

Pons

Meninges covering parietal lobe

Superior sagittal sinus

Corpus callosum

Fornix

Thalamus

Mamillary body

Corpora quadrigemina

Occipital lobe

Fourth ventricle

Cerebellum

Medulla oblongata

Margin of foramen magnum

Dura mater

Subarachnoid space

Spinal cord

3.1a Sagittal section

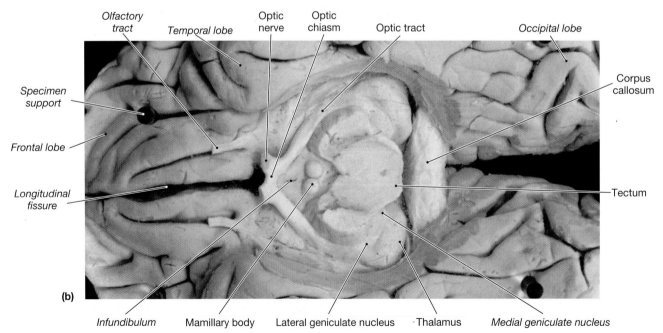

Olfactory tract

Temporal lobe

Optic nerve

Optic chiasm

Optic tract

Occipital lobe

Specimen support

Corpus callosum

Frontal lobe

Longitudinal fissure

Tectum

(b)

Infundibulum

Mamillary body

Lateral geniculate nucleus

Thalamus

Medial geniculate nucleus

3.1b Inferior sectional view of the brain

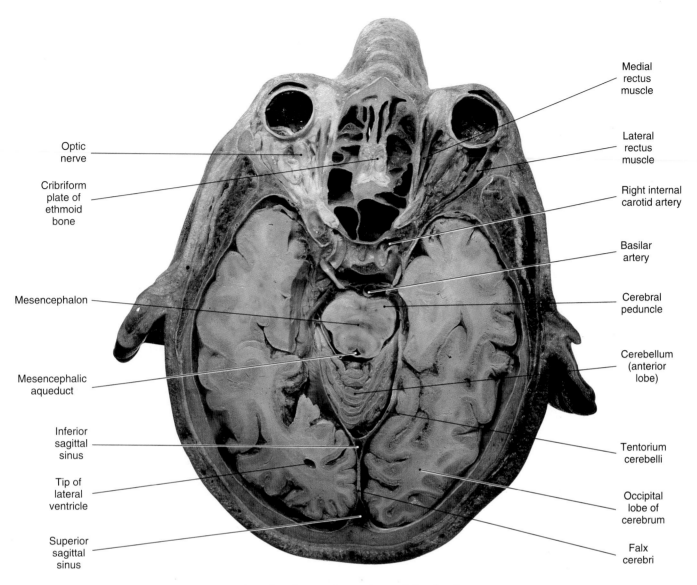

Optic nerve

Cribriform plate of ethmoid bone

Mesencephalon

Mesencephalic aqueduct

Inferior sagittal sinus

Tip of lateral ventricle

Superior sagittal sinus

Medial rectus muscle

Lateral rectus muscle

Right internal carotid artery

Basilar artery

Cerebral peduncle

Cerebellum (anterior lobe)

Tentorium cerebelli

Occipital lobe of cerebrum

Falx cerebri

3.2 Horizontal section of the brain

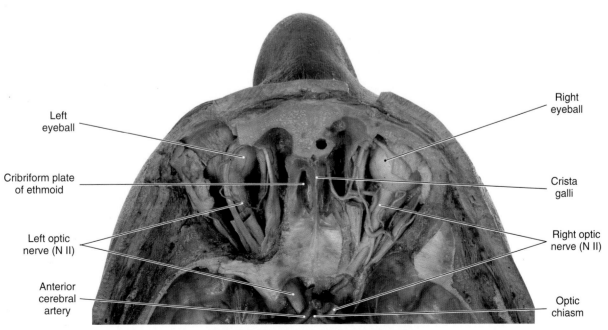

Left eyeball

Cribriform plate of ethmoid

Left optic nerve (N II)

Anterior cerebral artery

Right eyeball

Crista galli

Right optic nerve (N II)

Optic chiasm

3.3 Anatomy of the orbital region

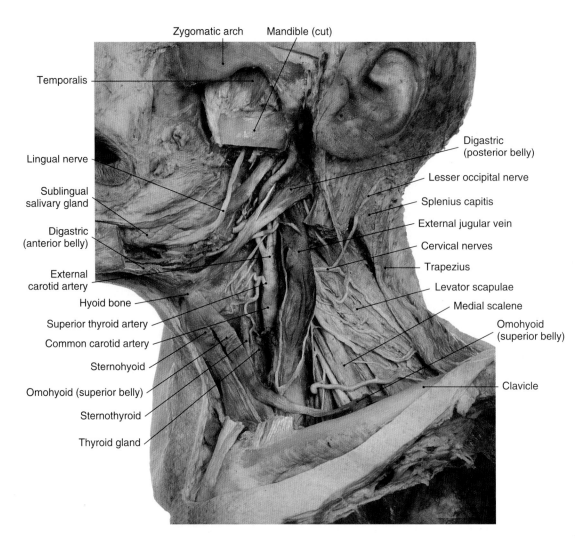

Zygomatic arch — Mandible (cut)

Temporalis

Lingual nerve

Sublingual salivary gland

Digastric (anterior belly)

External carotid artery

Hyoid bone

Superior thyroid artery

Common carotid artery

Sternohyoid

Omohyoid (superior belly)

Sternothyroid

Thyroid gland

Digastric (posterior belly)

Lesser occipital nerve

Splenius capitis

External jugular vein

Cervical nerves

Trapezius

Levator scapulae

Medial scalene

Omohyoid (superior belly)

Clavicle

3.4 Structures of the neck, lateral view

3.5a Cerebral arteries of the left hemisphere

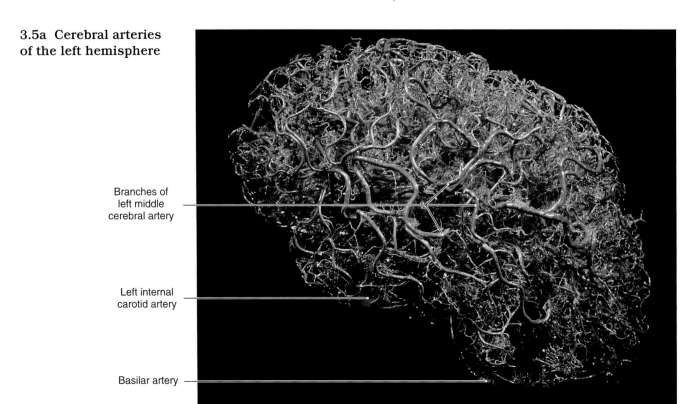

Branches of left middle cerebral artery

Left internal carotid artery

Basilar artery

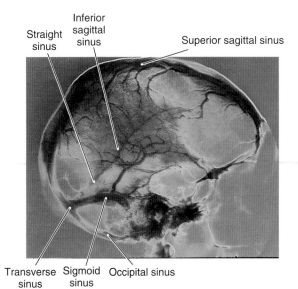

3.5b Veins draining the cranium

Straight sinus

Inferior sagittal sinus

Superior sagittal sinus

Transverse sinus

Sigmoid sinus

Occipital sinus

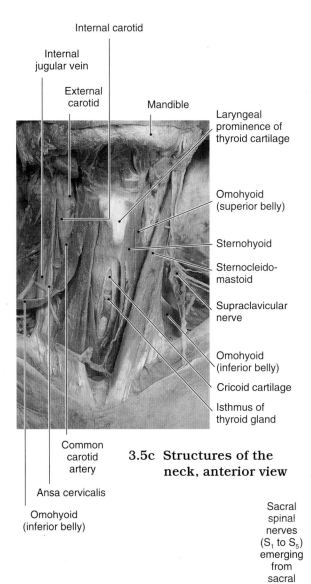

Internal carotid

Internal jugular vein

External carotid

Mandible

Laryngeal prominence of thyroid cartilage

Omohyoid (superior belly)

Sternohyoid

Sternocleido-mastoid

Supraclavicular nerve

Omohyoid (inferior belly)

Cricoid cartilage

Isthmus of thyroid gland

Common carotid artery

Ansa cervicalis

Omohyoid (inferior belly)

3.5c Structures of the neck, anterior view

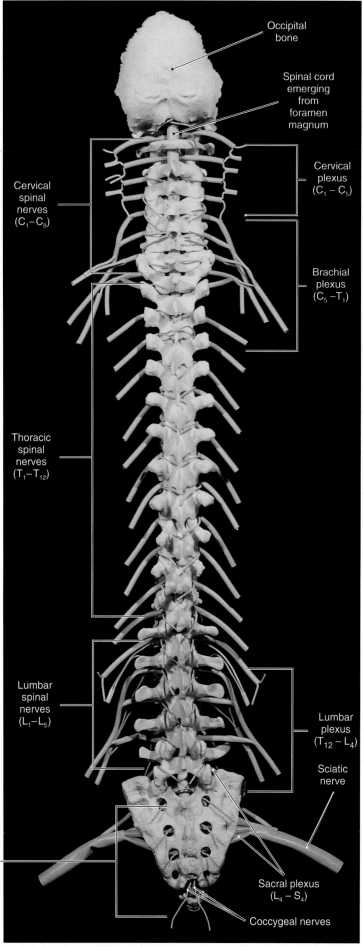

Occipital bone

Spinal cord emerging from foramen magnum

Cervical spinal nerves (C_1–C_8)

Cervical plexus (C_1–C_5)

Brachial plexus (C_5–T_1)

Thoracic spinal nerves (T_1–T_{12})

Lumbar spinal nerves (L_1–L_5)

Lumbar plexus (T_{12}–L_4)

Sciatic nerve

Sacral spinal nerves (S_1 to S_5) emerging from sacral foramina

Sacral plexus (L_4–S_4)

Coccygeal nerves

3.6 Posterior view of vertebral column and spinal nerves

C-11

PLATE 4   THE HEAD AND NECK
Surface Anatomy of the Head and Neck

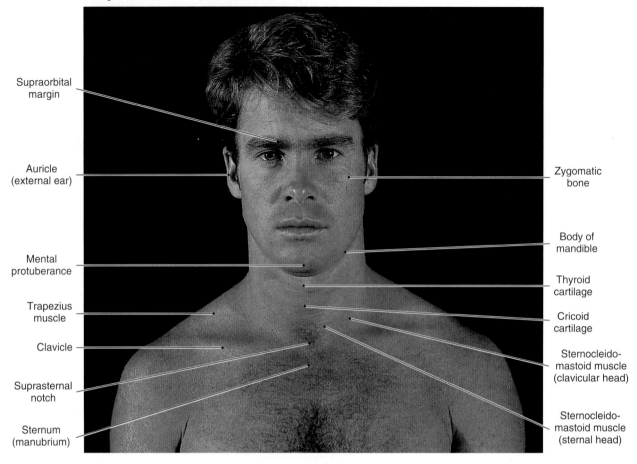

Supraorbital margin

Auricle (external ear)

Mental protuberance

Trapezius muscle

Clavicle

Suprasternal notch

Sternum (manubrium)

Zygomatic bone

Body of mandible

Thyroid cartilage

Cricoid cartilage

Sternocleido-mastoid muscle (clavicular head)

Sternocleido-mastoid muscle (sternal head)

4.1a Anterior view

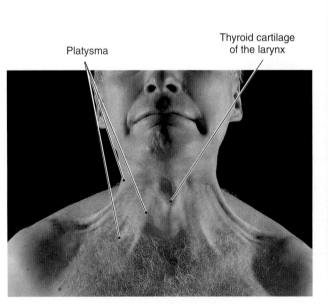

Platysma

Thyroid cartilage of the larynx

4.1b The anterior neck

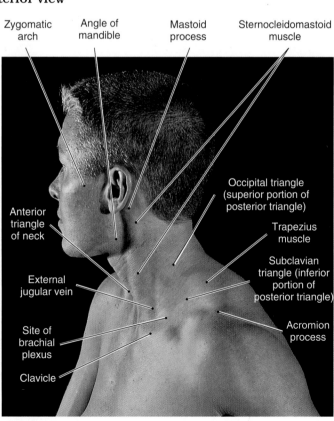

Zygomatic arch

Angle of mandible

Mastoid process

Sternocleidomastoid muscle

Anterior triangle of neck

External jugular vein

Site of brachial plexus

Clavicle

Occipital triangle (superior portion of posterior triangle)

Trapezius muscle

Subclavian triangle (inferior portion of posterior triangle)

Acromion process

4.2a Posterior cervical triangle

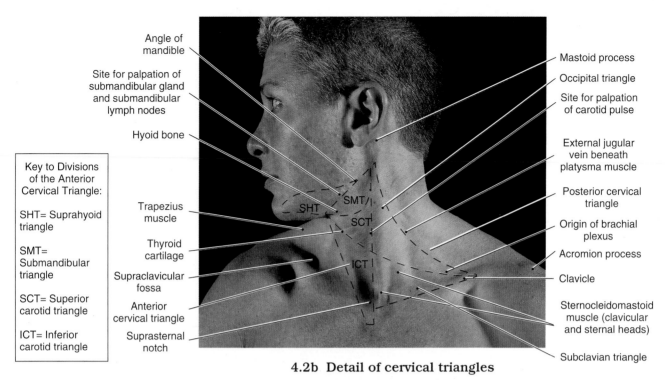

Key to Divisions
of the Anterior
Cervical Triangle:

SHT= Suprahyoid
triangle

SMT=
Submandibular
triangle

SCT= Superior
carotid triangle

ICT= Inferior
carotid triangle

Angle of
mandible

Site for palpation of
submandibular gland
and submandibular
lymph nodes

Hyoid bone

Trapezius
muscle

Thyroid
cartilage

Supraclavicular
fossa

Anterior
cervical triangle

Suprasternal
notch

Mastoid process

Occipital triangle

Site for palpation
of carotid pulse

External jugular
vein beneath
platysma muscle

Posterior cervical
triangle

Origin of brachial
plexus

Acromion process

Clavicle

Sternocleidomastoid
muscle (clavicular
and sternal heads)

Subclavian triangle

SMT

SHT

SCT

ICT

4.2b Detail of cervical triangles

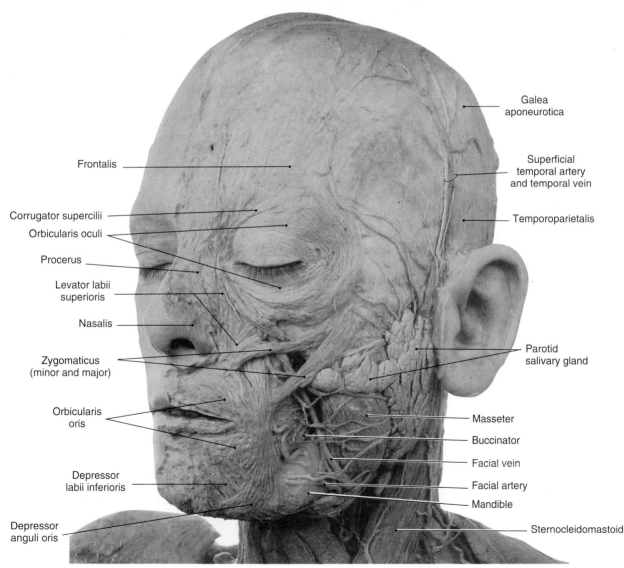

Frontalis

Corrugator supercilii

Orbicularis oculi

Procerus

Levator labii
superioris

Nasalis

Zygomaticus
(minor and major)

Orbicularis
oris

Depressor
labii inferioris

Depressor
anguli oris

Galea
aponeurotica

Superficial
temporal artery
and temporal vein

Temporoparietalis

Parotid
salivary gland

Masseter

Buccinator

Facial vein

Facial artery

Mandible

Sternocleidomastoid

4.3a Cadaver head and neck, lateral view

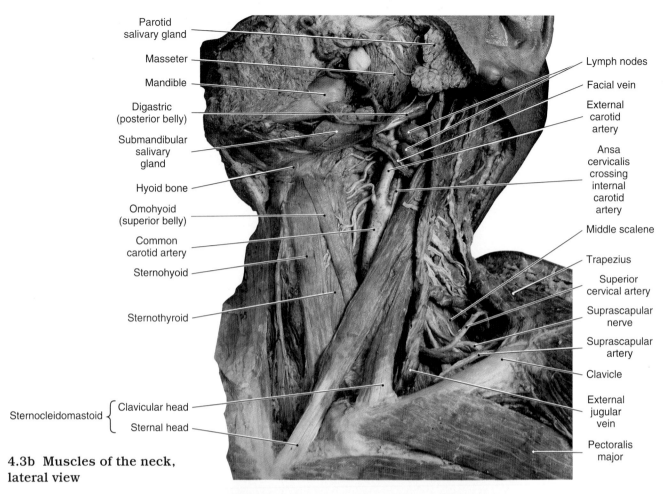

Parotid salivary gland
Masseter
Mandible
Digastric (posterior belly)
Submandibular salivary gland
Hyoid bone
Omohyoid (superior belly)
Common carotid artery
Sternohyoid
Sternothyroid
Sternocleidomastoid { Clavicular head / Sternal head

Lymph nodes
Facial vein
External carotid artery
Ansa cervicalis crossing internal carotid artery
Middle scalene
Trapezius
Superior cervical artery
Suprascapular nerve
Suprascapular artery
Clavicle
External jugular vein
Pectoralis major

4.3b Muscles of the neck, lateral view

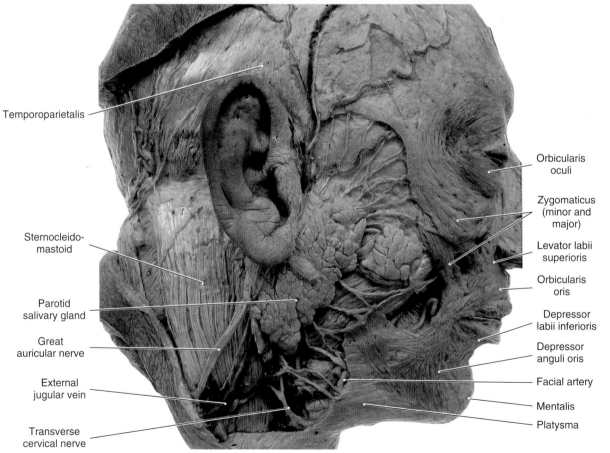

Temporoparietalis
Sternocleido-mastoid
Parotid salivary gland
Great auricular nerve
External jugular vein
Transverse cervical nerve

Orbicularis oculi
Zygomaticus (minor and major)
Levator labii superioris
Orbicularis oris
Depressor labii inferioris
Depressor anguli oris
Facial artery
Mentalis
Platysma

4.3c Muscles of the face

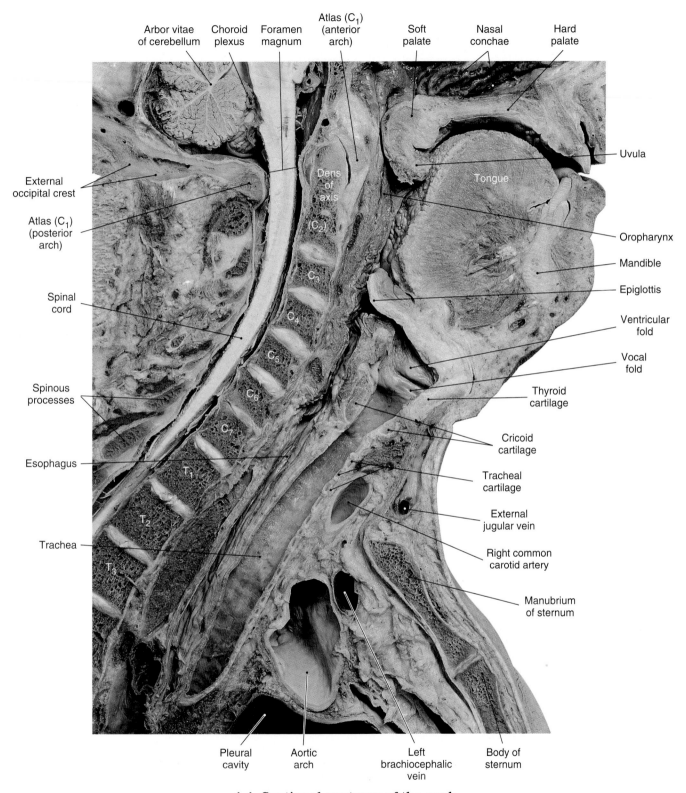

Arbor vitae of cerebellum — Choroid plexus — Foramen magnum — Atlas (C$_1$) (anterior arch) — Soft palate — Nasal conchae — Hard palate

External occipital crest

Atlas (C$_1$) (posterior arch)

Spinal cord

Spinous processes

Esophagus

Trachea

Dens of axis

(C$_2$)

C$_3$

C$_4$

C$_5$

C$_6$

C$_7$

T$_1$

T$_2$

T$_3$

Tongue

Uvula

Oropharynx

Mandible

Epiglottis

Ventricular fold

Vocal fold

Thyroid cartilage

Cricoid cartilage

Tracheal cartilage

External jugular vein

Right common carotid artery

Manubrium of sternum

Pleural cavity — Aortic arch — Left brachiocephalic vein — Body of sternum

4.4 Sectional anatomy of the neck

PLATE 5 THE UPPER LIMBS
The Right Upper Limb

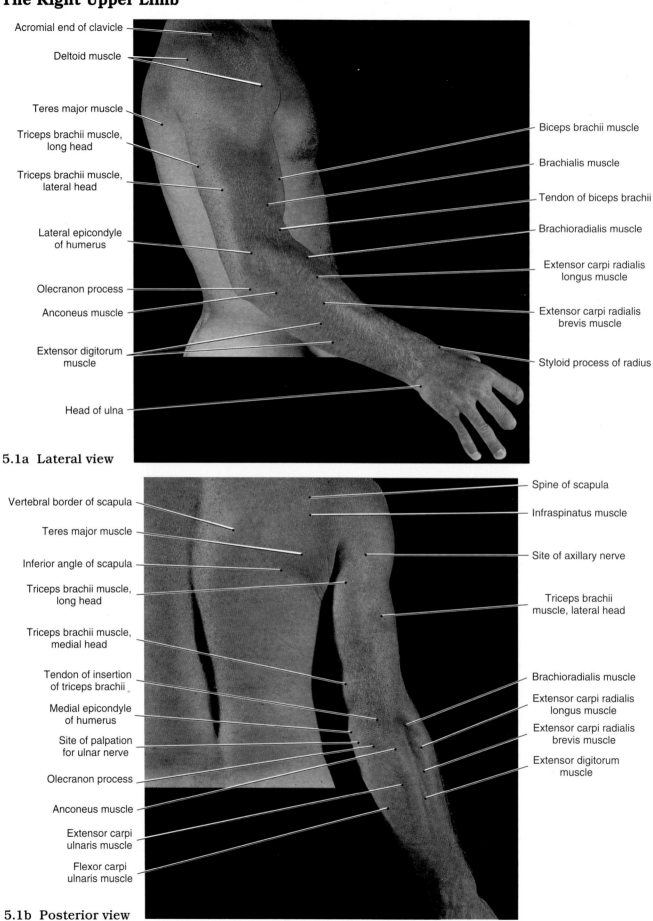

Acromial end of clavicle

Deltoid muscle

Teres major muscle

Triceps brachii muscle, long head

Triceps brachii muscle, lateral head

Lateral epicondyle of humerus

Olecranon process

Anconeus muscle

Extensor digitorum muscle

Head of ulna

Biceps brachii muscle

Brachialis muscle

Tendon of biceps brachii

Brachioradialis muscle

Extensor carpi radialis longus muscle

Extensor carpi radialis brevis muscle

Styloid process of radius

5.1a Lateral view

Vertebral border of scapula

Teres major muscle

Inferior angle of scapula

Triceps brachii muscle, long head

Triceps brachii muscle, medial head

Tendon of insertion of triceps brachii

Medial epicondyle of humerus

Site of palpation for ulnar nerve

Olecranon process

Anconeus muscle

Extensor carpi ulnaris muscle

Flexor carpi ulnaris muscle

Spine of scapula

Infraspinatus muscle

Site of axillary nerve

Triceps brachii muscle, lateral head

Brachioradialis muscle

Extensor carpi radialis longus muscle

Extensor carpi radialis brevis muscle

Extensor digitorum muscle

5.1b Posterior view

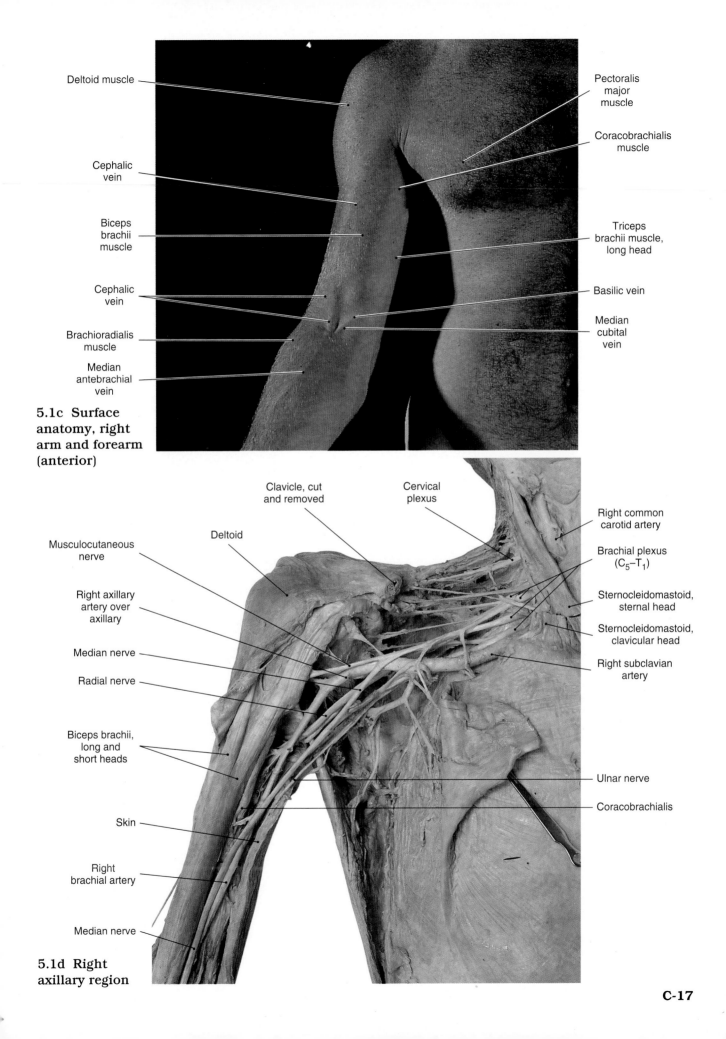

Deltoid muscle

Cephalic vein

Biceps brachii muscle

Cephalic vein

Brachioradialis muscle

Median antebrachial vein

Pectoralis major muscle

Coracobrachialis muscle

Triceps brachii muscle, long head

Basilic vein

Median cubital vein

5.1c Surface anatomy, right arm and forearm (anterior)

Clavicle, cut and removed

Cervical plexus

Deltoid

Musculocutaneous nerve

Right axillary artery over axillary

Median nerve

Radial nerve

Biceps brachii, long and short heads

Skin

Right brachial artery

Median nerve

Right common carotid artery

Brachial plexus (C$_5$–T$_1$)

Sternocleidomastoid, sternal head

Sternocleidomastoid, clavicular head

Right subclavian artery

Ulnar nerve

Coracobrachialis

5.1d Right axillary region

5.2a Right forearm, cadaver

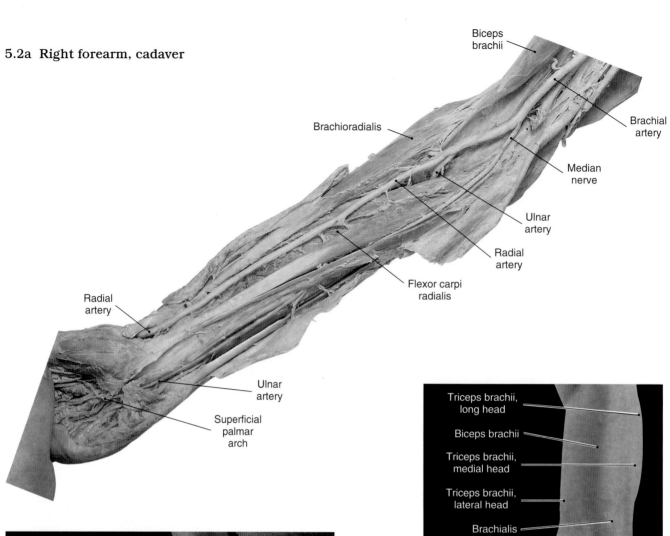

Biceps brachii

Brachioradialis

Brachial artery

Median nerve

Ulnar artery

Radial artery

Flexor carpi radialis

Radial artery

Ulnar artery

Superficial palmar arch

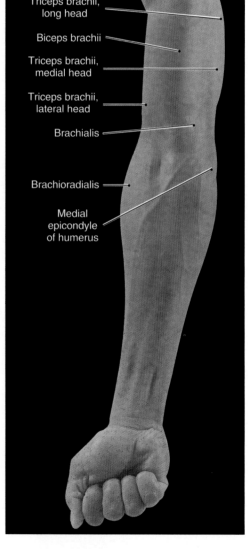

Triceps brachii, long head

Biceps brachii

Triceps brachii, medial head

Triceps brachii, lateral head

Brachialis

Brachioradialis

Medial epicondyle of humerus

5.3a Right arm and forearm, anterior view

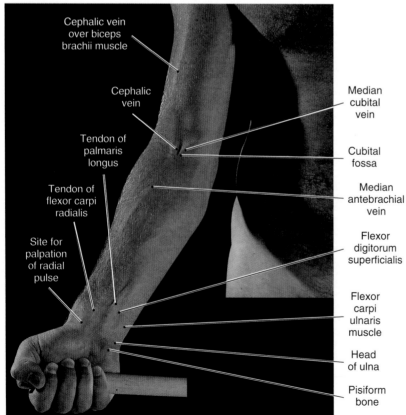

Cephalic vein over biceps brachii muscle

Cephalic vein

Tendon of palmaris longus

Tendon of flexor carpi radialis

Site for palpation of radial pulse

Median cubital vein

Cubital fossa

Median antebrachial vein

Flexor digitorum superficialis

Flexor carpi ulnaris muscle

Head of ulna

Pisiform bone

5.2b Right arm, forearm and wrist, anterior view
C-18

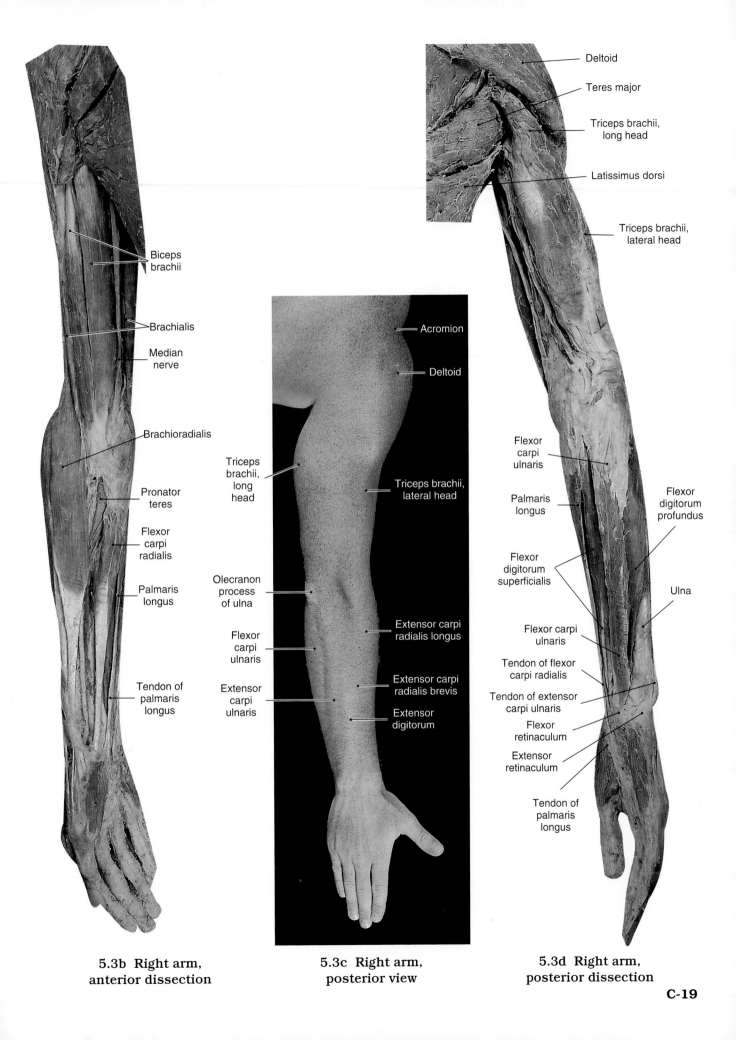

Deltoid

Teres major

Triceps brachii,
long head

Latissimus dorsi

Triceps brachii,
lateral head

Biceps
brachii

Brachialis

Median
nerve

Brachioradialis

Pronator
teres

Flexor
carpi
radialis

Palmaris
longus

Tendon of
palmaris
longus

Acromion

Deltoid

Triceps
brachii,
long
head

Triceps brachii,
lateral head

Olecranon
process
of ulna

Flexor
carpi
ulnaris

Extensor
carpi
ulnaris

Extensor carpi
radialis longus

Extensor carpi
radialis brevis

Extensor
digitorum

Flexor
carpi
ulnaris

Palmaris
longus

Flexor
digitorum
superficialis

Flexor carpi
ulnaris

Tendon of flexor
carpi radialis

Tendon of extensor
carpi ulnaris

Flexor
retinaculum

Extensor
retinaculum

Tendon of
palmaris
longus

Flexor
digitorum
profundus

Ulna

**5.3b Right arm,
anterior dissection**

**5.3c Right arm,
posterior view**

**5.3d Right arm,
posterior dissection**

C-19

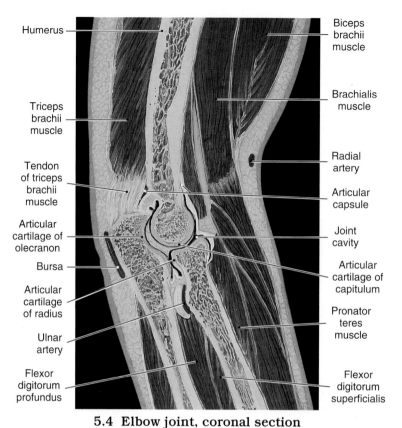

Humerus	Biceps brachii muscle
Triceps brachii muscle	Brachialis muscle
Tendon of triceps brachii muscle	Radial artery
	Articular capsule
Articular cartilage of olecranon	Joint cavity
Bursa	Articular cartilage of capitulum
Articular cartilage of radius	Pronator teres muscle
Ulnar artery	
Flexor digitorum profundus	Flexor digitorum superficialis

5.4 Elbow joint, coronal section

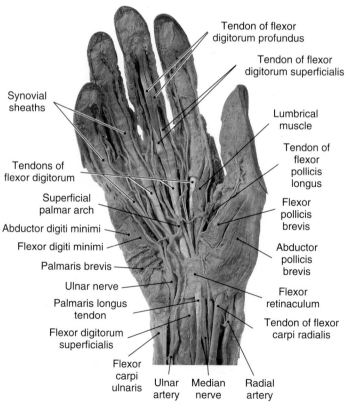

Tendon of extensor indicis

Tendon of extensor pollicis longus

Digital nerve

Dorsal digital arteries

Collateral ligament

Interphalangeal joint

Extensor digiti minimi tendon

Radial artery

Extensor digitorum tendons

Abductor digiti minimi

Tendon of extensor carpi radialis longus

Dorsal interossei

Tendon of extensor carpi radialis brevis

Extensor retinaculum

(a)

5.5a A model of the right wrist and hand, posterior view

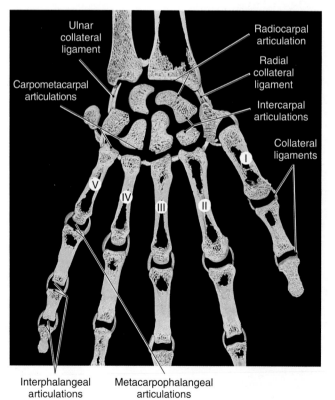

Ulnar collateral ligament

Radiocarpal articulation

Radial collateral ligament

Carpometacarpal articulations

Intercarpal articulations

Collateral ligaments

I

V

IV

III

II

Interphalangeal articulations

Metacarpophalangeal articulations

5.5b Joints of the right wrist and hand, sectional view

Tendon of flexor digitorum profundus

Tendon of flexor digitorum superficialis

Synovial sheaths

Lumbrical muscle

Tendon of flexor pollicis longus

Tendons of flexor digitorum

Flexor pollicis brevis

Superficial palmar arch

Abductor digiti minimi

Abductor pollicis brevis

Flexor digiti minimi

Palmaris brevis

Flexor retinaculum

Ulnar nerve

Tendon of flexor carpi radialis

Palmaris longus tendon

Flexor digitorum superficialis

Flexor carpi ulnaris

Ulnar artery

Median nerve

Radial artery

5.5c Muscles, tendons, and ligaments of the right wrist and hand, anterior view

PLATE 6   THE TRUNK
The Thorax

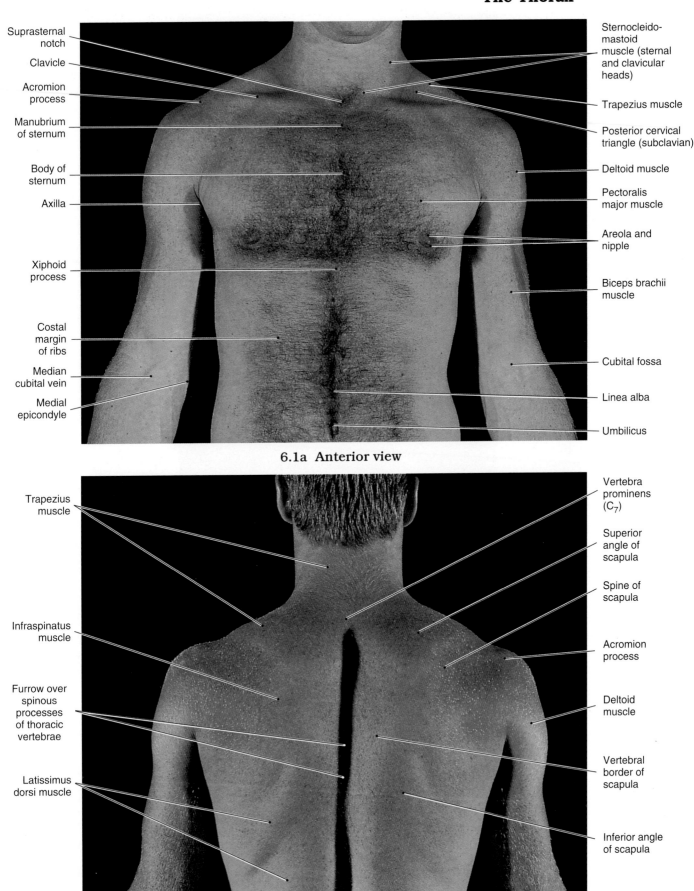

Suprasternal notch

Clavicle

Acromion process

Manubrium of sternum

Body of sternum

Axilla

Xiphoid process

Costal margin of ribs

Median cubital vein

Medial epicondyle

Sternocleido-mastoid muscle (sternal and clavicular heads)

Trapezius muscle

Posterior cervical triangle (subclavian)

Deltoid muscle

Pectoralis major muscle

Areola and nipple

Biceps brachii muscle

Cubital fossa

Linea alba

Umbilicus

6.1a Anterior view

Trapezius muscle

Infraspinatus muscle

Furrow over spinous processes of thoracic vertebrae

Latissimus dorsi muscle

Vertebra prominens (C$_7$)

Superior angle of scapula

Spine of scapula

Acromion process

Deltoid muscle

Vertebral border of scapula

Inferior angle of scapula

6.1b Posterior view

Muscles that Move the Arm

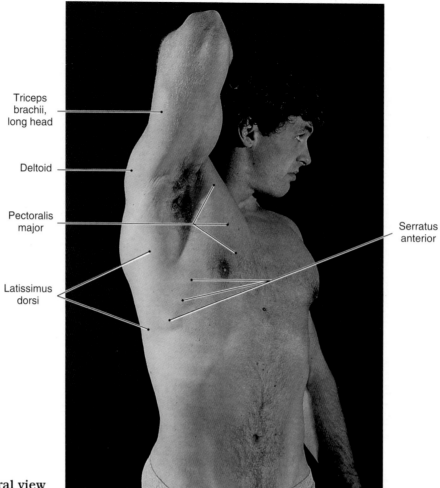

Triceps brachii, long head

Deltoid

Pectoralis major

Latissimus dorsi

Serratus anterior

6.2a Anterolateral view

Acromion

Triceps brachii, lateral head

Triceps brachii, long head

Infraspinatus

Teres major

Inferior angle of scapula

Deltoid

Trapezius

Latissimus dorsi

Erector spinae

6.2b Posterior view

6.3a The thoracic and abdominal cavities, anterior view (superficial)

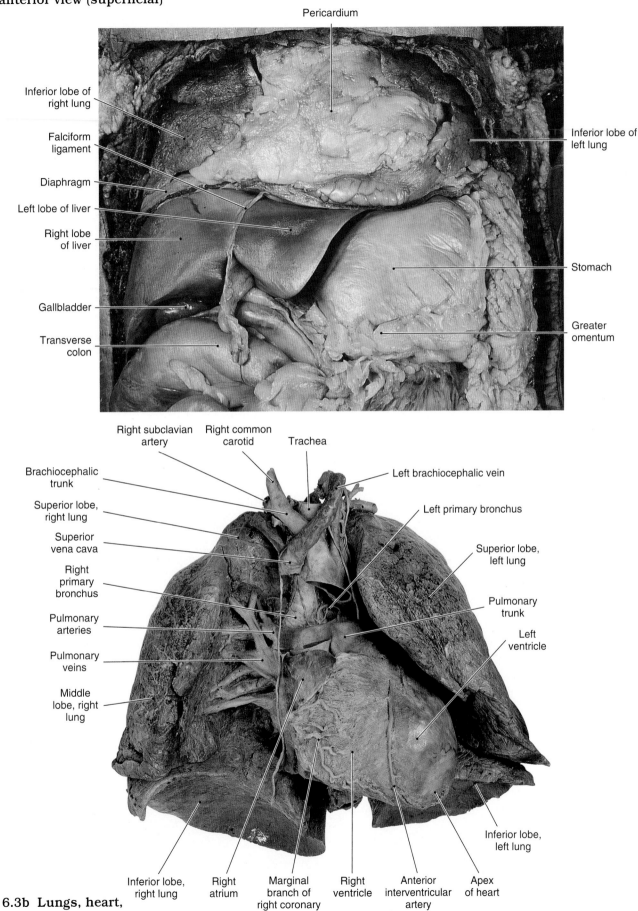

Pericardium

Inferior lobe of right lung

Falciform ligament

Diaphragm

Left lobe of liver

Right lobe of liver

Gallbladder

Transverse colon

Inferior lobe of left lung

Stomach

Greater omentum

Right subclavian artery

Right common carotid

Trachea

Left brachiocephalic vein

Brachiocephalic trunk

Superior lobe, right lung

Superior vena cava

Right primary bronchus

Pulmonary arteries

Pulmonary veins

Middle lobe, right lung

Left primary bronchus

Superior lobe, left lung

Pulmonary trunk

Left ventricle

Inferior lobe, left lung

Inferior lobe, right lung

Right atrium

Marginal branch of right coronary artery

Right ventricle

Anterior interventricular artery

Apex of heart

6.3b Lungs, heart, and great vessels

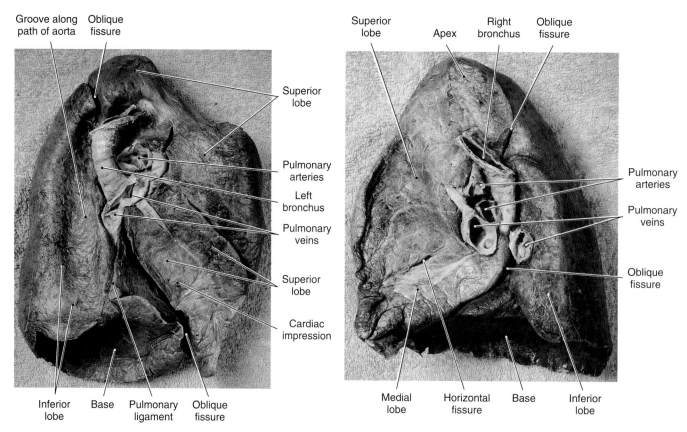

Groove along path of aorta   Oblique fissure

Superior lobe

Pulmonary arteries

Left bronchus

Pulmonary veins

Superior lobe

Cardiac impression

Inferior lobe   Base   Pulmonary ligament   Oblique fissure

6.3c Left lung, mediastinal surface

Superior lobe   Apex   Right bronchus   Oblique fissure

Pulmonary arteries

Pulmonary veins

Oblique fissure

Medial lobe   Horizontal fissure   Base   Inferior lobe

6.3d Right lung, mediastinal surface

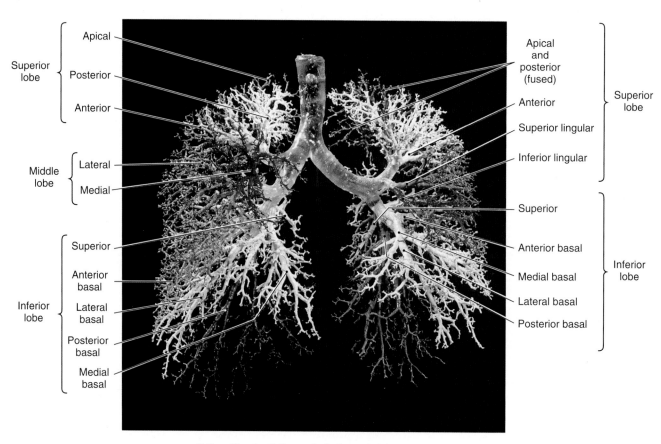

Superior lobe
Apical
Posterior
Anterior

Middle lobe
Lateral
Medial

Inferior lobe
Superior
Anterior basal
Lateral basal
Posterior basal
Medial basal

Apical and posterior (fused)
Anterior
Superior lingular
Inferior lingular
Superior lobe

Superior
Anterior basal
Medial basal
Lateral basal
Posterior basal
Inferior lobe

6.3e Cast of the adult bronchial tree

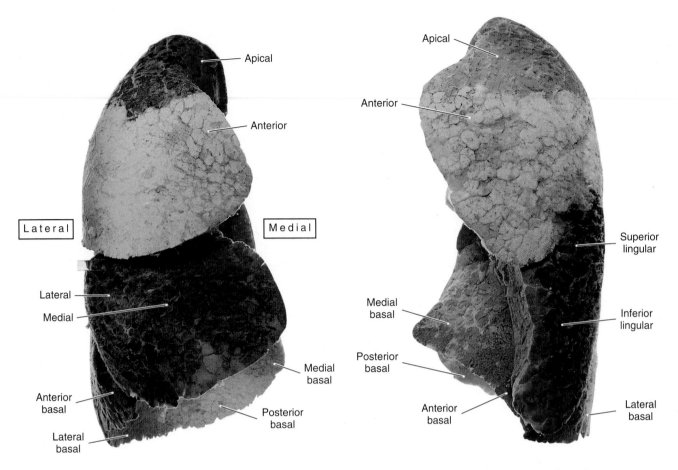

6.3f Right lung, anterior view

6.3g Left lung, anterior view

6.3h Right lung, lateral view

6.3i Left lung, lateral view

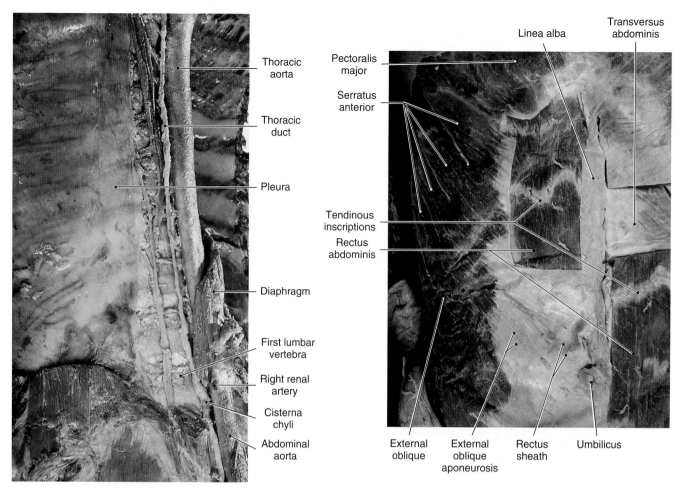

6.3j Major lymphatic vessels of the trunk

Thoracic aorta

Thoracic duct

Pleura

Diaphragm

First lumbar vertebra

Right renal artery

Cisterna chyli

Abdominal aorta

6.4a Cadaver, superficial anterior view of the abdominal wall

Linea alba

Transversus abdominis

Pectoralis major

Serratus anterior

Tendinous inscriptions

Rectus abdominis

External oblique

External oblique aponeurosis

Rectus sheath

Umbilicus

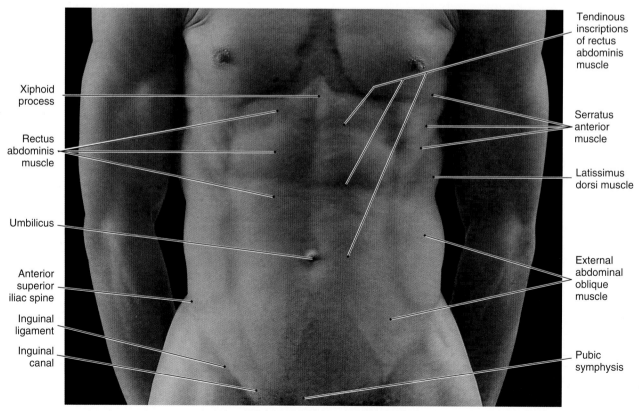

Xiphoid process

Rectus abdominis muscle

Umbilicus

Anterior superior iliac spine

Inguinal ligament

Inguinal canal

Tendinous inscriptions of rectus abdominis muscle

Serratus anterior muscle

Latissimus dorsi muscle

External abdominal oblique muscle

Pubic symphysis

6.4b Abdominal wall, anterior view

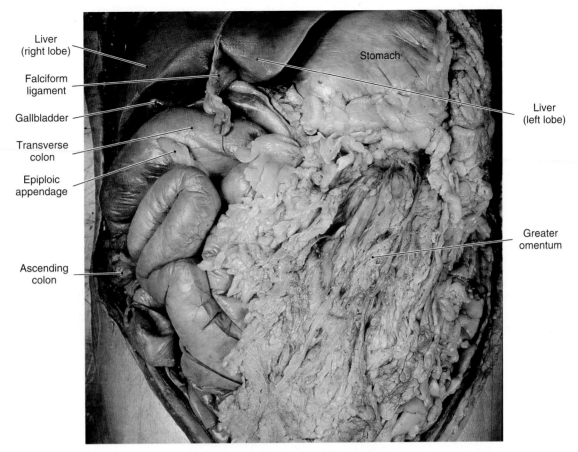

Liver (right lobe)

Falciform ligament

Gallbladder

Transverse colon

Epiploic appendage

Ascending colon

Stomach

Liver (left lobe)

Greater omentum

6.5a Abdominopelvic cavity

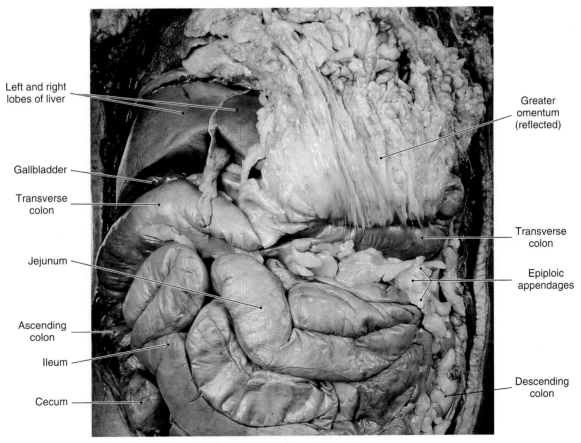

Left and right lobes of liver

Gallbladder

Transverse colon

Jejunum

Ascending colon

Ileum

Cecum

Greater omentum (reflected)

Transverse colon

Epiploic appendages

Descending colon

6.5b Abdominopelvic viscera

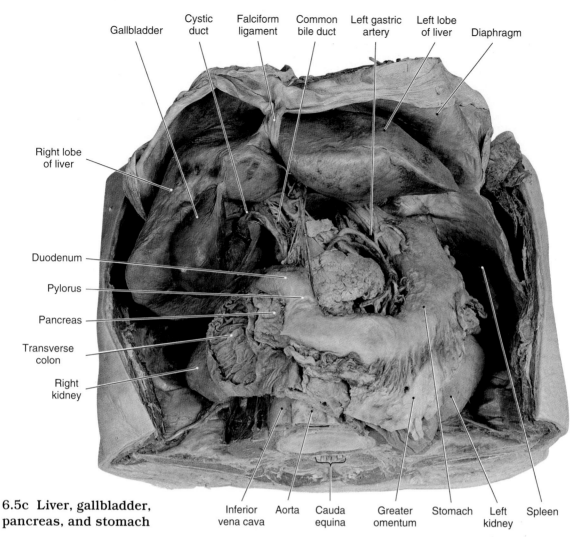

Gallbladder

Cystic duct

Falciform ligament

Common bile duct

Left gastric artery

Left lobe of liver

Diaphragm

Right lobe of liver

Duodenum

Pylorus

Pancreas

Transverse colon

Right kidney

Inferior vena cava

Aorta

Cauda equina

Greater omentum

Stomach

Left kidney

Spleen

6.5c Liver, gallbladder, pancreas, and stomach

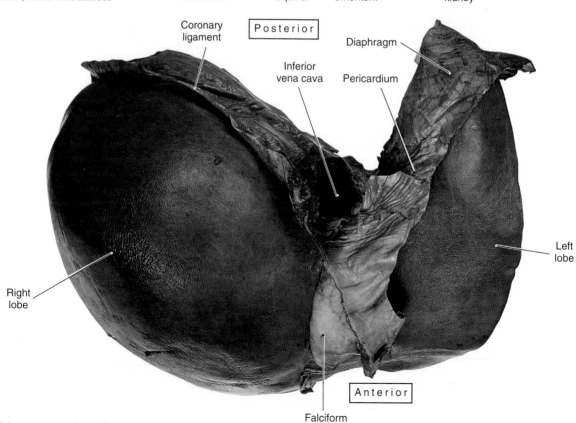

Coronary ligament

Posterior

Inferior vena cava

Diaphragm

Pericardium

Left lobe

Right lobe

Anterior

Falciform ligament

6.5d Liver, superior view
C-28

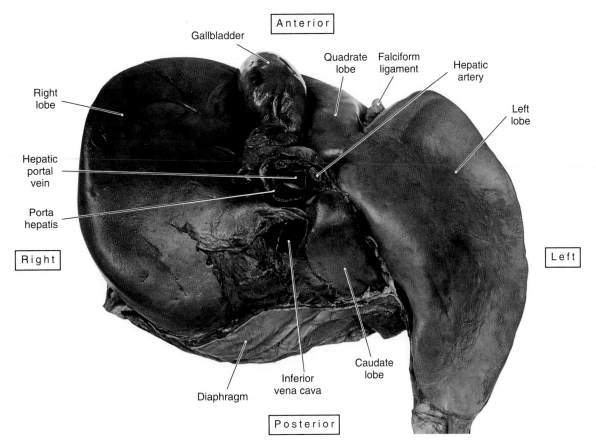

Anterior

Gallbladder

Quadrate lobe

Falciform ligament

Hepatic artery

Right lobe

Left lobe

Hepatic portal vein

Porta hepatis

Right

Left

Caudate lobe

Diaphragm

Inferior vena cava

Posterior

6.5e Liver, inferior view

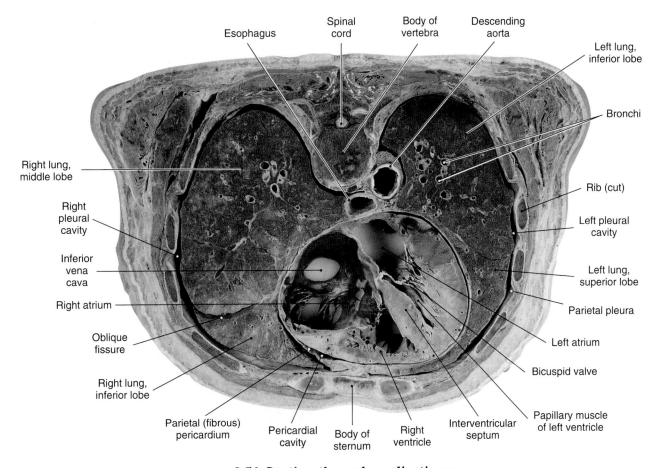

Esophagus

Spinal cord

Body of vertebra

Descending aorta

Left lung, inferior lobe

Bronchi

Right lung, middle lobe

Right pleural cavity

Rib (cut)

Left pleural cavity

Inferior vena cava

Left lung, superior lobe

Right atrium

Parietal pleura

Oblique fissure

Left atrium

Bicuspid valve

Right lung, inferior lobe

Papillary muscle of left ventricle

Parietal (fibrous) pericardium

Pericardial cavity

Body of sternum

Right ventricle

Interventricular septum

6.5f Section through mediastinum

C-29

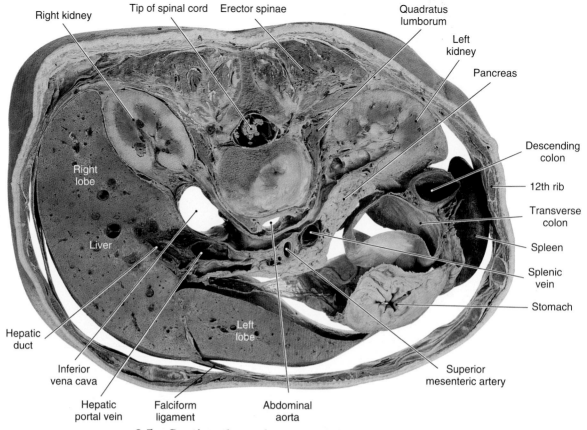

Right kidney · Tip of spinal cord · Erector spinae · Quadratus lumborum · Left kidney · Pancreas

Right lobe

Liver

Descending colon

12th rib

Transverse colon

Spleen

Splenic vein

Stomach

Left lobe

Hepatic duct

Inferior vena cava

Hepatic portal vein

Falciform ligament

Abdominal aorta

Superior mesenteric artery

6.5g Section through upper abdomen

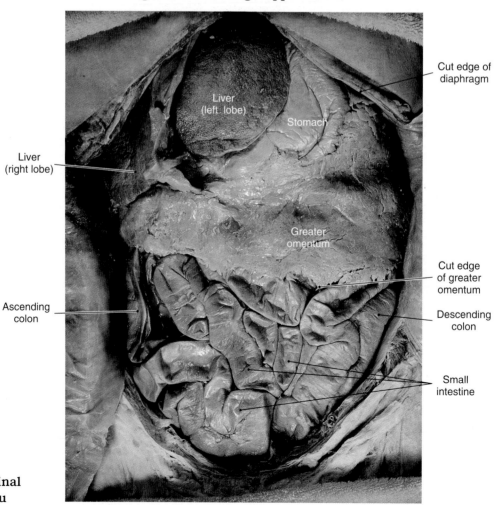

Liver (left lobe)

Stomach

Cut edge of diaphragm

Liver (right lobe)

Greater omentum

Cut edge of greater omentum

Descending colon

Ascending colon

Small intestine

6.5h Abdominal viscera in situ

C-30

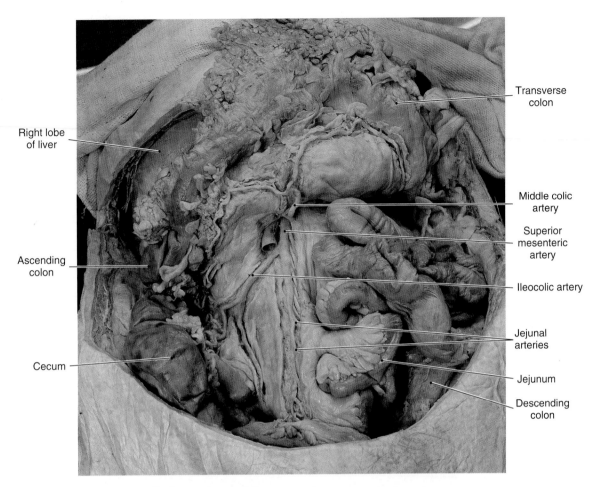

Right lobe
of liver

Ascending
colon

Cecum

Transverse
colon

Middle colic
artery

Superior
mesenteric
artery

Ileocolic artery

Jejunal
arteries

Jejunum

Descending
colon

6.5i Branches of the superior mesenteric artery

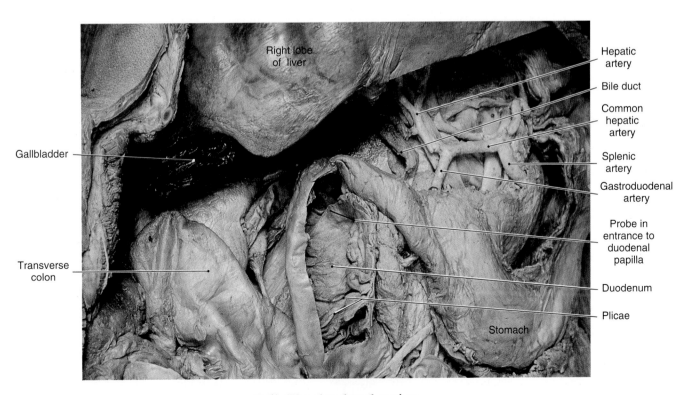

Right lobe
of liver

Gallbladder

Transverse
colon

Hepatic
artery

Bile duct

Common
hepatic
artery

Splenic
artery

Gastroduodenal
artery

Probe in
entrance to
duodenal
papilla

Duodenum

Plicae

Stomach

6.5j The duodenal region

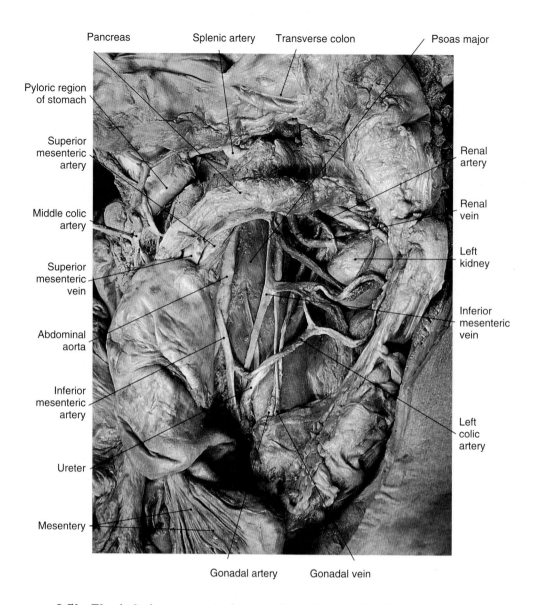

Pancreas · Splenic artery · Transverse colon · Psoas major

Pyloric region of stomach

Superior mesenteric artery

Middle colic artery

Superior mesenteric vein

Abdominal aorta

Inferior mesenteric artery

Ureter

Mesentery

Renal artery

Renal vein

Left kidney

Inferior mesenteric vein

Left colic artery

Gonadal artery · Gonadal vein

6.5k The inferior mesenteric vessels and associated structures

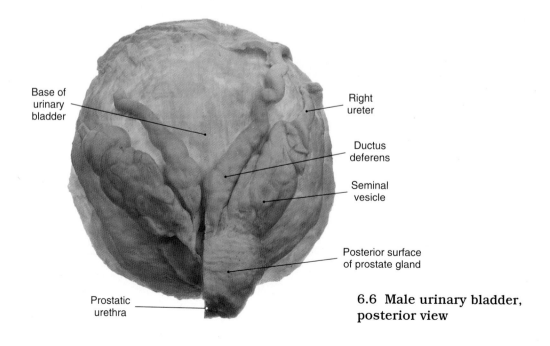

Base of urinary bladder

Right ureter

Ductus deferens

Seminal vesicle

Posterior surface of prostate gland

Prostatic urethra

6.6 Male urinary bladder, posterior view

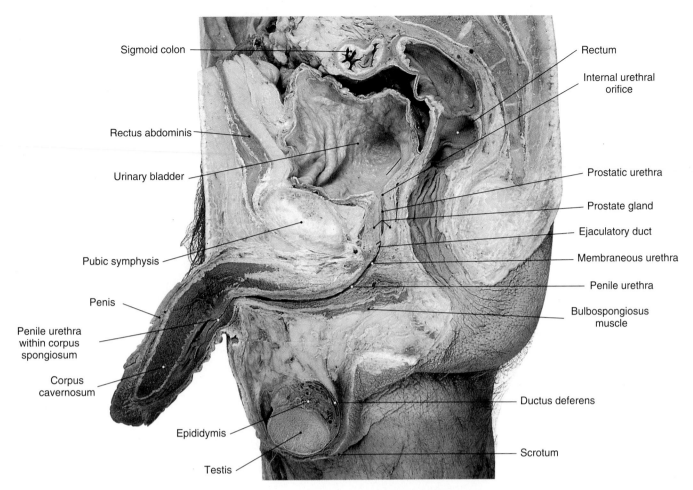

6.7a The male reproductive system, sagittal section

Sigmoid colon

Rectus abdominis

Urinary bladder

Pubic symphysis

Penis

Penile urethra within corpus spongiosum

Corpus cavernosum

Epididymis

Testis

Rectum

Internal urethral orifice

Prostatic urethra

Prostate gland

Ejaculatory duct

Membraneous urethra

Penile urethra

Bulbospongiosus muscle

Ductus deferens

Scrotum

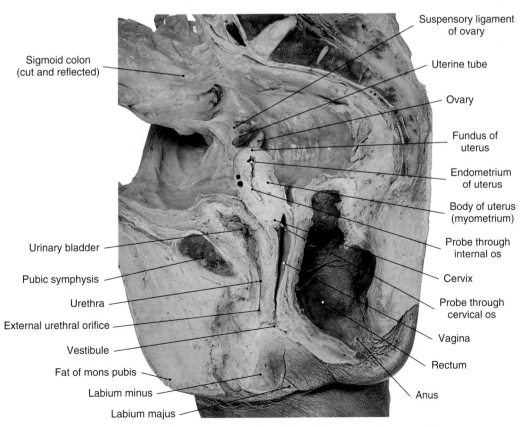

6.7b The female reproductive system, sagittal section

Sigmoid colon (cut and reflected)

Urinary bladder

Pubic symphysis

Urethra

External urethral orifice

Vestibule

Fat of mons pubis

Labium minus

Labium majus

Suspensory ligament of ovary

Uterine tube

Ovary

Fundus of uterus

Endometrium of uterus

Body of uterus (myometrium)

Probe through internal os

Cervix

Probe through cervical os

Vagina

Rectum

Anus

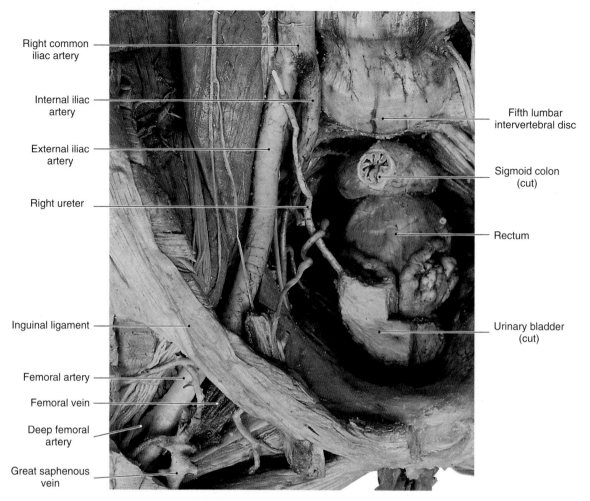

Right common iliac artery

Internal iliac artery

External iliac artery

Right ureter

Inguinal ligament

Femoral artery

Femoral vein

Deep femoral artery

Great saphenous vein

Fifth lumbar intervertebral disc

Sigmoid colon (cut)

Rectum

Urinary bladder (cut)

6.8 Major vessels of the pelvis

PLATE 7   THE LOWER LIMBS

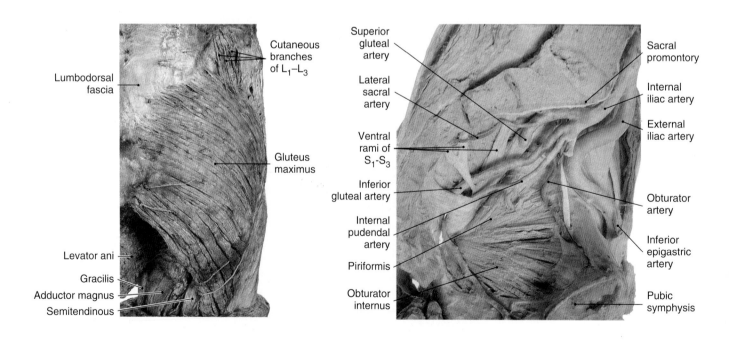

Lumbodorsal fascia

Cutaneous branches of L₁–L₃

Gluteus maximus

Levator ani

Gracilis

Adductor magnus

Semitendinous

Superior gluteal artery

Lateral sacral artery

Ventral rami of S₁-S₃

Inferior gluteal artery

Internal pudendal artery

Piriformis

Obturator internus

Sacral promontory

Internal iliac artery

External iliac artery

Obturator artery

Inferior epigastric artery

Pubic symphysis

7.1a The muscles of the hip and thigh, posterior view (superficial)

7.1b A sagittal section through the pelvis

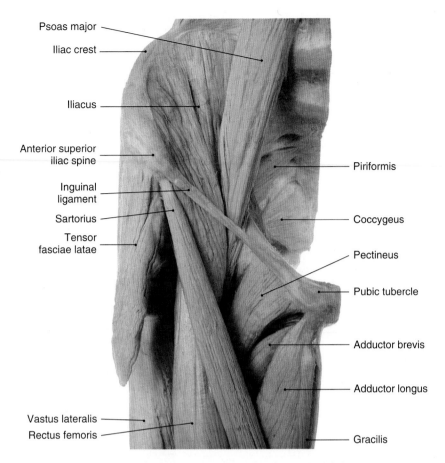

Psoas major

Iliac crest

Iliacus

Anterior superior
iliac spine

Inguinal
ligament

Sartorius

Tensor
fasciae latae

Piriformis

Coccygeus

Pectineus

Pubic tubercle

Adductor brevis

Adductor longus

Vastus lateralis

Rectus femoris

Gracilis

**7.1c The muscles of the hip and thigh,
anterior view (superficial)**

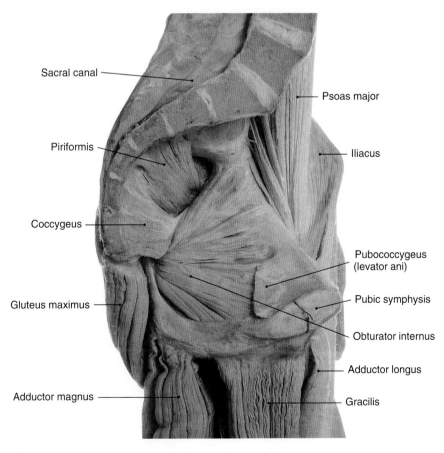

Sacral canal

Piriformis

Coccygeus

Gluteus maximus

Adductor magnus

Psoas major

Iliacus

Pubococcygeus
(levator ani)

Pubic symphysis

Obturator internus

Adductor longus

Gracilis

**7.1d A sagittal section through the pelvis, providing a medial
view of the muscles of the hip and thigh**

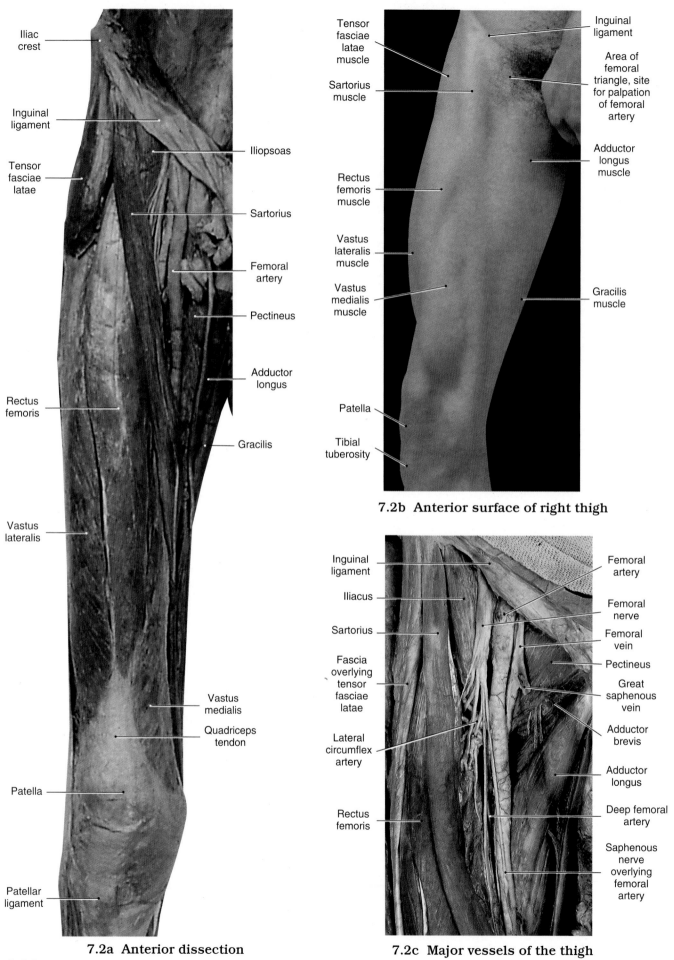

Iliac
crest

Inguinal
ligament

Tensor
fasciae
latae

Rectus
femoris

Vastus
lateralis

Patella

Patellar
ligament

Iliopsoas

Sartorius

Femoral
artery

Pectineus

Adductor
longus

Gracilis

Vastus
medialis

Quadriceps
tendon

7.2a Anterior dissection

Tensor
fasciae
latae
muscle

Sartorius
muscle

Rectus
femoris
muscle

Vastus
lateralis
muscle

Vastus
medialis
muscle

Patella

Tibial
tuberosity

Inguinal
ligament

Area of
femoral
triangle, site
for palpation
of femoral
artery

Adductor
longus
muscle

Gracilis
muscle

7.2b Anterior surface of right thigh

Inguinal
ligament

Iliacus

Sartorius

Fascia
overlying
tensor
fasciae
latae

Lateral
circumflex
artery

Rectus
femoris

Femoral
artery

Femoral
nerve

Femoral
vein

Pectineus

Great
saphenous
vein

Adductor
brevis

Adductor
longus

Deep femoral
artery

Saphenous
nerve
overlying
femoral
artery

7.2c Major vessels of the thigh

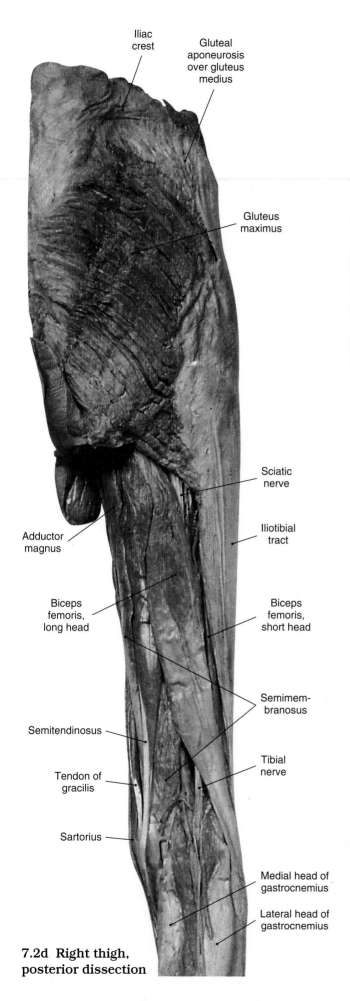

Iliac crest

Gluteal aponeurosis over gluteus medius

Gluteus maximus

Sciatic nerve

Iliotibial tract

Adductor magnus

Biceps femoris, long head

Biceps femoris, short head

Semimem-branosus

Semitendinosus

Tibial nerve

Tendon of gracilis

Sartorius

Medial head of gastrocnemius

Lateral head of gastrocnemius

7.2d Right thigh, posterior dissection

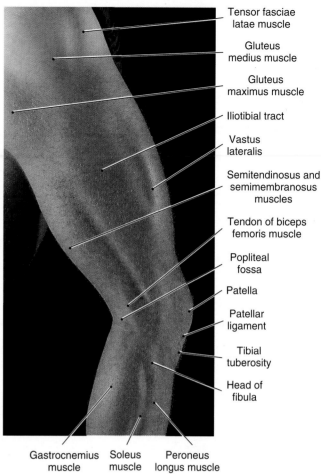

Tensor fasciae latae muscle

Gluteus medius muscle

Gluteus maximus muscle

Iliotibial tract

Vastus lateralis

Semitendinosus and semimembranosus muscles

Tendon of biceps femoris muscle

Popliteal fossa

Patella

Patellar ligament

Tibial tuberosity

Head of fibula

Gastrocnemius muscle   **Soleus muscle**   **Peroneus longus muscle**

7.2e Gluteal region and right thigh, lateral view

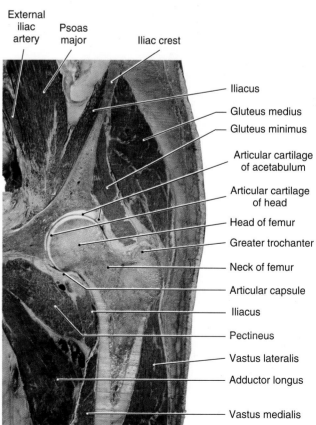

External iliac artery   **Psoas major**   **Iliac crest**

Iliacus

Gluteus medius

Gluteus minimus

Articular cartilage of acetabulum

Articular cartilage of head

Head of femur

Greater trochanter

Neck of femur

Articular capsule

Iliacus

Pectineus

Vastus lateralis

Adductor longus

Vastus medialis

7.3a Cadaver, coronal section of hip joint

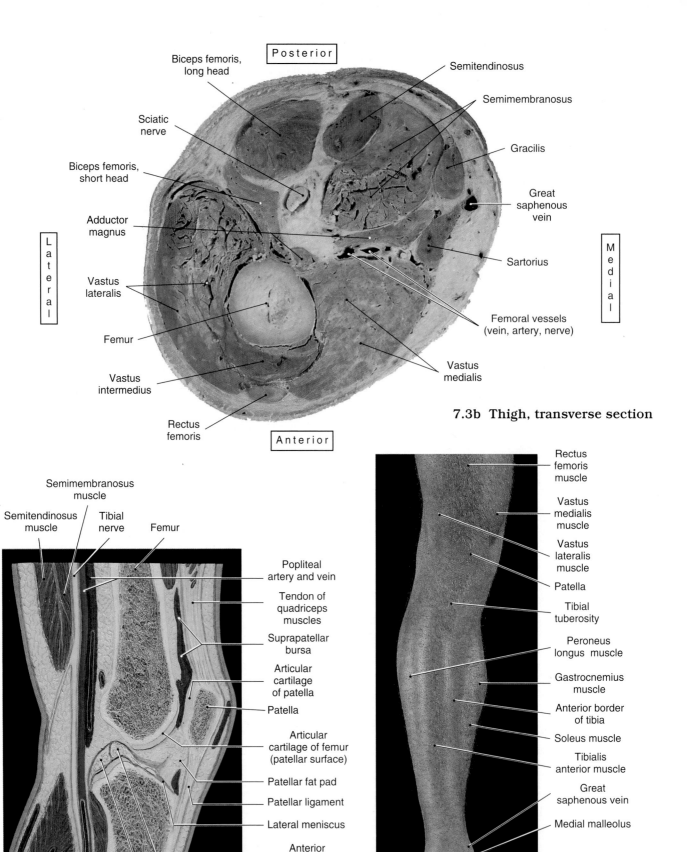

Posterior

Biceps femoris, long head

Sciatic nerve

Biceps femoris, short head

Adductor magnus

Vastus lateralis

Femur

Vastus intermedius

Rectus femoris

Anterior

Lateral

Medial

Semitendinosus

Semimembranosus

Gracilis

Great saphenous vein

Sartorius

Femoral vessels (vein, artery, nerve)

Vastus medialis

7.3b Thigh, transverse section

Semimembranosus muscle

Semitendinosus muscle

Tibial nerve

Femur

Gastrocnemius muscle

Soleus muscle

Popliteus muscle

Popliteal artery and vein

Tendon of quadriceps muscles

Suprapatellar bursa

Articular cartilage of patella

Patella

Articular cartilage of femur (patellar surface)

Patellar fat pad

Patellar ligament

Lateral meniscus

Anterior cruciate ligament

Posterior cruciate ligament

Tibia

7.3c Sagittal section of the knee

Rectus femoris muscle

Vastus medialis muscle

Vastus lateralis muscle

Patella

Tibial tuberosity

Peroneus longus muscle

Gastrocnemius muscle

Anterior border of tibia

Soleus muscle

Tibialis anterior muscle

Great saphenous vein

Medial malleolus

Lateral malleolus

Dorsal venous arch

Tendon of extensor hallucis longus

Tendons of extensor digitorum longus

7.4a Knee and leg, anterior view

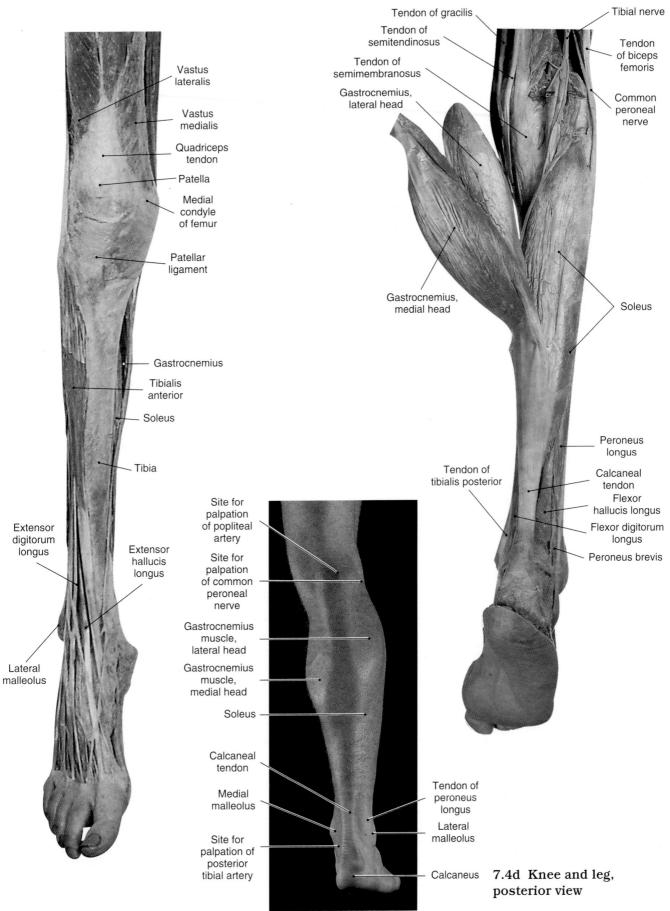

7.4b Superficial muscles of the right leg, cadaver, anterior view

Vastus lateralis

Vastus medialis

Quadriceps tendon

Patella

Medial condyle of femur

Patellar ligament

Gastrocnemius

Tibialis anterior

Soleus

Tibia

Extensor digitorum longus

Extensor hallucis longus

Lateral malleolus

7.4c Superficial muscles of the right leg, cadaver, posterior view

Tendon of gracilis

Tendon of semitendinosus

Tendon of semimembranosus

Gastrocnemius, lateral head

Tibial nerve

Tendon of biceps femoris

Common peroneal nerve

Gastrocnemius, medial head

Soleus

Tendon of tibialis posterior

Peroneus longus

Calcaneal tendon

Flexor hallucis longus

Flexor digitorum longus

Peroneus brevis

Site for palpation of popliteal artery

Site for palpation of common peroneal nerve

Gastrocnemius muscle, lateral head

Gastrocnemius muscle, medial head

Soleus

Calcaneal tendon

Medial malleolus

Site for palpation of posterior tibial artery

Tendon of peroneus longus

Lateral malleolus

Calcaneus

7.4d Knee and leg, posterior view

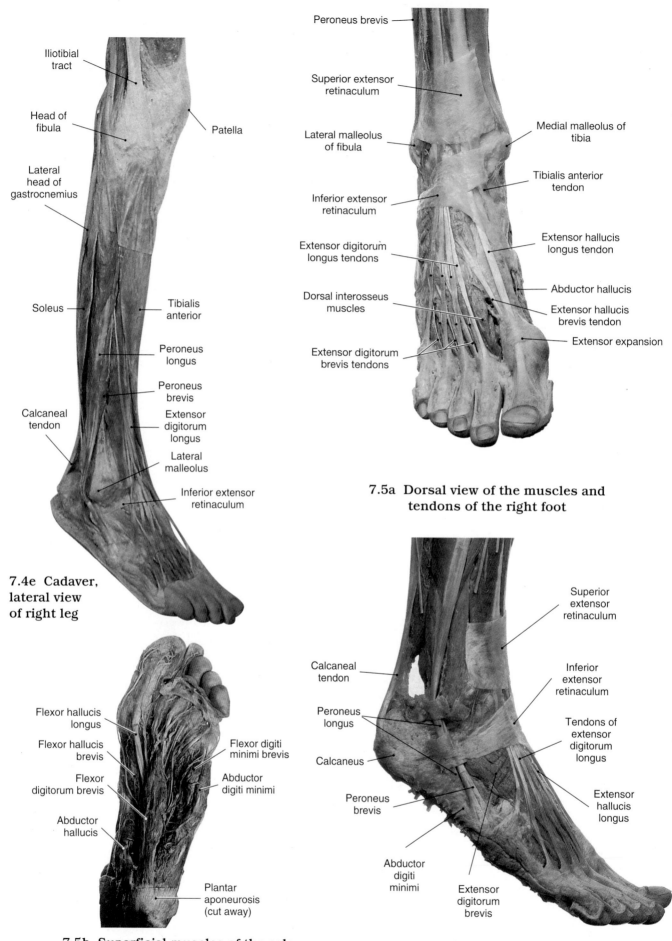

Iliotibial tract

Head of fibula

Patella

Lateral head of gastrocnemius

Soleus

Tibialis anterior

Peroneus longus

Peroneus brevis

Extensor digitorum longus

Lateral malleolus

Calcaneal tendon

Inferior extensor retinaculum

7.4e Cadaver, lateral view of right leg

Peroneus brevis

Superior extensor retinaculum

Lateral malleolus of fibula

Medial malleolus of tibia

Inferior extensor retinaculum

Tibialis anterior tendon

Extensor digitorum longus tendons

Extensor hallucis longus tendon

Dorsal interosseus muscles

Abductor hallucis

Extensor hallucis brevis tendon

Extensor digitorum brevis tendons

Extensor expansion

7.5a Dorsal view of the muscles and tendons of the right foot

Flexor hallucis longus

Flexor hallucis brevis

Flexor digitorum brevis

Abductor hallucis

Flexor digiti minimi brevis

Abductor digiti minimi

Plantar aponeurosis (cut away)

7.5b Superficial muscles of the sole, after removal of the plantar aponeurosis

Calcaneal tendon

Peroneus longus

Calcaneus

Peroneus brevis

Abductor digiti minimi

Extensor digitorum brevis

Superior extensor retinaculum

Inferior extensor retinaculum

Tendons of extensor digitorum longus

Extensor hallucis longus

7.5c Lateral aspect of the foot